STUDENT'S SOLUTIONS MANUAL

J. JACKSON BARNETTE
University of Alabama at Birmingham

IAN C. WALTERS, JR.
D'Youville College

BIOSTATISTICS FOR THE BIOLOGICAL AND HEALTH SCIENCES

Marc M. Triola, M.D.
New York University School of Medicine

Mario F. Triola

Boston San Francisco New York
London Toronto Sydney Tokyo Singapore Madrid
Mexico City Munich Paris Cape Town Hong Kong Montreal

Reproduced by Pearson Addison-Wesley from electronic files supplied by the author.

Publishing as Pearson Addison-Wesley, 75 Arlington Street, Boston, MA 02116.

ISBN 0-321-28689-8

Contents

Chapter 1. Introduction

1 – 2 Types of Data

In Exercises 1 & 3, determine whether the given value is a statistic or a parameter.

1. Statistic, since 12 males (the sample size of males) is a characteristic of a sample.
3. Statistic, since 226 m (the sample mean) is a characteristic of a sample of frigate birds.

In Exercises 5 & 7, determine whether the given values are from a discrete or continuous data set.

5. Discrete, since the number of birds must be in the form of whole numbers, a fraction of a bird is not possible.
7. Discrete, since the number of families having guns in their homes must be in the form of whole numbers; a fraction of a home is not possible.

In Exercises 9 – 15, determine which of the four levels of measurement (nominal, ordinal, interval ratio) is most appropriate.

9. Ratio, since there are equal intervals and a natural starting point of 0 inches.
11. Interval, since body temperatures, either measured on a Fahrenheit or Centigrade scale, have equal intervals, but do not have a natural starting point of 0.
13. Nominal, since the measure is used as a label for a category and the category does not have any ordinal property.
15. Ratio, since the number of manatees killed by boats has a natural 0 starting point.

In Exercises 17 & 19, identify the (a) sample and (b) population. Also determine whether the sample is likely to be representative of the population.

17. This would be a sample that is likely not representative of the population of rainbow trout, but it could be representative of rainbow trout in the stream where she is netting the trout.
19. This would be a sample that would likely be representative of a population from which the sample was drawn.

1 – 3 Design of Experiments

In Exercises 1 & 3, determine whether the given description corresponds to an observational study or an experiment.

1. **Drug Testing** Experiment, since the patients are being treated or exposed to a treatment, in this case Lipitor.
3. **Quality Control** Observation, since the amount of aspirin is observed in each tablet, they are not exposed to any treatment.

In Exercises 5 & 7, identify the type of observational study (cross-sectional, retrospective, prospective).

5. **Medical Research** Retrospective, since the study focus is on what has happened in a period of the past (five years) for head injury patients.
7. **Flu Incidence** Cross-sectional, since the study focus is on what happened at a specific time period.

In Exercises 9 – 19, identify which of these types of sampling is used: random, systematic, convenience, stratified, or cluster.

9. **Aspirin Usage** Convenience, since the researcher asks those who enter the clinic for treatment.

11. **Telephone Polls** Random, since telephone survey respondents were selected at random from a list of phone numbers; each phone number would have the same chance of being selected. However, we need to exercise some caution in calling this random. When a large majority of potential survey respondents had phones and researchers had access to the phone numbers, these types of surveys were often considered random. However, with changes like "no call lists", unlisted numbers, phones that can be programmed to screen out unrecognized calls, and individuals who have made cell phones their phone of choice and have dropped land line phones, random dialing of phone numbers is less likely to provide a representative sample than has been the case in the past.
13. **Student Drinking** Cluster, since classes form the clusters and clusters are randomly selected and all members of the selected clusters become study participants.
15. **Sobriety Checkpoint** Systematic, since every i^{th}, in this case every fifth, driver is stopped for a sobriety check sample.
17. **Education and Health** Stratified, since categories or strata are identified (three education level strata) and 150 survey participants are randomly selected from within each strata.
19. **Medical Research** Cluster, since each hospital represents a cluster and 30 clusters are randomly selected and all the cardiac patients in the cluster are surveyed.

Exercises 21 – 25 relate to random samples and simple random samples.

21. **Sampling Aspirin Tablets** Simple random sample, since every possible sample of size 50 has the same chance of being chosen.
23. **Convenience Sample** Neither a random nor simple random sample, since random selection is not done in any way to select the sample.
25. **Stratified Sample** Simple random sample, since we would assume that within each group, there is a random selection of possible survey respondents whereby every possible sample of size 100 men and 100 women has an equal chance of being selected.
27. **Cluster Sample** Random sample, since the blocks are randomly selected in such a way that all different combinations of 10 blocks would have an equal chance of being selected so relative to blocks this would be a simple random sample. However since the blocks are made up of subjects who must remain within the blocks, there is no equal chance for any of the possible samples of all the residents to be selected.

29. **Sample Design** The study is prospective since subjects will be observed over a period of time after treatment introduction. It is a randomized study if participants were randomly selected from a population and then randomly assigned into one of the three treatment conditions. The study is double-blind when neither the treatment recipients nor those who administer the treatments know what treatment any participant receives. The study is placebo-controlled when there is a control comparison group where the participants in that group receive a treatment that is expected to have no affect on levels of the outcome or dependent variable, but they are lead to believe they are receiving a treatment.
31. **Motorcycle Helmets** The very important group that was not able to testify would be those who did not survive an accident whether or not they were using a helmet.

Review Exercises

1. **Sampling** No, they cannot be considered representative of the population of the US since only those who were Internet subscribers were included, those without an Internet connection did not have the opportunity to respond. In addition, those who responded to the survey self-selected themselves to do so and thus were not even representative of all Internet subscribers.
3. **Level of Measurement**
   a. Ratio, since there is a natural zero point and each pound or kilogram is equivalent
   b. Ordinal, since the categories do have order, but there is no basis for determining meaningful differences between the categories.
   c. Nominal, since the observation is a category with no order to the categories.
   d. Interval, since the observation is in equal units of years, but the zero point does not represent the beginning of the scale.

5. **Smokers**
   a. Systematic sample, this sample would likely be representative of all who smoke.
   b. Convenience sample, this sample would only represent those who are runners, not the population of all who smoke, so it would not be representative.
   c. Cluster sample, this sample would likely not be representative since the grocery stores may not be representative of grocery stores where smokers live.
   d. Random sample, this sample would be likely be representative of all who smoke.
   e. Stratified sample, this sample would likely not be representative since there are probably different distributions of smokers in varying zip codes.

Cumulative Review Exercises

1. **Percentages**
   a. 57% of 1500 is 0.57 * 1500= 855
   b. 26% of 950 is 0.26 * 950= 247
3. **Percentages in a Study of Lipitor**
   a. 19 out of 270 is equal to 19/270 * 100%= 0.070 * 100%= 7% of patients who received a placebo reported headaches
   b. 3% of the 270 patients reported back pains, this is 3%/100% * 270= 0.03 * 270= 8 patients out of 270 reported back pains
5. **Percentages in the Media** This would relate to using a ratio scale where zero would represent the absence of plaque. For such a scale, the observation of the condition can only be reduced by 100%. Once that is done, there is no more to be reduced.

In Exercises 7 & 9, the given expressions are designed to yield results expressed in the form of scientific notation. Use a calculator or computer to perform the indicated operation and express the result as an ordinary number that is not in scientific notation.

7. 0.95^{500} = 7.2744916E^{-12}= 0.0000000000072744916 (moving the decimal point 12 places to the left)
9. 9^{12}= 2.8242954E^{11}= 282,429,540,000 (moving the decimal point 11 places to the right) left)

Chapter 2. Describing, Exploring, and Comparing Data

2-2 Frequency Distributions

In exercises 1 & 3, identify the class width, class midpoints, and class boundaries for the given frequency distribution based on Data Set 1 in Appendix B.

1. Class width is 10 units of systolic blood pressure of men. This is the difference between two consecutive lower or upper class boundaries such as 100-90 for the two lower class limits of the two lowest intervals or 159-149= 10 for the two upper limits of the two highest intervals.
Class midpoints are the middle points of the intervals. For each interval, they are determined by finding the average of the upper and lower limits of the interval For the 90 – 99 interval, the midpoint is (90 + 99)/2 = 189/2 = 94.5. For this example, the midpoints are: 94.5, 104.5, 114.5, 124.5, 134.5, 144.5, and 155.5. Note that the difference between adjacent midpoints is the class width of 10.
Class boundaries represent the point between each interval. These are determined by finding the average of the upper class limit of the lower interval and the lower class limit of the higher interval. In this example, the class boundary between the 90-99 and 100-109 intervals would be (99 + 100)/2= 199/2= 99.5 and the series of class boundaries for this example are: 89.5, 99.5, 109.5, 119.5, 129.5, 139.5, 149.5, and 159.5. Note that all the scores fall in the range of the lower class boundary of the lowest interval and the higher class interval of the highest interval (89.5 to 159.5) and they are all 10 (the class width) apart.

Systolic Blood Pressure of Men	Class Boundaries	Class Midpoints	Frequency
90 – 99	89.5 – 99.5	94.5	1
100 – 109	99.5 – 109.5	104.5	4
110 – 119	109.5 – 119.5	114.5	17
120 – 129	119.5 – 129.5	124.5	12
130 – 139	129.5 – 139.5	134.5	5
140 – 149	139.5 – 149.5	144.5	0
150 – 159	149.5 – 159.5	155.5	1

3. Class width is 200 units of cholesterol of men. This is the difference between two consecutive lower or upper class boundaries such as 0-200 for the two lower class limits of the two lowest intervals or 1200-1000= 200 for the two upper limits of the two highest intervals.
Class midpoints are the middle points of the intervals. For each interval, they are determined by finding the average of the upper and lower limits of the interval. For the 0 – 199 interval, the midpoint is (0 + 199)/2 = 199/2 = 99.5. For this example, the midpoints are: 99.5, 299.5, 499.5, 699.5, 899.5, 1099.5 and 1299.5. Note that the difference between adjacent midpoints is the class width of 200.
Class boundaries represent the point between each interval. These are determined by finding the average of the upper class limit of the lower interval and the lower class limit of the higher interval. In this example, the class boundary between the 0 – 199 and 200 – 399 intervals would be (199 + 200)/2= 399/2= 199.5 and the series of class boundaries for this example are: -0.5, 199.5, 399.5, 599.5, 799.5, 999.5, 1199.5, and 1399.5. Note that all the scores fall in the range of the lower class boundary of the lowest interval and the higher class interval of the highest interval (199.5 to 1399.5) and they are all 200 units of cholesterol (the class width) apart.

Cholesterol of Men	Class Boundaries	Class Midpoints	Frequency
0-199	0.5 – 199.5	99.5	13
200-399	199.5 – 399.5	299.5	11
400-599	399.5 – 599.5	499.5	5
600-799	599.5 – 799.5	699.5	8
800-999	799.5 – 999.5	899.5	2
1000-1199	999.5 – 1199.5	1099.5	0
1200-1399	1199.5 – 1399.5	1299.5	1

In exercises 5 & 7, construct the relative frequency distributions that correspond to the frequency distribution in the exercise indicated.

The relative frequency is the proportion, usually converted to a percent, of scores in each interval, the equation to be used is:

$$\text{Relative frequency} = \frac{\text{Class frequency}}{\text{Sum of all frequencies}}$$

5. For **Systolic Blood Pressure of Men** (Exercise 1), the sum of frequencies (n) is 40
For the 90 – 99 interval, relative frequency = 1/40 * 100% = 0.025 * 100%= 2.5%
For the 120 – 129 interval, relative frequency = 12/40 * 100% = 0.30 * 100%= 30.0%

Systolic Blood Pressure of Men	Frequency	Relative Frequency
90 – 99	1	2.5%
100 – 109	4	10.0%
110 – 119	17	42.5%
120 – 129	12	30.0%
130 – 139	5	12.5%
140 – 149	0	0.0%
150 – 159	1	2.5%

7. For **Cholesterol of Men** (Exercise 3), the sum of frequencies (n) is 40.
For the 200 – 399 interval, relative frequency is 11/40 * 100% = 0.275 * 100%= 27.5%
For the 1200 – 1399 interval, relative frequency is 1/40 * 100% = 0.025 * 100= 2.50%

Cholesterol of Men	Frequency	Relative Frequency
0 – 199	13	32.5%
200 – 399	11	27.5%
400 – 599	5	12.5%
600 – 799	8	20.0%
800 – 999	2	5.0%
1000 – 1199	0	0.0%
1200 – 1399	1	2.5%

In exercises 9 & 11, construct the cumulative frequency distribution that corresponds to the frequency distribution in the exercise indicated.

The cumulative frequency in each interval is the sum of the frequencies in the interval and all of the frequencies in the intervals below (in score terms) that interval. Cumulative frequency in the highest score interval must equal the sum of all frequencies (n).

9 For **Systolic Blood Pressure of Men** (Exercise 5)

Systolic Blood Pressure of Men	Frequency	Working column (not in final table) f in interval + cumulative f up to interval	Cumulative Frequency
90-99	1	1 + 0 = 1	1
100-109	4	4 + 1 = 5	5
110-119	17	17 + 5 = 22	22
120-129	12	22 + 12 = 34	34
130-139	5	5 + 34 = 39	39
140-149	0	0 + 39 = 39	39
150-159	1	1 + 39 = 40	40

11 For **Cholesterol of Men** (Exercise 7)

Cholesterol of Men	Frequency	Working column (not in final table) f in interval + cumulative f up to interval	Cumulative Frequency
0-199	13	13 + 0 = 13	13
200-399	11	11 + 13 = 24	24
400-599	5	5 + 24 = 29	29
600-799	8	8 + 29 = 37	37
800-999	2	2 + 37 = 39	39
1000-1199	0	0 + 39 = 39	39
1200-1399	1	1 + 39 = 40	40

13. **Bears** Frequency Distribution of Weight of Bears in Pounds (From Data Set 6)

Weight of Bears in Pounds	Frequency
0 – 49	6
50 – 99	10
100 –149	10
150 –199	7
200 – 249	8
250 – 299	2
301 – 349	4
350 – 399	3
400 – 449	3
450 – 499	0
500 – 549	1

15. **Head Circumferences** in cm. of Two-Month-Old Babies (From Data Set 4)

Head Circumference in cm.	Frequency of Males	Frequency of Females
34.0 – 35.9	2	1
36.0 – 37.9	0	3
38.0 – 39.9	5	14
40.0 – 41.9	29	27
42.0 – 43.9	14	5

From the comparison of the two frequency distributions, it seems that the head circumference for the male babies is higher than for the female babies. However, we cannot conclude that the difference is statistically significant until we conduct specific tests to test these differences.

17. Yeast Cell Counts (From Data Set 11)

Yeast Cell Counts	Frequency
1	20
2	43
3	53
4	86
5	70
6	54
7	37
8	18
9	10
10	5
11	2
12	2

The most common cell count is 4 with a frequency of 86.

2 – 3 Visualizing Data

In Exercises 1 & 3, answer the questions by referring to the SPSS-generated histogram given below The histogram represents the lengths (mm) of cuckoo eggs found in the nests of other birds (based on data from O.M. Latter and Data and Story Library).

1. **Center** Center is a very general term, but basically it looks for a central value that could be used to represent the entire distribution. One value that could be used is the center of the scale which would be at the average of the low (19.75) and high (24.76) points, which would be (19.75 + 24.76)/2= 22.25 mm of egg length. Another possible value for center would be the value that occurs most often which would be taken as the midpoint of the 22.0 interval. One statistic that is reported in the table is the mean of 22.46 mm. This is one of the descriptive statistics that is often used to represent the center of a continuous distribution.

3. **Percentage** Note the midpoints of the intervals are not all labeled. There are three intervals that have midpoints of 20.75, 21.00, and 21.25 (21.00 is not printed, but it is there). The lower and upper limits of these three intervals are: 20.625 – 20.875, 20.875 – 21.125, and 21.125 – 21.37521.125 is the upper limit of the 20.875 to 21.125 interval. Therefore, all of the eggs in that interval and below are less than 21.125 mm. Counting them from the start up though that interval, we have 2 + 2 + 1 + 0 + 6 + 6 = 17 and that equates to a percentage of 17/120 (100%) = 14.2%.

In Exercise 5, refer to the accompanying pie chart of blood groups for a large sample of people (based on data from the Greater New York Blood Program).

5. **Interpreting Pie Chart** It would appear that about 40% of the observed group has Group A blood and out of 500 in the sample, this would be about 200 people.

7. **Bears** Histogram of Weights in Pounds

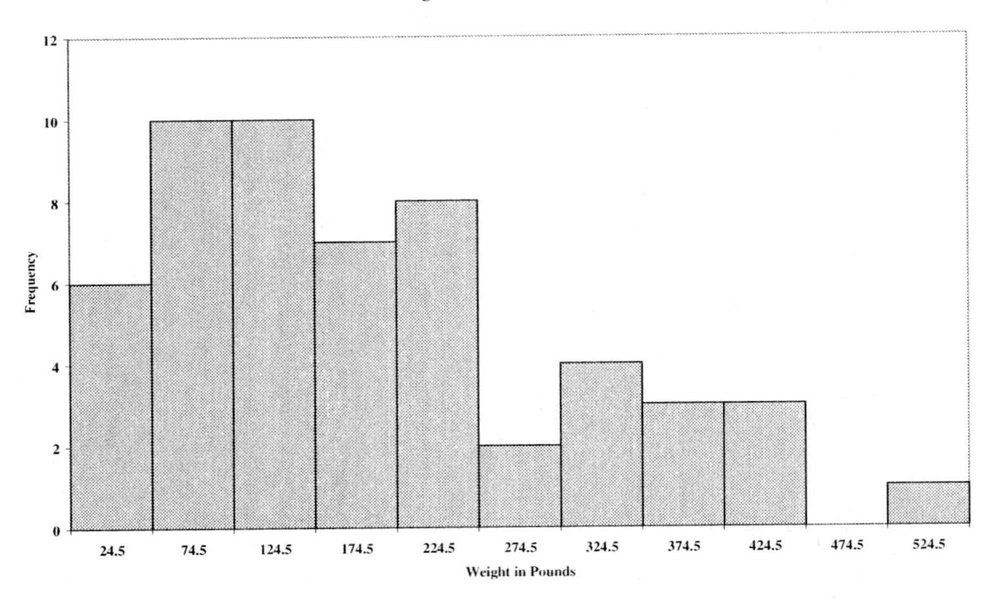

The approximate weight at the center seems to be between 174.5 and 224.5 or about 200 pounds. We could compute the mean and find it is 183 pounds, which would be a useful measure of centrality of the distribution.

9. **Yeast Cell Counts** Histogram (Data Set 11)

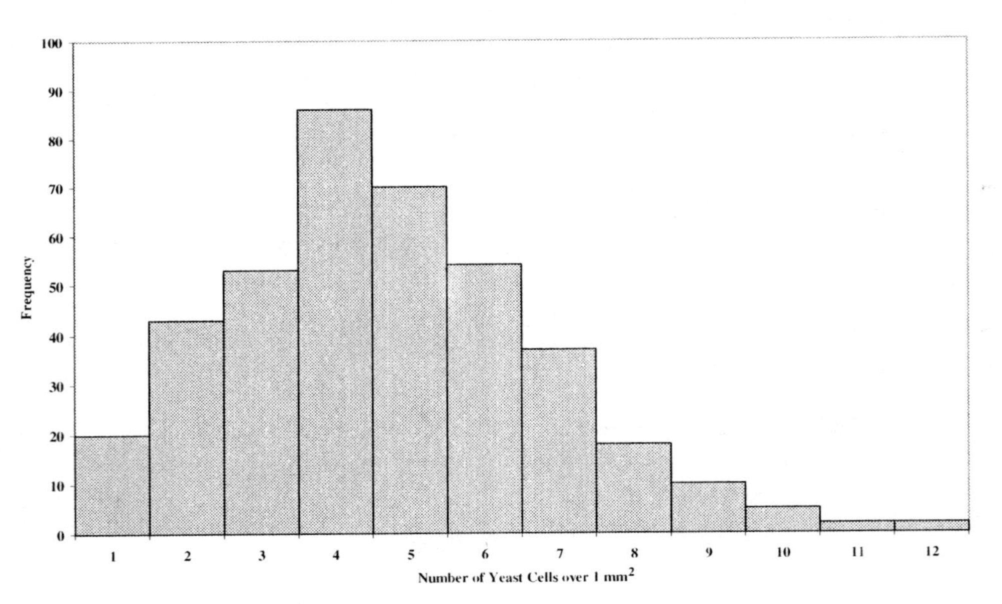

The distribution seems to depart somewhat from being bell-shaped It seems to have a reasonable degree of positive skewness. Other specific measures can be used to assess the exact degree of skewness.

In Exercise 11, use graphs to compare the two indicated data sets.

11. Poplar Trees Histogram comparing Heights (in meters) of Control Group and Irrigation Only Trees (From Data Set 9)
The two distributions do not seem to have much departure from each other and this would not seem to be enough for us to conclude that the distributions are significantly different. There is no sound evidence to conclude that the irrigation treatment had any effect on poplar tree height.

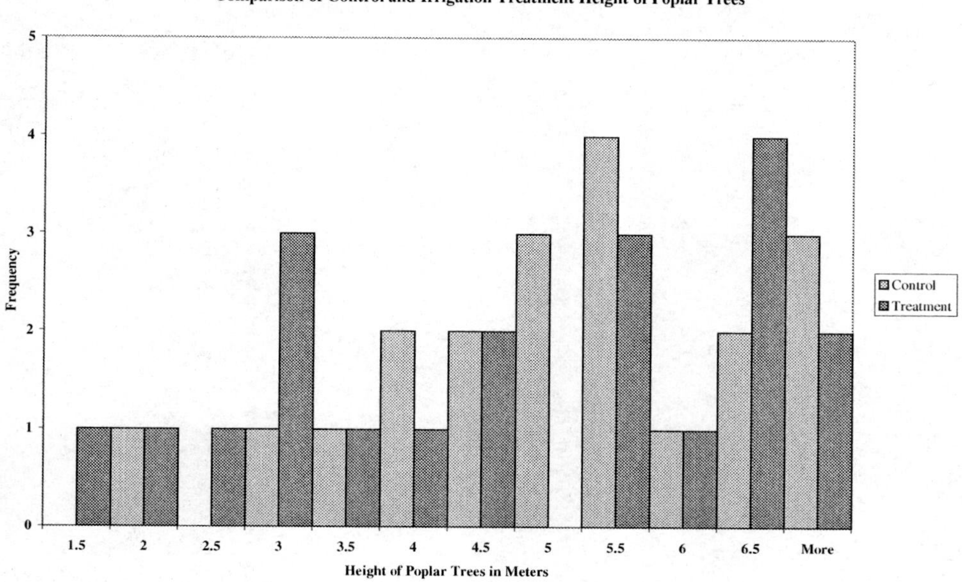

In Exercise 13, list the original data represented by the given stem-and-leaf plots.

13. To find the original scores, the leaf in units of tens and ones is added to the stem which is in units of hundreds.
5012 5012 5012 5055 5200 5200 5200 5200 5327 5327 5335 5472

In Exercise 15, construct the dotplot for the data represented by the stem-and-leaf plot in the given exercise.

15. Dotplot of Exercise 13 Data

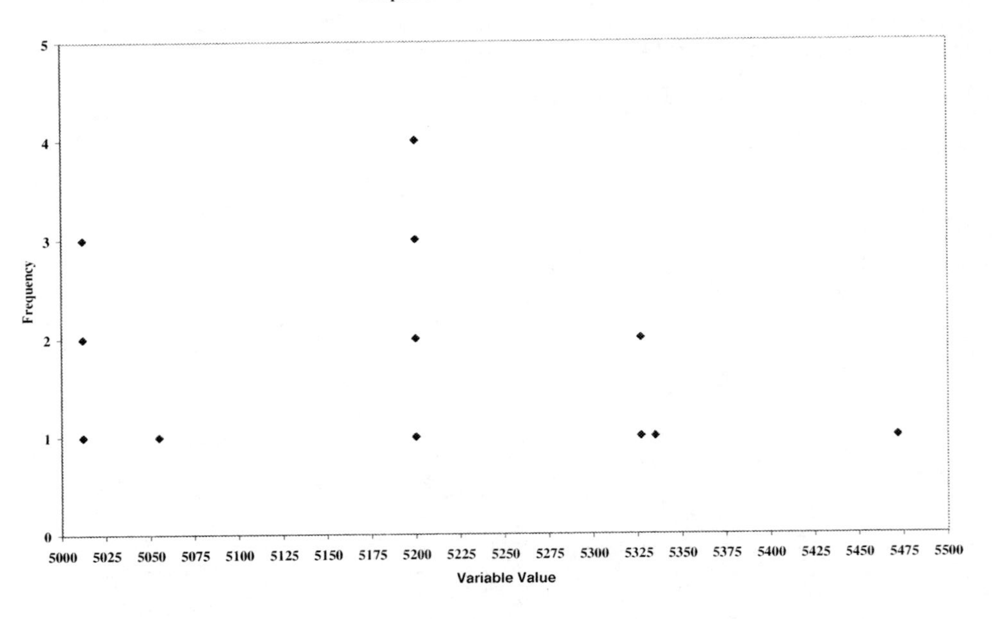

In Exercises 17 & 19, use the given paired data from Appendix B to construct a scatter diagram.

17. Parent/Child Height (Data Set 3)
To determine a relationship graphically, we will use a scatter diagram with mother's height on the horizontal (x-axis) scale and the child's height on the vertical (y-axis) scale.

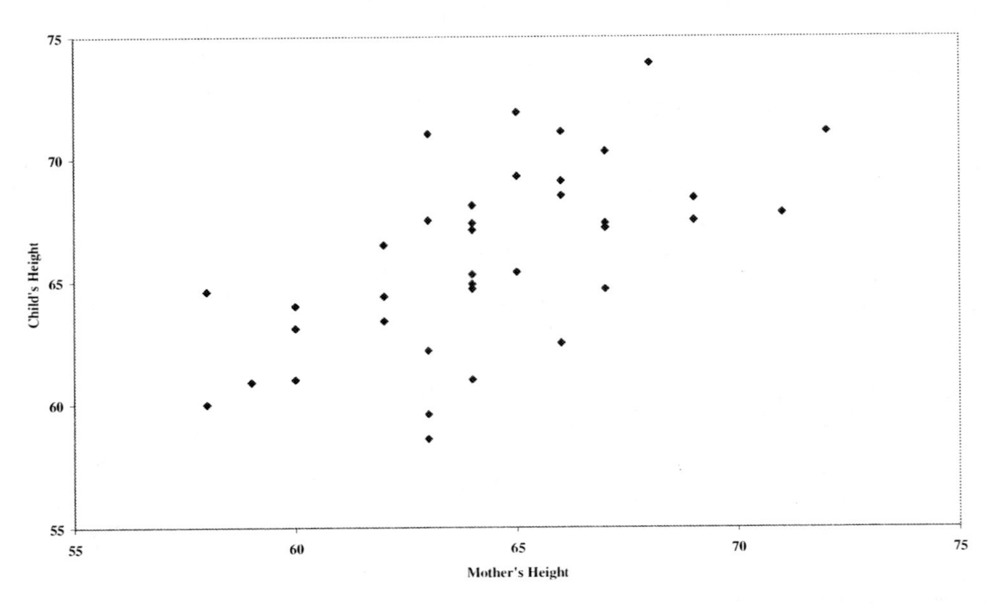

There appears to be a relationship and the nature of the relationship is that low heights on both cluster together and high heights on both cluster together or another way of saying this is that as height on one increases we observe height on the other increasing, also called a direct or positive relationship.

19. Sepal/Petal Length (Data Set 7)

Scatter Diagram of Setosa Class Sepal and Petal Lengths

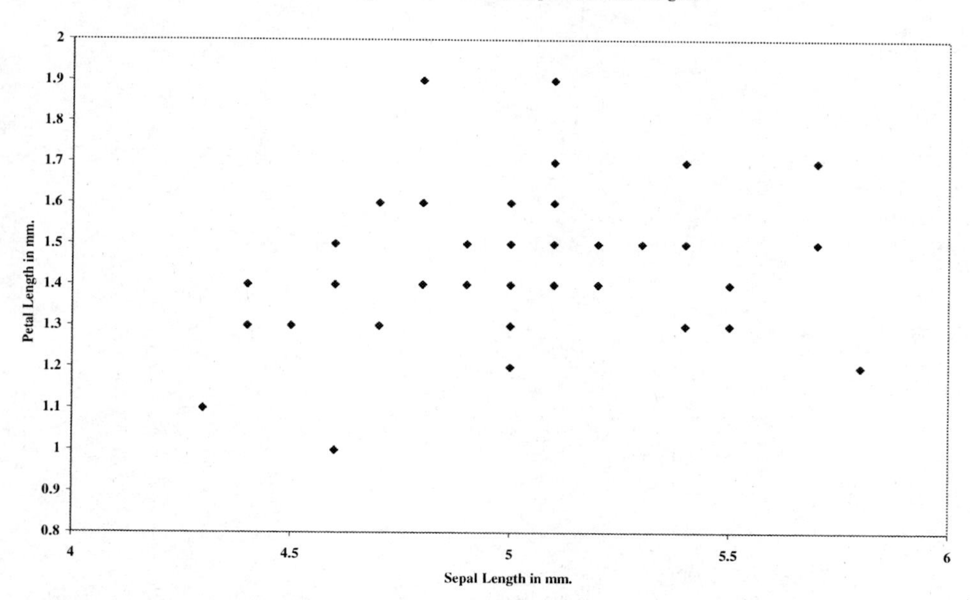

There does not appear to be a relationship between sepal length and petal length for the Setosa Class. As one of the variables changes in a given direction, there seems to be no corresponding change in the other variable in either direction.

2 – 4 Measures of Center

In Exercises 1 – 7, find the (a) mean, (b) median, (c) mode, and (d) midrange for the given sample data.

The four measures of center used below are:

- **a.** the mean ($\bar{x}$) which we find using the following equation:
  Mean: $\bar{x} =$ x/n where x represents the sum of scores
- **b.** the median ($\tilde{x}$) which is determined with the following rules:
  If there are an odd number of scores, the median is the score that is exactly in the middle of the ordered set of scores.
  If there are an even number of scores, the median is the point exactly between or the average of the two middle scores.
- **c.** the mode, which is the score that occurs most often; if there are two that occur most often that are tied, the distribution is bimodal and two modes are reported; if there are three that occur most often that are tied, the distribution is trimodal and three modes are reported; if no score occurs more often than any others or if there are more than three that occur most often and they are tied, no mode is reported.
- **d.** the midrange is the point exactly between or the average of the lowest and highest scores

1. Tobacco Use in Children's Movies Arranged in order, the scores are:

#	1	2	3	4	5	6
Score	0	0	0	176	223	548

a. Mean: $\bar{x} = \quad x/n = 947/6 = 157.8$

b. Median: $\tilde{x}$ is the midpoint between the two middle scores (3rd and 4th scores) if an even number of scores This would be (0 + 176)/2= 176/2= 88.0

c. Mode: Mode is score that occurs most often, this is 0 for these data

d. Midrange: Midpoint (or average) between lowest and highest scores, (548 + 0)/2 = 548/2= 274.0

3. Body Mass Index Arranged in order, the scores are:

#	1	2	3	4	5	6	7	8	9	10	11	12
Score	17.7	19.6	19.6	20.6	21.4	22.0	23.8	24.0	25.2	27.5	28.9	29.1
#	13	14	15									
Score	29.9	33.5	37.7									

a. Mean: $\bar{x} = \quad x/n = 380.5/15 = 25.37$

b. Median: $\tilde{x}$ is the middle score (8th score) if an odd number of scores This would be 24.0

c. Mode: Mode is score that occurs most often, the only score that occurs more than once is 19.6 so that is the mode

d. Midrange: Midpoint (or average) between lowest and highest scores, (37.7 + 17.7)/2 = 20.0/2= 27.7
Is the mean of this sample reasonably close to the mean of 25.74, which is the mean for all 40 women included in Data Set 1? Yes, they are very close, especially when on considers the range of values to be from about 17 to 38.

5. **Motorcycle Fatalities** Arranged in order, the scores are:

#	1	2	3	4	5	6	7	8	9	10	11	12	13	14	15
Score	14	16	17	18	20	21	23	24	25	27	28	30	31	34	37
#	16	17	18												
Score	38	40	42												

a. Mean: $\bar{x} = \quad x/n = 485/18 = 26.9$

b. Median: $\tilde{x}$ is the average of the middle scores (9th and 10th scores) if an even number of scores This would be (25 + 27)/2= 52/2= 26.0

c. Mode: Mode is score that occurs most often. In this case no score occurs more than once so there is no mode

d. Midrange: Midpoint (or average) between lowest and highest scores, (14 + 42)/2 = 56/2= 28.0

Do the results support the common belief that such fatalities are incurred by a greater proportion of younger drivers? Following is a histogram of the frequencies:

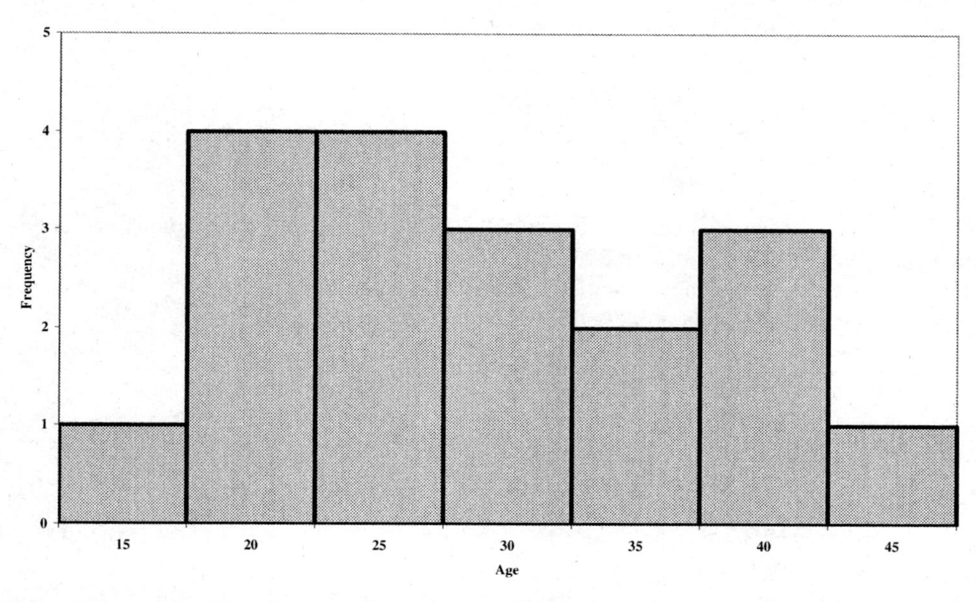

In the range of ages of fatalities for these data, the distribution is fairly even from ages 14 to 42, with a slightly higher distribution toward the lower range, but not such a distinct difference that would warrant concluding that such fatalities are incurred by a greater proportion of younger drivers. It is difficult to answer this question since we don't know and might reasonable question whether there would be an equal proportion of motorcycle riders across all the age groups. Also, this data set is relatively small to permit making such a conclusion.

7. Blood Pressure Measurements Arranged in order, the scores are:

#	1	2	3	4	5	6	7	8	9	10	11	12	13	14
Score	120	120	125	130	130	130	130	135	138	140	140	143	144	150

a. Mean: $\bar{x} = \quad x/n = 1875/14 = 133.9$

b. Median: $\tilde{x}$ is the midpoint between the two middle scores (7th and 8th scores) if an even number of scores This would be (130 + 135)/2= 265/2= 132.5

c. Mode: Mode is score that occurs most often. In this case the mode is 130 since that occurs most often (4 times)

d. Midrange: Midpoint (or average) between lowest and highest scores, (120 + 150)/2 = 270/2= 135
What is notable about this data set? The measures of centrality are relatively close, which would indicate the distribution seems relatively symmetrical.

In Exercises 9 & 11, find the mean, median, mode, and midrange for each of the samples then compare the two sets of results.

9. Patient Waiting Times
Single Line Central Statistics: Arranged in order, the scores are:

#	1	2	3	4	5	6	7	8	9	10
Score	65	66	67	68	71	73	74	77	77	77

a. Mean: $\bar{x} = \quad x/n = 715/10 = 71.5$

b. Median: $\tilde{x}$ is the midpoint between the two middle scores (5th and 6th scores) if an even number of scores This would be (71 + 73)/2= 144/2= 72.0

c. Mode: Mode is score that occurs most often. In this case the mode is 77 since that occurs most often

d. Midrange: Midpoint (or average) between lowest and highest scores, (65 + 77)/2 = 142/2= 71.0

Multiple Line Central Statistics: Arranged in order, the scores are:

#	1	2	3	4	5	6	7	8	9	10
Score	42	54	58	62	67	77	77	85	93	100

a Mean: $\bar{x} = \quad x/n = 715/10 = 71.5$

b Median: $\tilde{x}$ is the midpoint between the two middle scores (5th and 6th scores) if an even number of scores. This would be (67 + 77)/2= 144/2= 72.0

c Mode: Mode is score that occurs most often. In this case the mode is 77 since that occurs most often

d Midrange: Midpoint (or average) between lowest and highest scores, (42 + 100)/2 = 142/2= 71

Comparison of the two distributions relative to central measures: The central measures are exactly the same for both groups. However, there is more spread of the scores from each other for the multiple line group (42 to 100) compared with the single line group (65 to 77).

11. Poplar Trees

Control Group Central Statistics: Arranged in order, the scores are:

#	1	2	3	4	5	6	7	8	9	10	11	12	13	14	15
Score	1.9	2.9	3.2	3.6	3.9	4.1	4.1	4.6	4.8	4.9	5.1	5.5	5.5	5.5	6.0
#	16	17	18	19	20										
Score	6.3	6.5	6.8	6.9	6.9										

a Mean: $\bar{x} = \quad x/n = 99.0/20 = 4.95$

b Median: $\tilde{x}$ is the midpoint between the two middle scores (10th and 11th scores) if an even number of scores. This would be (4.9 + 5.1)/2= 10.0/2= 5.0

c Mode: Mode is score that occurs most often. In this case, 5.5 occurs more than any other score (3 times) so the mode is 5.5

d Midrange: Midpoint (or average) between lowest and highest scores, (1.9 + 6.9)/2 = 8.8/2= 4.4

Irrigation Treatment Group Central Statistics: Arranged in order, the scores are:

#	1	2	3	4	5	6	7	8	9	10	11	12	13	14	15
Score	1.2	1.6	2.5	2.8	2.9	2.9	3.3	3.6	4.1	4.5	5.1	5.1	5.5	5.8	6.1
#	16	17	18	19	20										
Score	6.1	6.4	6.5	6.8	6.8										

a Mean: $\bar{x} = \quad x/n = 89.6/20 = 4.48$

b Median: $\tilde{x}$ is the midpoint between the two middle scores (10th and 11th scores) if an even number of scores. This would be (4.5 + 5.1)/2= 9.6/2= 4.80

c Mode: Mode is score that occurs most often. In this case, 4 scores occur the same number of times (twice) so the distribution has four modes with modes at: 2.9, 5.1, 6.1 and 6.8 occurs more than any other score (3 times) so there is no distinct mode

d Midrange: Midpoint (or average) between lowest and highest scores, (1.2 + 6.8)/2 = 8.0/2= 4.0

Comparison of the two distributions relative to central measures: The central measures are slightly higher on the mean and the median for the control group on the heights (in meters) of the trees.

In Exercises 13 & 15, refer to the data set in Appendix B. Use computer software or a calculator to find the means and medians, then compare the results as indicated.

13. Head Circumference Comparison head circumference in cm. of males and females on central measures (Data Set 4)

Statistic	Males	Females
Mean	41.10	40.05
Median	41.10	40.20

There does appear to be a slight difference since both central measures are higher for the males than for the females with males having a 1.05 cm. higher mean circumference.

15. Petal Lengths of Irises Comparison of petal lengths of the three classes (Data Set 7)

Statistic	Setosa	Versicolor	Virginica
Mean	1.46	4.26	5.55
Median	1.50	4.35	5.55

No, they do not appear to have the same petal lengths. It is very clear that there are differences in central measures among these three classes of Iris petal lengths. Petal length is much higher for the Versicolor and Virginica classes compared with the Setosa class while the length of the Virginica class seems to be higher than the length of petals for the Versicolor class They may all be different from each other, but this would need to be tested with more precise methods to decide

In Exercise 17, find the mean of the data summarized in the given frequency distribution.

17. Mean from a Frequency Distribution The following frequency distribution will be used to find the grouped mean:

Frequency Distribution of Cotinine Levels of Smokers

Serum Cotinine Level (mg/ml)	Frequency (f)	Interval Midpoint (x)	$f * x$
0 – 99	11	49.5	544.5
100 – 199	12	149.5	1794.0
200 – 299	14	249.5	3493.0
300 – 399	1	349.5	349.5
400 - 499	2	449.5	899.0
Sum	$\Sigma f = 40$		$\Sigma(f * x) = 7080.0$

$$\text{Grouped Mean} = \frac{(f * x)}{f} = \frac{7080.0}{40} = 177.0$$

The mean computed using the actual raw values was 172.5.

The Grouped Mean is close to the actual mean. The grouped mean is an estimate of the actual mean. With the availability of computing equipment, the actual mean would be easy to compute and would always be preferred. However, if the raw scores were not available, this procedure could provide a reasonable estimate.

19. Mean of Means No, this process would not result in an accurate computation of the mean salary of physicians for the country. This process would provide a single mean entry per state in such a way that it would assume the same number of physicians were in each state. Of course this is not true. There would be many more physicians in states like New York and California compared with states like New Hampshire and Wyoming. The actual number of physicians within each state would need to be accounted for in this computation.

21. Censored Data The mean of the five values, including the one that is censored, would be 3.42. Since the two values of 5 are minimal values and their actual values would be higher than 5, we would conclude that the mean would be 3.42 or higher.

2 – 5 Measures of Variation

In Exercises 1 – 7, find the range, variance, and standard deviation for the given sample data (The same data were used in Section 2 – 4 where we found measures of center Here we find measures of variation).

In the following computations of variance and standard deviation, the equation used to compute the variance includes the following terms:

n represents the sample size

x represents the sum of the scores (the scores are added together)

$(\ x)^2$ represents the squared value for the sum of scores (the scores are added together and then that value is squared)

x^2 represents the sum of the squared scores (each score is squared and then these are all added together)

1. Tobacco Use in Children's Movies

a. Range = highest score – lowest score = 548.0 – 0 = 548.0

b. Variance $s^2 = \dfrac{n(\ x^2)-(\ x)^2}{n(n-1)} = \dfrac{6(381009)-(947)^2}{6(6-1)} = \dfrac{1389245}{30} = 46308.2$

c. Standard Deviation $s = \sqrt{s^2} = \sqrt{46308.2} = 215.2$

3. Body Mass Index

a. Range = highest score – lowest score = 37.7 – 17.7 = 20.0

b. Variance $s^2 = \dfrac{n(\ x^2)-(\ x)^2}{n(n-1)} = \dfrac{15(10101.23)-(380.5)^2}{15(15-1)} = \dfrac{6738.20}{210} = 32.09$

c. Standard Deviation $s = \sqrt{s^2} = \sqrt{32.0867} = 5.66$

Would a body mass index of 34.0 be considered "unusual"?
Need to use the mean. From before it is 25.37 (Exercise 4.3).
A body mass index of 34 is not lower than the minimum usual value of
mean – 2 s or 25.37 – 2 (5.66) = 25.37 – 11.32 = 14.04,
AND a body max index of 34 is not higher than the maximum usual value of
mean + 2 s or 25.37 + 2 (5.66) = 25.37 + 11.32 = 36.69
A body mass index of 34 is between the usual minimum and usual maximum values of 20.68 and 36.69. Thus, a body mass of 34 would not be considered unusual.

5. Motorcycle Fatalities

a. Range = highest score – lowest score = 42 – 14 = 28.0

b. Variance $s^2 = \dfrac{n(\ x^2)-(\ x)^2}{n(n-1)} = \dfrac{18(14343)-(485)^2}{18(18-1)} = \dfrac{22949}{306} = 75.00$

c. Standard Deviation $s = \sqrt{s^2} = \sqrt{75.00} = 8.66$

How does the variation of these ages compare to the variation of ages of all licensed drivers in the general population? The variation of this group will be lower since the range of 14 – 42 is lower than the typical range of the age of licensed drivers which would probably be from 15 to over 80. One would expect the variance and standard deviation of this group to be lower as well.

7. Blood Pressure Measurements

a. Range = highest score – lowest score = 150 – 120 = 30.0

b. Variance $s^2 = \dfrac{n(\ x^2)-(\ x)^2}{n(n-1)} = \dfrac{14(252179)-(1875)^2}{14(14-1)} = \dfrac{14881}{182} = 81.76$

c. Standard Deviation $s = \sqrt{s^2} = \sqrt{81.76} = 9.04$

What does this suggest about the accuracy of the readings? They do not appear to be very accurate when one considers that all of these measures were taken on the same patient.

In Exercises 9 & 11, find the range, variance, and standard deviation for each of the two samples, then compare the two sets of results. (The same data were used in Section 2 – 4.)

9. Patient Waiting Times

Single line:

a. Range = highest score – lowest score = 77 – 65 = 12.0

b. Variance $s^2 = \dfrac{n(\ x^2)-(\ x)^2}{n(n-1)} = \dfrac{10(51327)-(715)^2}{10(10-1)} = \dfrac{2045}{90} = 22.72$

c. Standard Deviation $s = \sqrt{s^2} = \sqrt{22.72} = 4.77$

Multiple line:

a. Range = highest score – lowest score = 100 – 42 = 58.0

b. Variance $s^2 = \dfrac{n(\ x^2)-(\ x)^2}{n(n-1)} = \dfrac{10(54109)-(715)^2}{10(10-1)} = \dfrac{29865}{90} = 331.83$

c. Standard Deviation $s = \sqrt{s^2} = \sqrt{331.83} = 18.22$

Compare the results: There is clearly higher variation for the multiple line as compared with the single line.

11. Poplar Trees

Control Group:

a. Range = highest score – lowest score = 6.9 – 1.9 = 5.0

b. Variance $s^2 = \dfrac{n(\ x^2)-(\ x)^2}{n(n-1)} = \dfrac{20(528.42)-(99)^2}{20(20-1)} = \dfrac{767.40}{380} = 2.019$

c. Standard Deviation $s = \sqrt{s^2} = \sqrt{2.019} = 1.42$

Irrigation Group:

a. Range = highest score – lowest score = 6.8 – 1.2 = 5.6

b. Variance $s^2 = \dfrac{n(\ x^2)-(\ x)^2}{n(n-1)} = \dfrac{20(461.84)-(89.6)^2}{20(20-1)} = \dfrac{1208.64}{380} = 3.181$

c. Standard Deviation $s = \sqrt{s^2} = \sqrt{3.181} = 1.78$

Compare the results. There is slightly higher variation in the heights of the Irrigation Group (s^2= 3.18) compared with the Control Group (s^2= 2.02).

In Exercises 13 & 15, refer to the data set in Appendix B. Use computer software or a calculator to find the standard deviations, then compare the results.

13. Head Circumference

s_F = 1.64 s_M = 1.50

There does not appear to be a substantial difference in the variation.

15. Petal Length of Irises

$s_{Setosa} = 0.174$ $s_{Versicolor} = 0.470$ $s_{Virginica} = 0.552$

No, they do not appear to have the same variation of petal lengths. It is very clear that there are differences in variances among these three classes of Iris petal lengths. Petal length is much more variable for the Versicolor and Virginica classes when compared with the Setosa class while the variation for the Virginica class seems to be higher than the variation of length of petals for the Versicolor class.

17. Finding Standard Deviation from a Frequency Distribution

Frequency Distribution of Cotinine Levels of Smokers

Serum Cotinine Level (mg/ml)	Frequency (f)	Interval Midpoint (x)	Interval Midpoint Squared (x^2)	$f * x$	$f * x^2$
0 – 99	11	49.5	2450.25	544.5	26952.75
100 – 199	12	149.5	22350.25	1794.0	268203.00
200 – 299	14	249.5	62250.25	3493.0	871503.50
300 – 399	1	349.5	122150.25	349.5	122150.25
400 – 499	2	449.5	202050.25	899.0	404100.50
Sum	$\Sigma f = 40$	---	---	$\Sigma(f * x) = 7080.0$	$\Sigma(f * x^2) = 1692910.00$

Find the variance first and then the standard deviation

$$s^2 = \frac{n[\ (f * x^2)] - [(\ (f * x)^2]}{n(n-1)} = \frac{40(1692910.00) - (7080)^2}{40(40-1)} = \frac{17590000}{1560} = 11275.64$$

Standard Deviation $s = \sqrt{s^2} = \sqrt{11275.64} = 106.2$

Comparison: In this case, the standard deviation (106.2) is lower when computed as a grouped standard deviation compared with when computed with the original 40 scores (119.5). We make the assumption that the midpoint of the interval is representative of all the scores within the interval. This is not likely to be exactly the case so to the extent that it is not, there will be errors in the estimate and those errors could result in estimates that are lower or higher than the ungrouped estimate. The ungrouped estimate should be used whenever possible, but if the original scores are not available, yet the grouped frequency distribution is, this approach could be used to get an estimate.

19. Range Rule of Thumb Estimate the standard deviation of all faculty member ages at your college. I work at the University of Alabama at Birmingham. I would estimate the youngest faculty member to be about 25 and the oldest to be about 75. The range is 50 years. Using the range rule of thumb, the estimated standard deviation would be:

$$s \approx \frac{Range}{4} = \frac{50}{4} = 12.5 \text{ years of age}$$

21. Empirical Rule Mean of 176 cm and standard deviation of 7 cm in the population.

Approximate percentage of men between:

a. 169 cm and 183 cm

169 cm is one standard deviation (176 – 7= 169) below the mean and 183 cm is one standard deviation (176 + 7= 173) above the mean The percentage of scores between one standard deviation below the mean and one standard deviation above the mean ($\overline{x} \pm 1s$) is 68% in a bell-shaped distribution.

b. 155 cm and 197 cm
155 cm is three standard deviations below the mean [176 – (3 * 7) = 155] and 197 cm is three standard deviations above the mean [176 + (3 * 7) = 197] The percentage of scores between three standard deviation below the mean and three standard deviation above the mean ($\bar{x} \pm 3s$) is 99.7% in a bell-shaped distribution.

23. Coefficient of Variation
Heights of men: $n = 6$ $\bar{x} = 69.17$ in $s = 2.14$ in

$$CV = \frac{s}{\bar{x}} \bullet 100\% = \frac{2.14}{69.17} \bullet 100\% = 3.09\%$$

Lengths (mm) of cuckoo eggs: $n = 9$ $\bar{x} = 22.14$ mm $s = 1.13$ mm

$$CV = \frac{s}{\bar{x}} \bullet 100\% = \frac{1.13}{22.14} \bullet 100\% = 5.10\%$$

Comparison: The relative variation of the cuckoo eggs lengths is greater by a factor of about 1.7 times, than the relative variation of the men's heights, but this does not appear to be a substantial difference.

25. Understanding Units of Measurement Data consisting of longevity times (in days) of fruit flies
Units of standard deviation are days while units of variance are in days squared. This is the primary reason we convert variance to standard deviation since finding the square root also affects the unit of measurement and this puts our measure of variability back on the same continuum as the scores. It's hard for us to comprehend squared days, squared blood pressure, squared minutes, etc.

27. Why Divide by n – 1? Let the population be: 3, 6, 9
Population Parameters: $n = 3$ $\mu = 6$ $\sigma^2 = 6.0$ $\sigma = 2.45$

a. Population variance (using n rather than n – 1) for denominator, $\sigma^2 = 6.00$
Nine possible samples of $n = 2$ and their variances as samples:

b.

Possible samples of size 2, with replacement*	Sample Mean	**b.** Sample Variances Division by(n – 1)	**c.** Population Variances Division by n	Sample Standard Deviation
3,3	3.0	0.0	0.00	0.00
3,6	4.5	4.5	2.25	2.12
3,9	6.0	18.0	9.00	4.24
6,3	4.5	4.5	2.25	2.12
6,6	6.0	0.0	0.00	0.00
6,9	7.5	4.5	2.25	2.12
9,3	6.0	18.0	9.00	4.24
9,6	7.5	4.5	2.25	2.12
9,9	9.0	0.0	0.00	0.00
Mean	6.00	6.00	3.00	1.88

With replacement is a critical aspect of sampling. When we pull a score from a distribution, we record it and then return it back to the population for the next sample or when we pull out a sample we record the sample values and then return them back to the population before selecting the next sample. The reason this is done is so that every sampling is done from the same population. Otherwise, we could not have had the samples with scores 3 and 3, 6 and 6, and 9 and 9.

d. The average variance for those samples where n – 1 was used exactly matched the population variance. Thus, the sample variance, on average, is an unbiased estimator of the population variance. An unbiased estimator is one where the average of the sampled statistics will be the population parameter it is designed to estimate. However, notice that the variation is higher for the n – 1 situation. This difference would decrease as the sample size increases.

e. If we find the standard deviation of the sample variances (column e) and then find their average of 1.88, we see that this average is not the same as the population standard deviation of 2.45. Thus, the sample standard deviation is not an unbiased estimator of the population standard deviation. In this case, we would say that the sample standard deviation is negatively biased meaning that the average estimator is lower than the parameter it is being used to estimate.

2 – 6 Measures of Relative Standing

In Exercises 1 & 3, express all z scores with two decimal places

1 **Darwin's Height** Darwin's height was 182 cm, population values: $\mu = 176$ cm, $\sigma = 7$ cm

a. Difference between Darwin's height and mean $x = 182 - 176 = 6$ cm

b. Number of standard deviations is 6/7 = 0.86

c. z score $z = \frac{x-\mu}{\sigma} = \frac{182-176}{7} = 0.86$

d. Usual heights are between -2 and +2 z values Is Darwin's height usual or unusual?
Darwin's z value of +0.86 is between -2 and +2 z values, so his height would be considered <u>usual.</u>

3 **Pulse Rates of Adults** Population pulse rate is: $\mu = 72.9$ bpm, $\sigma = 12.3$ bpm

a. Difference between pulse rate of 48 and mean, $x = 48 - 72.9 = -24.9$

b. How many standard deviations is that? $-24.9/12.3 = -2.02$

c. z score $z = \frac{x-\mu}{\sigma} = \frac{48-72.9}{12.3} = \frac{-24.9}{12.3} = -2.02$

d. If usual is between $\pm 2z$, then the range of pulse rates that are between -2 and +2 z values would be: $\mu - 2\sigma = 72.9 - 2(12.3) = 72.9 - 24.6 = 48.3$ and $\mu + 2\sigma = 72.9 + 2(12.3) = 72.9 + 24.6 = 97.5$. A pulse rate of 48 would be considered to be unusual since it is outside the range of 48.3 to 97.5. Among the reasons the pulse rate could be this low would be: heredity, young in age, non-smoker, regular exerciser, good diet, low stress job, etc.

In Exercises 5 & 7, Express all z scores with two decimal places Consider a score to be unusual if its z score is less than –2.00 or greater than +2.00.

5. Heights of Women – Women's height parameters: $\mu = 63.6$ in , $\sigma = 2.5$ in

z score for height of 70 in: $z = \frac{x-\mu}{\sigma} = \frac{70-63.6}{2.5} = \frac{6.4}{2.5} = 2.56$

A height of 70 in. is 2.56 standard deviations above the mean and is higher than the +2.00 criterion point for being "usual." Thus, this would be unusual and the Beanstalk Club members may be aptly named.

7. **Body Temperature** Body temperature parameters: $\mu = 98.20$°F, $\sigma = 0.62$°F

z score for temperature of 101 °F: $z = \frac{x-\mu}{\sigma} = \frac{101-98.2}{0.62} = \frac{2.80}{0.62} = 4.52$

A body temperature of 101 °F is 4.52 standard deviations above the mean and is higher than the +2.00 criterion point for being "usual." Thus, this would be an unusually high temperature. This would suggest that immediate actions are called for to reduce the temperature for this patient.

9. Comparing Test Scores Which is better?

Biology: $x = 85$ $z = \frac{x-\bar{x}}{s} = \frac{85-90}{10} = \frac{-5}{10} = -0.50$

Economics: $x = 45$ $z = \frac{x-\bar{x}}{s} = \frac{45-55}{5} = \frac{-10}{5} = -2.00$

The range of z values is about –3.00 to +3.00 and the order is such that as scores move from the low part of the range (–3.00) through 0 to the upper part of the range (up to about +3.00), scores are increasing. Thus, a z of –0.5 is higher than a z of –2.0 on that –3.00 to +3.00 continuum. The student performed better, relative to the other students taking the tests, on the biology test than on the economics test.

In Exercises 11 & 13, use the sorted continine levels of smokers listed in Table 2-11 Find the percentile corresponding to the given cotinine level.

It might make this a little easier if we put the order number with the score, as below (we would not usually go to all of this work, we would usually let a computer do all of this, especially if there were a large number of scores):

Score #	1	2	3	4	5	6	7	8	9	10	11	12	13	14
Score	0	1	1	3	17	32	35	44	48	86	87	103	112	121

Score#	15	16	17	18	19	20	21	22	23	24	25	26	27	28
Score	123	130	131	149	164	167	173	173	198	208	210	222	227	234

Score #	29	30	31	32	33	34	35	36	37	38	39	40
Score	245	250	253	265	266	277	284	289	290	313	477	491

The percentile is the percentage of the scores that fall at or below a given score, which is found using:

$$\text{Percentile}=\frac{\text{number of values less than the given score}}{\text{total number of values}}*100\%$$

11. Value of 149

$$\text{Percentile}=\frac{\text{number of values less than the given score}}{\text{total number of values}}*100\%=\frac{17}{40}*100\%=0.425*100\%=43rd \text{ percentile}$$

13. Value of 35

$$\text{Percentile}=\frac{\text{number of values less than the given score}}{\text{total number of values}}*100\%=\frac{6}{40}*100\%=0.15*100\%=15th \text{ percentile}$$

In Exercises 15-21, use the sorted cotinine levels of smokers listed in Table 2-11. Find the indicated percentile or quartile.

In these examples, we find L, which is the number that represents the order position of the score when the scores are ordered low to high. k is the percentile value and n is the number of scores

$L = \frac{k}{100} * n$ provides the order Locator

If the value of L is a whole number, the percentile is the average of the score at L and the score at $L + 1$

If the value of L is not a whole number, the percentile is the value of L rounded up to a whole number.

15. Find P_{20}

$L = \frac{k}{100} * n = \frac{20}{100} * 40 = 0.20 * 40 = 8$ 8 is a whole number, thus:

P_{20} is the average of the 8th and the 9th score or (44 + 48)/2 = 46.0, thus P_{20}= 46.0

17 Find P_{75}

$L = \frac{k}{100} * n = \frac{75}{100} * 40 = 0.75 * 40 = 30$ 30 is a whole number, thus:

P_{75} is the average of the 30th and 31st scores or (250 + 253)/2= 251.5, thus P_{75}= 251.5 (also Q_3)

19. Find P_{33}

$L = \frac{k}{100} * n = \frac{33}{100} * 40 = 0.33 * 40 = 13.2$ 13.2 is not a whole number so L is the next highest whole number or the 14th number in the order The 14th number is 121, thus P_{33} is 121.

21. Find P_1

$L = \frac{k}{100} * n = \frac{1}{100} * 40 = 0.01 * 40 = 0.4$

0.4 is not a whole number so L is the next highest whole number or the 1st number in the order. The 1st number is 0, thus P_1 is 0.

23. Continine Levels of Smokers

a. Interquartile Range, distance between Q_1 and Q_3

The interquartile range is $Q_3 - Q_1$

Finding Q_3 which is the same as P_{75}

$L = \frac{k}{100} * n = \frac{75}{100} * 40 = 0.75 * 40 = 30$

30 is a whole number, thus:

Q_3 is the average of the 30th and 31st scores or (250 + 253)/2= 251.5, thus Q_3= 251.5 (also P_{75})

Finding Q_1 which is the same as P_{25}

$L = \frac{k}{100} * n = \frac{25}{100} * 40 = 0.25 * 40 = 10$

10 is a whole number, thus Q_1 is the average of the 10th and 11th scores or (86 + 87)/2= 86.5, thus Q_1= 86.5

The interquartile range, IQR= $Q_3 - Q_1$= 251.5 – 86.5= 165.0

b. Midquartile, the point exactly between Q_1 and Q_3

The midquartile is defined as: $Midquartile = \frac{Q_3 + Q_1}{2}$

We have found that Q_3= 251.5 and Q_1= 86.5, thus:

$$Midquartile = \frac{Q_3 + Q_1}{2} = \frac{251.5 + 86.5}{2} = \frac{338.0}{2} = 169.0$$

c. 10 – 90 percentile range
Finding P_{10}
$$L = \frac{k}{100} * n = \frac{10}{100} * 40 = 0.10 * 40 = 4$$
P_{10}= Avg. of 4th and 5th scores= (3 + 17)/2= 10.0
Finding P_{90}
$$L = \frac{k}{100} * n = \frac{90}{100} * 40 = 0.90 * 40 = 36$$
P_{90}= Avg. of 36th and 37th scores = (289 + 290)/2 = 579/2 = 289.5
10 – 90 percentile range = $P_{90} - P_{10}$= 289.5 – 10.0= 279.5

d. Does $P_{50} = Q_2$? If so, does P_{50} always equal Q_2?
In order to find Q_2, we have to find P_{50}, since 50 must be used for the value of k in finding the locator L value Thus P_{50} always equals Q_2 Yes, $P_{50} = Q_2$ and yes, P_{50} always equals Q_2.

e. Does $Q_2 = (Q_1 + Q_3)/2$? If so, does Q_2 always equal $(Q_1 + Q_3)/2$?
In this case Q_1= 86.5 and Q_3= 251.5, $(Q_1 + Q_3)/2$ = (86.5 + 251.5)/2= 169.0 (this is also the value of the midquartile). Q_2 (the median) is 170.0. These values are very close, but they are not the same. They (the midquartile and the median) will be exactly the same if the distribution of scores is symmetrical around the median. To the extent the distribution is skewed, these values will be more different from each other. Actually, because these are so close in this example, we could infer that there is relative symmetry of this distribution.

2 – 7 Exploratory Data Analysis

The five-number summary referred to in the following Exercises are:

1. Minimum score
2. Q_1, the first quartile, also known as P_{25}
3. The median or Q_2, the second quartile, also known as P_{50}
4. Q_3, the third quartile, also known as P_{75}
5. Maximum score

These are used to develop boxplots for exploratory data analysis. Boxplots, also referred to as Box and Whisker Plots, were proposed by a famous modern statistician by the name of John Tukey. They have only recently been incorporated into statistical software packages. As indicated in your text, there is quite a variation on how these are presented by the various packages. Some have the plot going in a vertical direction and some have the plot going in a horizontal direction and there seems to be no evidence that one way is better than another. Some include the mean as a point, usually with the + symbol for the mean on the continuum. Some include points that would be considered outliers with marks to indicate these points, based on rules such as those described in the text. Some boxplots, use variations of the lines that represent the box, such as the boxplots produced by the STATDISK software. All are reasonable ways of presenting the data displayed.

There is no boxplot option in the current version of Excel. However, the boxplots presented in this workbook have been generated using Excel by setting the 5-number plots as a variable and a constant such as 1 for the other variable and then generating a scatterplot. If there is more than one to be plotted on the same graph, the other(s) would be given a different constant than 1. The Excel scatterplot provides the points and the line draw function of Excel is used to draw the box and the whiskers. Outliers are not identified in these boxplots.

1. Testing Corn Seeds Regular Corn, n= 11, Scores in rank order:

Score #	1	2	3	4	5	6	7	8	9	10	11
Score	1316	1444	1511	1612	1903	1910	1935	1961	2060	2108	2496

Minimum score is 1316
Q_1, the first quartile, also known as P_{25} is the 3rd score, which is 1511
The median or Q_2, the second quartile, also known as P_{50} is the 6th score or 1910 (Mean is 1841)
Q_3, the third quartile, also known as P_{75} is the 9th score, which is 2060
Maximum score is 2496
Following is the boxplot. For the first few of these boxplots, the values will be entered for the 5-number summary so you can see where they are plotted. In addition, the mean is indicated with the + sign. Some boxploting programs include the mean and others don't. It is often useful to see where the mean lies relative to the median which is the middle line of the box. The mean is included in the boxplots presented in this workbook.

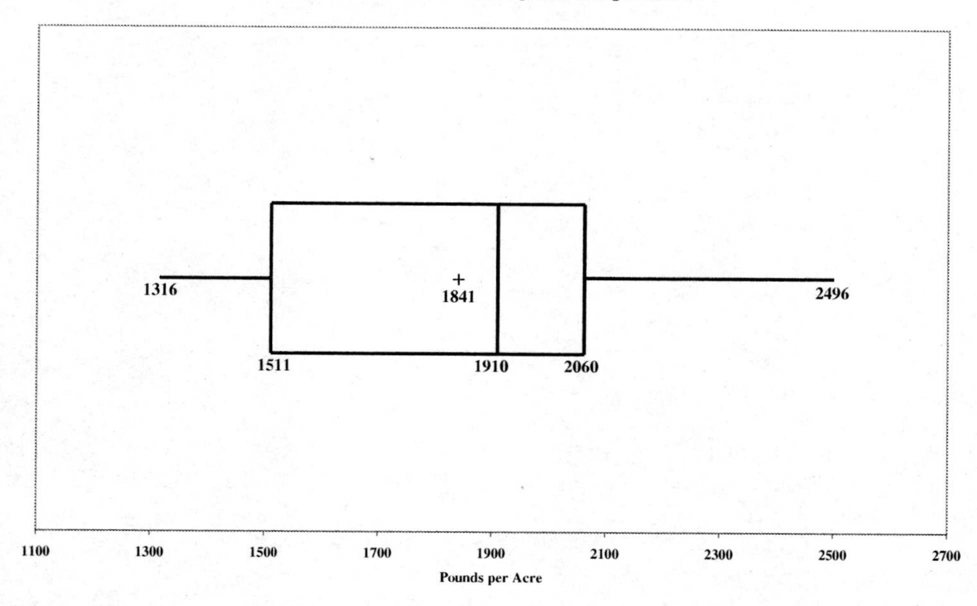

3. Fruit Flies $n = 11$, Scores in rank order:

Score #	1	2	3	4	5	6	7	8	9	10	11
Score	0.64	0.68	0.72	0.76	0.84	0.84	0.84	0.84	0.90	0.90	0.92

Minimum score is 0.64
Q_1, the first quartile, also known as P_{25} is the 3rd score, which is 0.72
The median or Q_2, the second quartile, also known as P_{50} is the 6th score or 0.84 (Mean is 0.81)
Q_3, the third quartile, also known as P_{75} is the 9th score, which is 0.90
Maximum score is 0.92
Following is the boxplot:

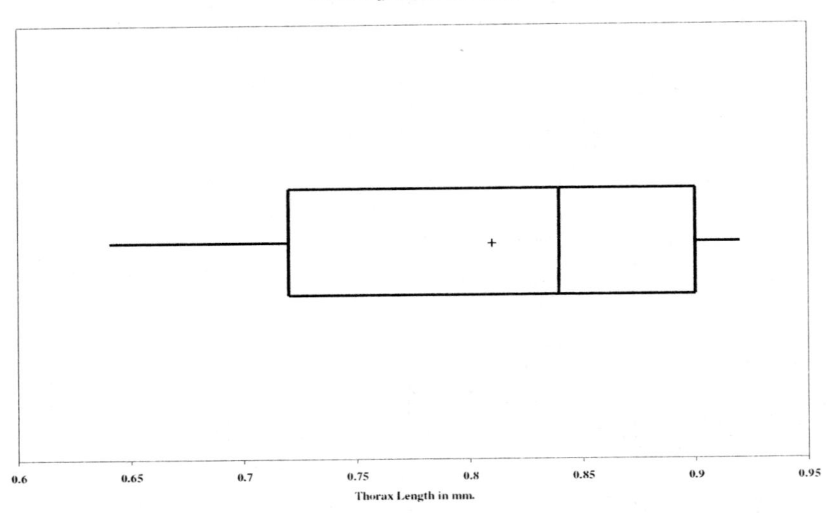

5. Body Temperatures (Data Set 2), $n = 106$
Minimum score is 96.5
Q_1, the first quartile, also known as P_{25} is the 27th score, which is 97.8
The median or Q_2, the second quartile, also known as P_{50} is the average of the 53rd and 54th scores or (98.4 + 98.4)/2= 98.4 (Mean is 98.2)
Q_3, the third quartile, also known as P_{75} is the 80th score, which is 98.6
Maximum score is 99.6
These data do not support the common belief that the mean body temperature is 98.6°F. In this distribution 98.6°F is at P_{75}. However, there is no indication that this sample is intended to be representative of the adult population and also these are temperatures taken at midnight. How many subjects needed to be awakened to have their temperatures taken? These factors might make a difference.

Following is the boxplot:

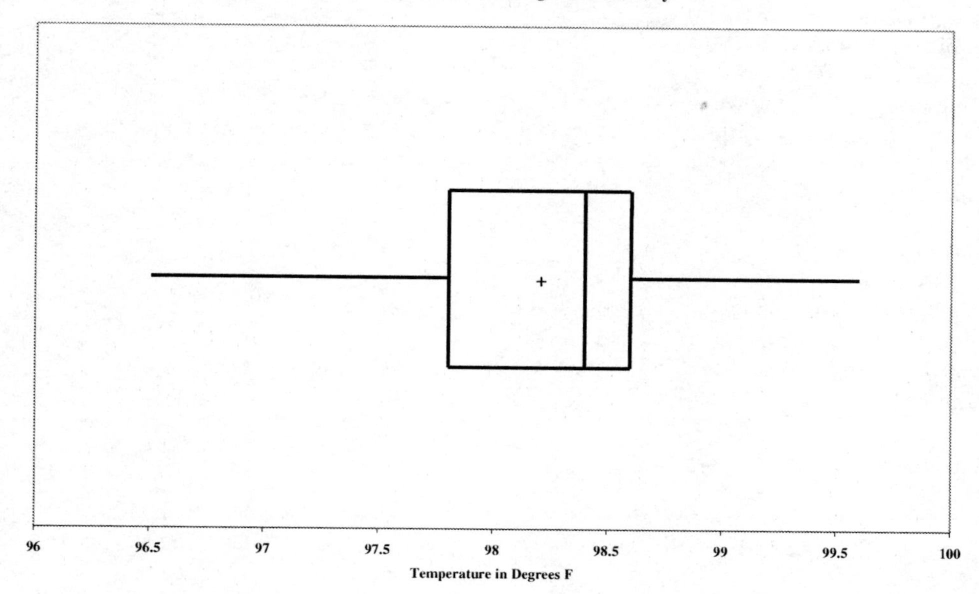

In Exercises 7 – 15, find the 5-numbr summaries, construct boxplots and compare the data sets.

7. Cuckoo Egg Lengths (Data Set 8)

Meadow pipits, n= 45
Minimum score is 19.65
Q_1, the first quartile, also known as P_{25} is the 12th score, which is 22.05
The median or Q_2, the second quartile, also known as P_{50} is the 23rd score or 22.25 (Mean is 22.30)
Q_3, the third quartile, also known as P_{75} is the 34th score, which is 22.85
Maximum score is 24.45

Tree pipits, n= 15
Minimum score is 21.05
Q_1, the first quartile, also known as P_{25} is the 4th score, which is 22.45
The median or Q_2, the second quartile, also known as P_{50} is the 8th score or 23.25 (Mean is 23.09)
Q_3, the third quartile, also known as P_{75} is the 12th score, which is 23.85
Maximum score is 24.05

Robins, n= 16
Minimum score is 21.05
Q_1, the first quartile, also known as P_{25} is the average of the 4th and 5th scores, which is 22.05
The median or Q_2, the second quartile, also known as P_{50} is the average of the 8th and 9th scores or 22.55 (Mean is 22.58)
Q_3, the third quartile, also known as P_{75} is the average of the 12th and 13th scores, which is 23.05
Maximum score is 23.85
Boxplots including Meadow Pipits from Exercise 6

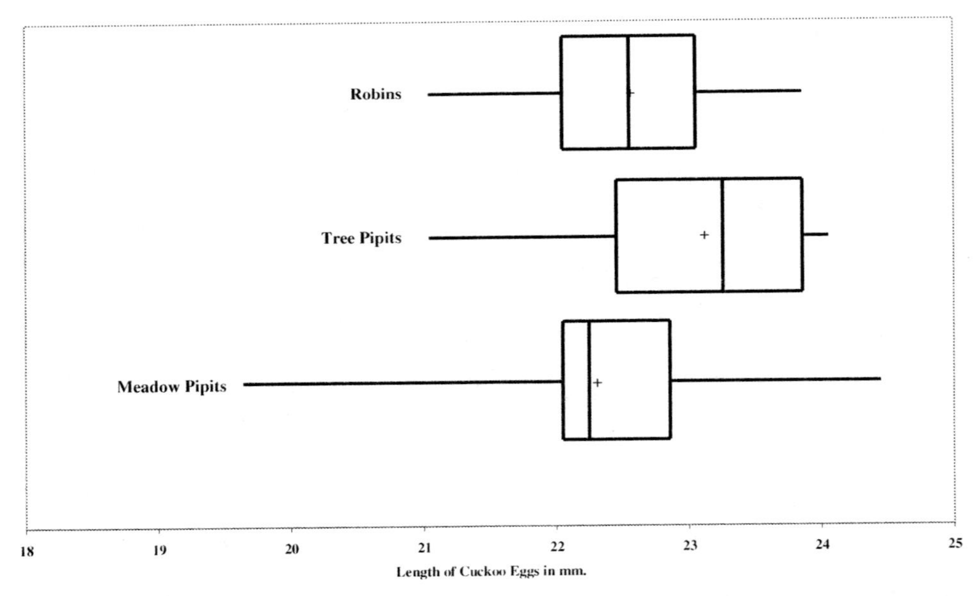

The length of cuckoo eggs is higher for the Tree Pipits nests and there is more variability of cuckoo eggs found in Meadow Pipits nests The distribution shape is relatively symmetric for the Robins but is negatively skewed for both the Tree Pipits and the Meadow Pipits.

9. **Head Circumference** (Data Set 4)

Boys, n= 50
Minimum score is 35.5
Q_1, the first quartile, also known as P_{25} is the 13th score, which is 40.4
The median or Q_2, the second quartile, also known as P_{50} is the average of the 25th and 26th scores or 41.1
(Mean is 41.1)
Q_3, the third quartile, also known as P_{75} is the 38th score, which is 42.0
Maximum score is 43.2

Girls, n= 50
Minimum score is 34.3
Q_1, the first quartile, also known as P_{25} is the 13th score, which is 39.5
The median or Q_2, the second quartile, also known as P_{50} is the average of the 25th and 26th scores or 40.2
(Mean is 40.1)
Q_3, the third quartile, also known as P_{75} is the 38th score, which is 40.9
Maximum score is 43.7
Boxplots:

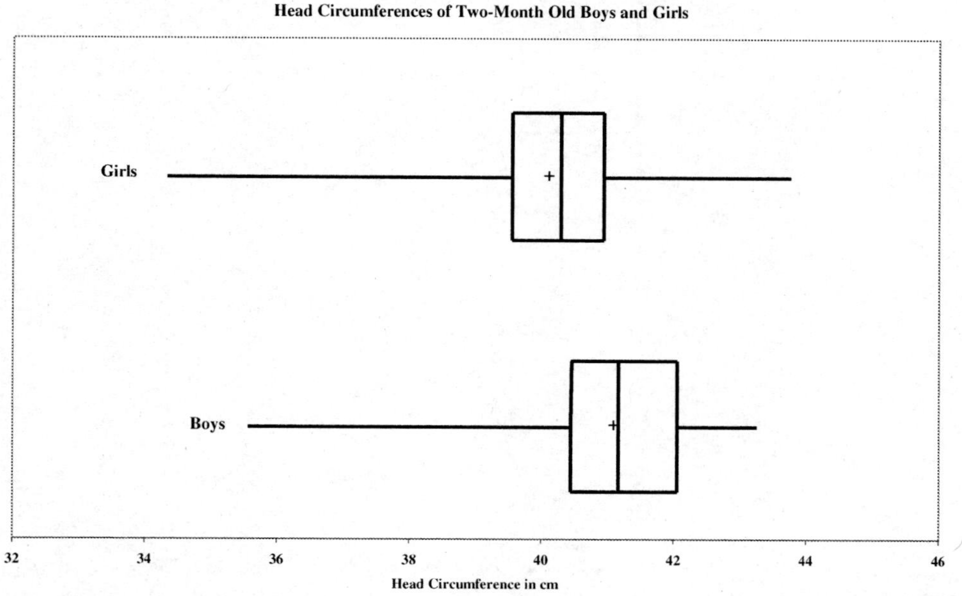

Head Circumferences of Two-Month Old Boys and Girls

Girls

Boys

32 34 36 38 40 42 44 46

Head Circumference in cm

The head circumference of the two-year old boys appears to be higher than that for the girls The variability for the girls appears to be slightly higher than for the boys

11. Ages of Oscar-Winning Actors and Actresses

Actors, $n = 39$
Minimum score is 31
Q_1, the first quartile, also known as P_{25} is the 10th score, which is 37
The median or Q_2, the second quartile, also known as P_{50} is the 20th score or 43 (Mean is 44.8)
Q_3, the third quartile, also known as P_{75} is the 30th score, which is 51
Maximum score is 76

Actresses, $n = 39$
Minimum score is 21
Q_1, the first quartile, also known as P_{25} is the 10th score, which is 30
The median or Q_2, the second quartile, also known as P_{50} is the 20th score or 34 (Mean is 38.1)
Q_3, the third quartile, also known as P_{75} is the 30th score, which is 41
Maximum score is 80

Boxplots:

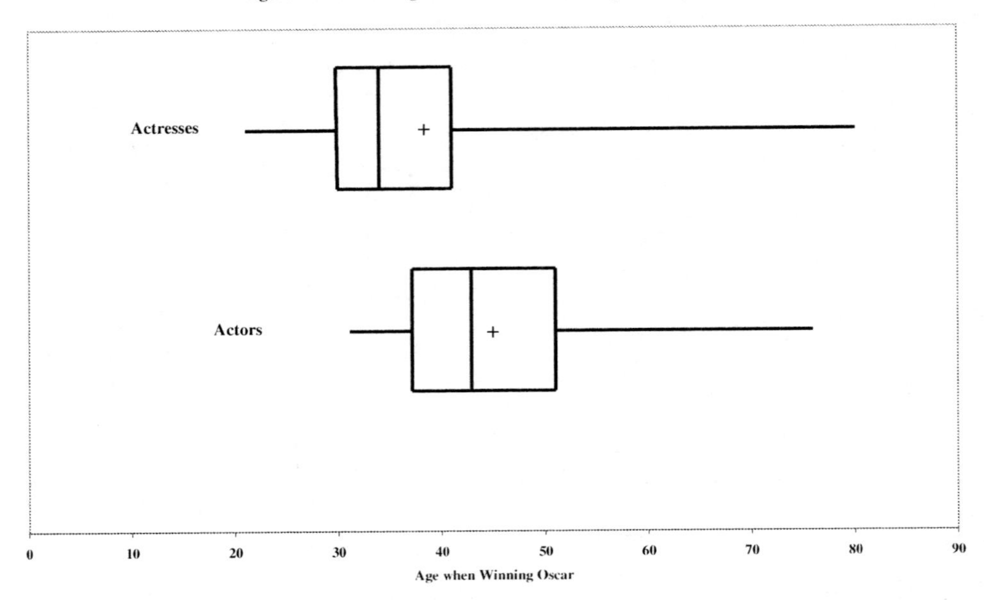

The average age for actresses when they won the Oscar was lower than the age of actors. There appears to be more variability of the actress ages compared with the actor ages and both distributions appear to have a degree of positive skewness.

13. Petal Widths of Irises (Data Set 7)

Setosa, n= 50
Minimum score is 0.1
Q_1, the first quartile, also known as P_{25} is the 13th score, which is 0.2
The median or Q_2, the second quartile, also known as P_{50} is the average of the 25th and 26th scores or 0.2
(Mean is 0.24)
Q_3, the third quartile, also known as P_{75} is the 38th score, which is 0.3
Maximum score is 0.6

Versicolor, n= 50
Minimum score is 1.0
Q_1, the first quartile, also known as P_{25} is the 13th score, which is 1.2
The median or Q_2, the second quartile, also known as P_{50} is the average of the 25th and 26th scores or 1.3
(Mean is 1.326)
Q_3, the third quartile, also known as P_{75} is the 38th score, which is 1.5
Maximum score is 1.8

Virginica, n= 50
Minimum score is 1.4
Q_1, the first quartile, also known as P_{25} is the 13th score, which is 1.8
The median or Q_2, the second quartile, also known as P_{50} is the average of the 25th and 26th scores or 2.0
(Mean is 2.03)
Q_3, the third quartile, also known as P_{75} is the 38th score, which is 2.3
Maximum score is 2.5

Boxplots:

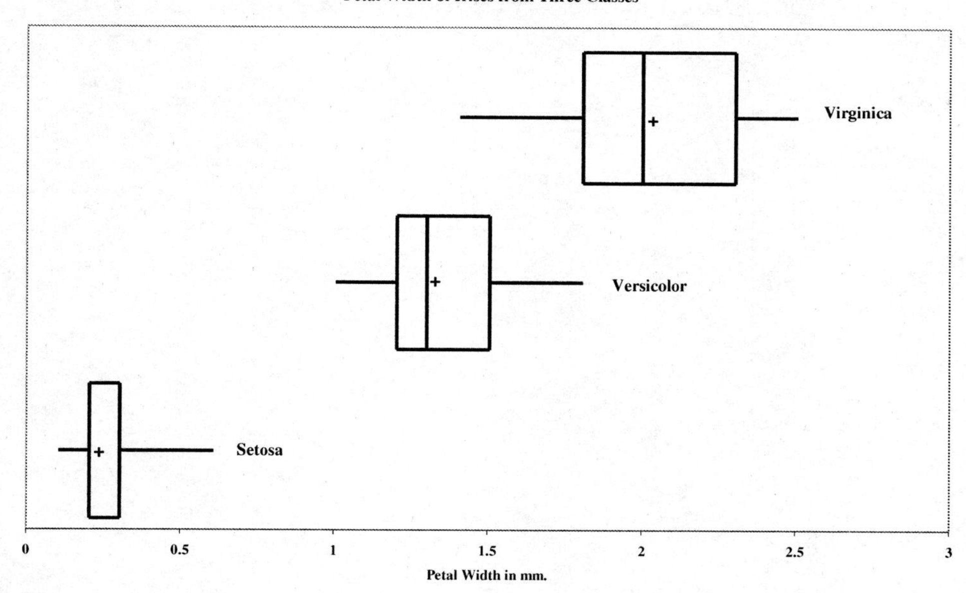

There are very clear differences in the average length of the petal widths among these three types of irises. The highest length is for the Virginica, the second highest length is for the Versicolor, and the lowest length is for the Setosa. The lowest variation of petal width was for Setosa, Veriscolor had the next highest variation, and the Virginica had the highest variation of petal lengths.

Review Exercises

1. **Tree Heights**, n= 20

a. Mean: $\bar{x} = \quad x/n = 90.7/20 = 4.54$

b. Median: $\tilde{x}$ is the midpoint between the two middle scores (10th and 11th scores) if an even number of scores This would be (3.9 + 4.0)/2= 7.9/2= 3.95

c. Mode: Mode is score that occurs most often. In this case, 3 scores occur more than any others (1.8, 3.7, and 5.1) so the distribution would be called trimodal with modes at 1.8, 3.7, and 5.1

d. Midrange: Midpoint (or average) between lowest and highest scores, (1.8 + 13.7)/2 = 15.5/2= 7.75

e. Range = highest score – lowest score = 13.7 – 1.8 = 11.9

f. Standard deviation

$$s = \sqrt{\frac{n(\quad x^2) - (\quad x)^2}{n(n-1)}} = \sqrt{\frac{20(544.85) - (90.7)^2}{20(20-1)}} = \sqrt{\frac{2670.51}{380}} = \sqrt{7.027} = 2.65$$

g. Variance $s^2 = 2.65^2 = 7.03$

h. Q_1 is the first quartile, also known as P_{25}, $L = \frac{k}{100} * n = \frac{25}{100} * 20 = 0.25 * 20 = 5$

Q_1 is the average of the 5th and 6th scores, which is (3.1 + 3.4)/2= 6.5/2= 3.25

i. Q_3, the third quartile, also known as P_{75}, $L = \frac{k}{100} * n = \frac{75}{100} * 20 = 0.75 * 20 = 15$

Q_3, is the average of the 15th and 16th scores, which is (5.1 + 5.2)/2= 10.3/2= 5.15

j. P_{10} is the 10th percentile, k is the percentile (10), $L = \frac{k}{100} * n = \frac{10}{100} * 20 = 0.10 * 20 = 2$

Since L is a whole number, in this case 2, P_{10} is taken as the average of the 2nd and 3rd scores or (1.8 + 1.9)/2= 1.85

3. Frequency Distribution

Tree Circumference in ft.	Frequency
1.0 – 2.9	4
3.0 – 4.9	9
5.0 – 6.9	5
7.0 – 8.9	1
9.0 – 10.9	0
11.0 – 12.9	0
13.0 – 14.9	1

5. Boxplot
5-number summary
Minimum score is 1.80
Q_1, the first quartile, also known as P_{25} is the average of the 5th and 6th scores, which is 3.25
The median or Q_2, the second quartile, also known as P_{50} is the average of the 10th and 11th scores or 3.95
(Mean is 4.54)
Q_3, the third quartile, also known as P_{75} is the average of the 15th and 16th scores, which is 5.15
Maximum score is 13.70
Boxplot:

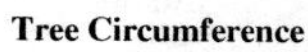

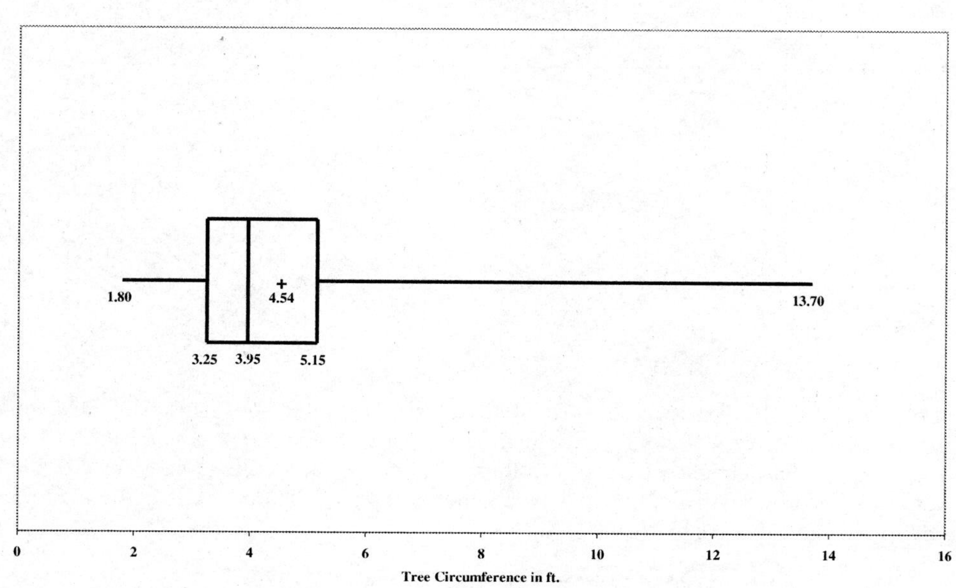

Cumulative Review Exercises

1. **Tree Measurements**
   a. The measures are continuous since the unit of measurement of feet can take on fractional values.
   b. The level of measurement is ratio since there is a natural zero or beginning point on the scale. There are no negative values on a ratio scale.

Critical Thinking

Age at Fatality	Relative Frequency of Licensed Drivers (millions)	Relative Frequency of Age of Drivers Killed in Car Crashes (Randomly selected)
16 – 19	9.2	30.0
20 – 29	33.6	24.0
30 – 39	40.8	14.0
40 – 49	37.0	8.0
50 – 59	24.2	8.0
60 – 69	17.5	3.0
70 – 79	12.7	8.0
80 – 89	4.3	5.0

The age category of 16 – 19 clearly has a substantially higher percentage of fatalities in the sample of 100 than in the group of licensed drivers It would seem justified to charge higher premiums for the 16 – 19 age group. The 80 – 89 age group has a slightly higher percentage for the ages at fatality. When one considers the absolute number of drivers involved, this may be another group that should be charged higher premiums. The following graph displays this comparison:

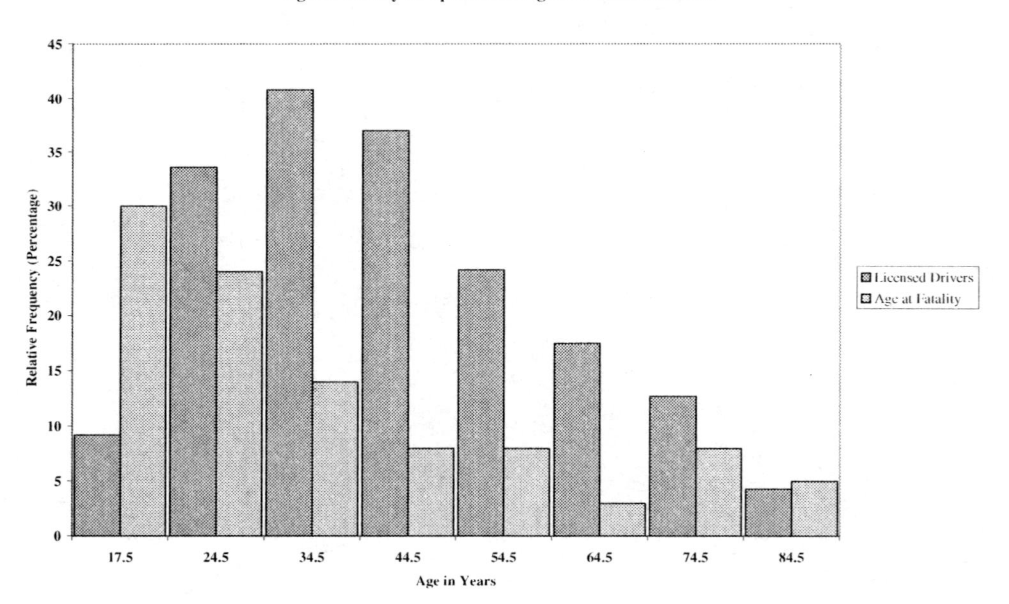

Chapter 3. Probability

3 – 1 Exercises

In exercise 1, express the indicated degree of likelihood as a probability value.

1. Identifying Probability Values

a. *P*(choosing correct answer)= 0.500

b. *P*(rain tomorrow)= 0.200

c. *P*(marrying my daughter)= 0.000

3. Identifying Probability Values

Which cannot be probabilities? Probabilities can only be in the range of 0 to 1.

-1, 2, 5/3, $\sqrt{2}$

5. Gender of Children

a. *P*(one girl out of three children)= 3/8= 0.375 (bbg, bgb, and gbb)

b. *P*(two girls out of three children)= 3/8= 0.375 (bgg, gbg, and ggb)

c. *P*(all girls out of three children)= 1/8= 0.125 (ggg)

7. Mendelian Genetics

$$P(\text{greenpea}) = \frac{\text{number of green peas}}{\text{number of trials or observations}} = \frac{428}{428+152} = \frac{428}{580} = 0.738$$

Yes, it is very close to the expected value of ¾ or 0.750.

In exercises 9 & 11, consider an event to be "unusual" if its probability is less than or equal to 0.05. (This is equivalent to the same criterion commonly used in inferential statistics, but the value of 0.05 is not absolutely rigid, and other values such as 0.01 are sometimes used instead.)

9. Probability of a Wrong Event

a. $P(\text{wrong}) = \frac{\text{number of wrong events}}{\text{number of trials or observations}} = \frac{5}{85} = 0.0588$

b. Since *P*= 0.0588 is greater than 0.05, it is not "unusual" for this test conclusion to be wrong for women who are pregnant.

11. Cholesterol-Reducing Drug

a. $P(\text{flu symptoms}) = \frac{\text{number of flu symptom events}}{\text{number of trials or observations}} = \frac{19}{19+844} = \frac{19}{863} = 0.0220$

b. Since *P*= 0.0220 is less than 0.05, it is considered "unusual" for patients taking this drug to experience flu symptoms.

13. Probability of a Birthday

a. $P(\text{correct}) = \frac{\text{number of ways correct birthday can occur}}{\text{number of different days}} = \frac{1}{365} = 0.00274$

b. $P(\text{correct}) = \frac{\text{number of ways correct October birthday can occur}}{\text{number of different days}} = \frac{31}{365} = 0.0849$

c. $P(\text{correct}) = \frac{\text{Number of ways birthday ends in y}}{\text{Number of different days in week}} = \frac{7}{7} = 1.000$

15. Probability of an Adverse Drug Reaction

$$P(\text{headache}) = \frac{\text{number of times headache occurred}}{\text{number of times patients observed}} = \frac{117}{117+617} = \frac{117}{734} = 0.159$$

Yes, this probability of about 16 times out of 100 patients would be of concern to many Viagra patients.

17. **Genetics: Constructing Sample Space**, both with blue-brown, brown dominant
   a. Possible Outcomes, blue-blue= blue, blue-brown= brown, brown-blue= brown, brown- brown= brown
   b. $P(\text{blue}) = \dfrac{\text{number of ways blue eyes can occur}}{\text{number of different possible outcomes}} = \dfrac{1}{4} = 0.250$
   c. $P(\text{brown}) = \dfrac{\text{number of ways brown eyes can occur}}{\text{number of different possible outcomes}} = \dfrac{3}{4} = 0.750$

19. **Genetics: Sample Space**, one with brown-brown and one with blue-blue, brown dominant
   a. Possible Outcomes, brown-blue= brown, brown-blue= brown, brown-blue= brown, brown-blue= brown
   b. $P(\text{blue}) = \dfrac{\text{number of ways blue eyes can occur}}{\text{number of different possible outcomes}} = \dfrac{0}{4} = 0.000$
   c. $P(\text{brown}) = \dfrac{\text{number of ways brown eyes can occur}}{\text{number of different possible outcomes}} = \dfrac{4}{4} = 1.000$

3 – 3 Addition Rule

For each part of exercise 1, are the two events disjoint for a single trial? (Hint: Consider "disjoint" to be equivalent to "separate" or "not overlapping.")

This is also referred as being "mutually exclusive" where both events cannot occur at the same time.

1. a. Not disjoint, since a cardiac surgeon could also be a female physician.
   b. Not disjoint, since a female college student could also drive a motorcycle.
   c. Disjoint, since if a study subject received Lipitor, he/she could not also be in a control group that received no medication.

3. **Finding Complements**
   The complement of the probability of an event is the probability that that even will not occur.
   The sum of a probability and its complement is equal to 1.
   a. If *P*(*A*)= 0.05, then $P(\overline{A})$= 1 – 0.05= 0.95
   b. If probability of having red/green blindness is 0.25% or 0.0025, then the probability of not having red/green blindness is 1 – 0.0025= 0.9975.

5. **Peas on Earth**, based on Table 3-2
   P(green pod or white flower)= *P*(green pod) + *P*(white flower) – *P*(both green pod and white flower) = 8/14 + 5/14 – 3/14= 10/14= 0.714

7. **National Statistics Day**
   P(birthday on Oct. 18)= 1/365, *P*(birthday not on Oct. 18)= 364/365= 0.997

9. ***Titanic* Passengers** (n= 2223), this is a disjoint probability
   P(women or child)= *P*(women) + *P*(child)= 422/2223 + (64 + 45)/2223=
   0.1898 + 109/2223= 0.1898 + 0.0490= 0.239

11. ***Titanic* Passengers** (n= 2223), this is not a disjoint probability
   P(child or survived)= *P*(child) + *P*(survived) – *P*(child and survived)=
   109/2223 + 706/2223 – (29 + 27)/2223= 0.0490 + 0.3176 – 56/2223=
   0.3660 – 0.0252= 0.341

In Exercises 13-19, use the data in the following table, which summarizes blood groups and Rh types for 100 typical people. These values may vary in different regions according to the ethnicity of the population.

		O	A	B	AB	Total
Rh Type	**Positive**	**39**	**35**	**8**	**4**	**86**
	Negative	**6**	**5**	**2**	**1**	**14**
	Total	**45**	**40**	**10**	**5**	**100**

13. Blood Groups and Types (n= 100), complement
$P(\bar{A}) = 1 - P(A) = 1 - 40/100 = 1 - 0.400 = 0.600$

15. Blood Groups and Types (n= 100), not a disjoint probability
$P(A \text{ or } Rh^-) = P(A) + P(Rh^-) - P(A \text{ and } Rh^-) = 40/100 + 14/100 - 5/100 =$
$0.400 + 0.140 - 0.050 = 0.490$

17. Blood Groups and Types (n= 100), complement
$P(\text{not } Rh^+) = 1 - P(Rh^-) = 1 - 14/100 = 1 - 0.140 = 0.860$

19. Blood Groups and Types (n= 100), not a disjoint probability
$P(AB \text{ or } Rh^+) = P(AB) + P(Rh^+) = P(AB \text{ and } Rh^+) = 5/100 + 86/100 - 4/100 =$
$0.050 + 0.860 - 0.040 = 0.870$

21. Poll Resistance. A disjoint probability

		Responded	Didn't Respond	Total
Age Group	**18-21**	**73**	**11**	**84**
	22-29	**255**	**20**	**275**
	Total	**328**	**31**	**359**

$P(\text{18-21 or didn't respond}) = P(\text{18-21}) + P(\text{didn't respond}) - P(\text{18-21 and didn't respond}) =$
$84/359 + 31/359 - 11/359 = 0.2340 + 0.0864 - 0.0306 = 0.290$

23. Disjoint Events
No. Take the events of: A is Pregnant, B is Male and C is Female. A (Pregnant) and B (Male) is disjoint, B(Male) and C(Female) is disjoint, but A(Pregnant) and C (Female) are not disjoint.

3 – 4 Multiplication Rule: Basics

In exercise 1, for each given pair of events, classify the events as independent or dependent.

1. **a.** Randomly selecting a grizzly bear and randomly selecting a male mammal
Independent, since the outcome of selecting a grizzly bear is not related to being a male mammal since the second selected Grizzly bear could be a female mammal

b. Randomly selecting a TV viewer who is watching *The Living Planet* and randomly selecting a second TV viewer who is watching *The Living Planet*
Dependent, assuming that both TV viewers are in the same room with only 1 TV. Otherwise, they would be independent if they were not in the same room with only 1 TV since the second person selected could be watching something else

c. Wearing plaid shorts with black socks and sandals and asking someone on a date and getting a positive response. Probably dependent since wearing such attire might reduce the likelihood of getting a positive response when asking for a date, but it could still happen.

3. Coin and Die, independent events
P(tail on coin and 3 on die)= *P*(tail) * *P*(3)= 1/2 * 1/6= 0.5000 * 0.1667= 0.0833

In Exercises 5 & 7, use the data in Table 3-1.

Table 3-1. Pregnancy Test Results

	Positive Test Result (Pregnancy is indicated)	Negative Test Result (Pregnancy is not indicated)	Total
Subject is pregnant	**80**	**5**	**85**
Subject is not pregnant	**3**	**11**	**14**
Total	**83**	**16**	**99**

5. Positive Test Result
P(first test +)= 83/99= 0.8384 (one positive test is removed from data set)
P(second test+)= 82/98= 0.8367
P(first and second tests are both +)= *P*(first test +) * *P*(second test +)=
0.8384 * 0.8367= 0.701

7. Pregnant
P(first subject is pregnant)= 85/99= 0.8586 (one pregnant subject is removed from data set)
P(second subject is pregnant)= 84/98= 0.8571
P(first and second subjects are both pregnant)=
P(first is pregnant) * *P*(second is pregnant)= 0.8586 * 0.8571= 0.736

In Exercises 9-13, use the following table, which summarizes blood groups and Rh types for 100 typical people. These values may vary in different regions according to the ethnicity of the population.

		O	A	B	AB	Total
Rh Type	**Positive**	**39**	**35**	**8**	**4**	**86**
	Negative	**6**	**5**	**2**	**1**	**14**
	Total	**45**	**40**	**10**	**5**	**100**

9. Blood Groups and Types *P*(both of two randomly selected have O blood)
 a. with first selection replaced back into same data pool
 P(both O)= *P*(first O) * *P*(second O from same pool)= 45/100 * 45/100= 0.450 * 0.450= 0.203
 b. with first selection not replaced in the same data pool
 P(both O)= *P*(first O) * *P*(second O from adjusted pool)= 45/100 * 44/99= 0.4500 * 0.4444= 0.200

11. Blood Groups and Types *P*(three of three randomly selected have B blood)
 a. with first and second selections replaced back into same data pool
 P(all B)= *P*(first B) * *P*(second B from same pool) * *P*(third B from same pool)=
 10/100 * 10/100 * 10/100= 0.100 * 0.100 * 0.100= 0.00100
 b. with first and second selections not replaced in the same data pool
 P(all B)= *P*(first B) * *P*(second B from adjusted pool) * *P*(third B from adjusted pool)=
 10/100 * 9/99 * 8/98= 0.1000 * 0.0909 * 0.0816= 0.000742

13. Blood Groups and Types
P(10 randomly selected will have all have Type A blood)= 0.400^{10}= 0.000105

15. Probability and Guessing
10 true/false questions, thus P(correct by chance per item)= 0.5

a. P(first 7 out of 10 correct, next three incorrect)=

$P(C) * P(C) * P(C) * P(C) * P(C) * P(C) * P(C) * P(\overline{C}) * P(\overline{C}) * P(\overline{C})=$
$0.5 * 0.5 * 0.5 * 0.5 * 0.5 * 0.5 * 0.5 * 0.5 * 0.5 * 0.5 = 0.5^{10} = 0.000977$

b. No, since there would be more than just this pattern of getting 7 out of 10 correct and these other patterns would have to be included in the determination of the overall probability of getting 7 correct out of 10

17. Testing Effectiveness of Gender-Selection Method
P(all ten babies are girls)= 0.5^{10}=0.000977
Yes, since the probability of getting all ten babies as girls is 0.000977 (or about 1 time out of 1000) if chance alone is operating, the actual outcome of getting all ten girls would indicate the gender selection method is effective since chance clearly would not be a viable reason for this outcome.

19. Quality Control
Would expect 2% of the monitors to have a defect.
P(none are defective out of 15 sampled monitors)= 0.98^{15}= 0.739
No, since this result could be attributed to chance or random variation; would happen almost three out of every four tests of 15 sampled monitors.

21. Stocking Fish
Identify all possible events: MMM, MMF, MFM, MFF, FFF, FMM, FFM, FMF
Of these combinations, 6/8 have both Male and Female, P= 0.750

3 – 5 Multiplication Rule: Beyond the Basics

In exercises 1 & 3, provide a written description of the complement of the given event.

1. Blood Testing
If it is not true that at least one of the 10 students has Group A blood, then none of them has Group A blood. The complement of an event where least one student out of 10 has group A blood is 0 or none.

3. X-Linked Disorder
When none of 12 males have a particular X-linked recessive gene, the probability of not having the gene is 1.000, thus the probability of at least one having the gene would be 1 minus the probability of not having the gene, or 1 – 0= 1.000.

5. Subjective Conditional Probability
It would be very unusual that the same credit card would be used in several different countries on the same day, assuming the person had to use it in person and the countries were not close to each other. Therefore, it is likely that it was being fraudulently used, thus the subjective conditional probability is close to 1.

7. Probability of At Least One Girl out of five children
Find probability of the complement of A where A is a girl:
$P(\overline{A})= P$(boy and boy and boy and boy and boy)= 0.5^5= 0.0313
P(of at least one girl)= $1.0000 - P(\overline{A}) = 1.0000 - 0.0313 = 0.9687$
The parents should be pretty confident that out of five children, at least one of them would be a girl.

9. Probability of a Girl as third child after two boys
These are independent events so the probability of getting a girl as the third child is 0.5 and there is no influence on that probability from the sex of the first two children.
No, the probability of getting three girls out of three children would be:
P(three girls)= (girl and girl and girl)= 0.5^3= 0.125
This is not the same probability as getting a girl as the third child after having two boys.

11. Clinical Trials of Pregnancy Test

$$P(\text{negative test} \mid \text{not pregnant}) = \frac{P(\text{not pregnant and negative test})}{P(\text{not pregnant})} = \frac{11/99}{14/99} = \frac{0.1111}{0.1414} = 0.786$$

She should probably be somewhat concerned that the test may not have accurately described the situation. There is a 0.786 probability that she will get a negative test when she is not pregnant.

13. Redundancy in Alarm Clocks, failure rate per clock is 0.01

P(at least one would work)= 1 – P(none of the three would work)

P(none of the three would work)= P(first will fail) * P(second will fail) * P(third will fail) = (0.01 * 0.01* 0.01)= 0.000001

P(at least one would work)= 1 – 0.000001= 0.999999

The surgeon certainly does gain by using three alarm clocks compared with one. If she used one, we would expect a failure one day out of every 100 days, but if she used three, we would expect a failure one out of every 1,000,000 days.

15. Using Composite Blood Sampling, 0.1 are positive for HIV

P(at least one positive)= 1 – P(none are positive)=

1 – [P(first is not positive) * P(second is not positive) * P(third is not positive)=

1 – (0.9 * 0.9 * 0.9)= 1 – 0.729= 0.271

***Conditional Probabilities.* In Exercises 17 & 19, use the Titanic mortality data in the accompanying table.**

	Men	Women	Boys	Girls	Total
Survived	**322**	**318**	**29**	**27**	**696**
Died	**1360**	**104**	**35**	**18**	**1517**
Total	**1682**	**422**	**64**	**45**	**2213**

17. $$P(\text{man} \mid \text{died}) = \frac{\text{number of men who died}}{\text{number who died}} = \frac{1360}{1517} = 0.897$$

19. $$P(\text{boy or girl} \mid \text{survived}) = \frac{\text{number of boys and girls who survived}}{\text{number who survived}} = \frac{56}{696} = 0.0805$$

In Exercises 21 & 23, assume that the Altigauge Manufacturing Company makes 80% of all electrocardiograph machines, the Bryant Company makes 15% of them, and the Cardioid Company makes the other 5%. The electrocardiograph machines made by Altigauge have a 4% rate of defects, the Bryant machines have a 6% rate of defect, and the Cardioid machines have a 9% rate of defects.

We'll make a table that displays these data had 10000 electrocardiograph machines been made.

	Altigauge	Bryant	Cartioid	Total
Not Defective	**7680**	**1410**	**455**	**9545**
Defective	**320**	**90**	**45**	**455**
Total	**8000**	**1500**	**500**	**10000**

21. a. $$P(\text{made by Altigauge}) = \frac{\text{number made by Altigauge}}{\text{total number made}} = \frac{8000}{10000} = 0.800$$

b. $$P(\text{Altiguage} \mid \text{defective}) = \frac{\text{number of defective Altigauge electrocardiographs}}{\text{number of defective electrocardiographs}} = \frac{320}{455} = 0.703$$

23. a. $P(\text{made by Cardioid}) = \dfrac{\text{number made by Cardioid}}{\text{total number made}} = \dfrac{500}{10000} = 0.050$

b. $P(\text{Cardioid} \mid \text{defective}) = \dfrac{\text{number of defective Cartioid electrocardiographs}}{\text{number of defective electrocardiographs}} = \dfrac{45}{455} = 0.0989$

25. HIV. We'll make a table that displays these results if 1000 subjects were in the study so we can deal with whole numbers:

	HIV Present	HIV Absent	Total
Screening Test Correct	**95**	**855**	**950**
Screening Test Incorrect	**5**	**45**	**50**
Total	**100**	**900**	**1000**

a. $P(\text{HIV} \mid \text{positive test}) = \dfrac{\text{number of HIV with positive test}}{\text{number of accurate tests}} = \dfrac{95}{95+45} = \dfrac{95}{140} = 0.679$

b. $P(\text{positive test} \mid \text{HIV}) = \dfrac{\text{number of HIV with positive test}}{\text{number with HIV}} = \dfrac{95}{100} = 0.950$

3 – 6 Risks and Odds

In Exercises 1 – 9, use the data in the accompanying table (based on data from Pfizer, Inc.). That table describes results from a clinical trial of the well-known drug Viagra.

	Viagra Treatment	Placebo	Total
Headache	**117**	**29**	**146**
No Headache	**617**	**696**	**1313**
Total	**734**	**725**	**1459**

1. Study is prospective since the data are collected directly during the conduct of the study by observing what happens to the study participants after treatment.

3. Compare $P(\text{headache} \mid \text{Viagra})$ with $P(\text{headache} \mid \text{placebo})$

$P(\text{headache} \mid \text{placebo}) = \dfrac{29}{725} = 0.040$

Risk of getting a headache is higher in Viagra treatment group than placebo group.

5. Number needed to treat to stop headache

First we find the absolute relative risk reduction for no headache

$$\text{Absolute risk reduction} = \left|P(\text{event occurring in treatment}) - P(\text{event occurring in control})\right|$$

$$= \left|\frac{617}{734} - \frac{696}{725}\right| = |0.8406 - 0.9600| = 0.119$$

$$\text{Number needed to treat} = \frac{1}{0.119} = 8.4 \quad \text{(rounded up to 9)}$$

7. Relative Risk of headache in treatment group compared with placebo group, interpret

$$\text{Relative Risk} = \frac{P_t}{P_c} = \frac{\text{Proportion in Treatment Group}}{\text{Proportion in Control Group}} = \frac{117/734}{29/725} = \frac{0.1594}{0.0400} = 3.985$$

The incidence or risk of getting a headache for the Viagra treatment group is almost four times as high as it is for the placebo group.

9. Should Viagra users be concerned about headache as an adverse reaction?
There is some basis for being concerned since headaches occur much more often for the Viagra treatment group than for the placebo group.

In Exercises 11-19, use the data in the accompanying table (based on data from "A Case-Control Study of the Effectiveness of Bicycle Safety Helmets in Preventing Facial Injury," by Thompson, Rivara, and Wolf, American Journal of Public Health, Vol. 80, No. 12).

	Helmet Worn	No Helmet	Total
Facial Injuries Received	30	182	212
All Injuries nonfacial	83	236	319
Total	113	418	531

11. $P(\text{facial injuries} \mid \text{no helmet}) = \dfrac{\text{number with facial injuries in no helmet group}}{\text{number in no helmet group}} = \dfrac{182}{418} = 0.435$

No, this probability should not be used as an estimate of incidence since subjects were not selected at random from the population of all bicycle riders.

13. Absolute Risk Reduction for helmet group and no helmet group
Absolute risk reduction $= \left|P(\text{event occurring in treatment}) - P(\text{event occurring in control})\right|$

$$= \left|\frac{30}{113} - \frac{182}{418}\right| = |0.2654 - 0.4354| = 0.170$$

15. Odds ratio for facial injuries for no helmet group compared with helmet group, interpret

$$\text{Odds ratio} = \frac{\text{odds in favor of event for no helmet group}}{\text{odds in favor of event for helmet group}} = \frac{182/236}{30/83} = \frac{0.7712}{0.3614} = 2.134$$

The odds of receiving a facial injury compared with not receiving a facial injury are 2.134 times higher for the no helmet group than for the helmet group.

17. Yes, wearing a helmet appears to decrease the risk of facial injuries. The odds of having a facial injury when wearing a helmet are 0.361 while the odds of having a facial injury when not wearing a helmet are 0.771, more than twice as high.

19. Tripling frequencies in first row of original table

$$\text{Odds ratio} = \frac{\text{odds in favor of event for no helmet group}}{\text{odds in favor of event for helmet group}} = \frac{546/236}{90/83} = \frac{2.3136}{1.0843} = 2.134$$

$$\text{Relative Risk} = \frac{P_{nh}}{P_h} = \frac{\text{Proportion in no helmet group}}{\text{Proportion in helmet group}} = \frac{546/782}{90/173} = \frac{0.6982}{0.5202} = 1.342$$

After tripling the first row frequencies or the event occurrence frequencies, there is no change in the odds ratio, but there is a change in the relative risk. The odds ratio is consistent which ever application there is (retrospective or prospective), but the relative risk changes depending on the application. Since the definition of the treatment and outcomes are applied to data already collected (retrospective) rather than for data directly observed in a controlled setting, the relative risk should be used only in prospective studies rather than in either.

3 – 7 Rates of Mortality, Fertility and Morbidity

In Exercises 1 – 7, use the data given below (based on data from the U.S. Census Bureau) to find the indicated rates. Round results to one decimal place, and use a multiplying factor of k= 1000 unless indicated otherwise.

1. Neonatal mortality rate

$$\text{Neonatal mortality rate} = \frac{\text{number of deaths of infants under 28 days of age}}{\text{total number of live births}} k = \frac{18{,}000}{4{,}026{,}000} * 1000$$

$$= 0.00447 * 1000 = 4.5 \text{ (rounded)}$$

3. Perinatal mortality rate

$$\text{Perinatal mortality rate} = \frac{\text{number of fetal deaths} + \text{number of neonatal deaths}}{\text{number of live births} + \text{number of fetal deaths}} k$$

$$= \frac{39{,}000 + 18{,}000}{4{,}026{,}000 + 39{,}900} * 1000 = \frac{57{,}900}{4{,}065{,}900} * 1000 = 0.0142 * 1000 = 14.2$$

5. General fertility rate

$$\text{General fertility rate} = \frac{\text{number of live births}}{\text{number in women aged 15-44}} k = \frac{4{,}026{,}000}{61{,}811{,}000} * 1000$$

$$= 0.0651 * 1000 = 65.1 \text{ (rounded)}$$

7. HIV incidence rate

$$\text{HIV incidence rate} = \frac{\text{number of reported cases of HIV}}{\text{total number in population}} k$$

$$= \frac{900{,}000}{285{,}318{,}000} * 1000 = 0.00315 * 1000 = 3.2 \text{ (rounded)}$$

9. Finding Probability
Crude mortality rate= 8.5 per 1000
P(died during year)= 8.5/1000= 0.0085
The advantage of the crude mortality rate is that it is in the form of a specific number out of 1000 rather than in a fractional value with several decimal places. Thus, it is a little easier to comprehend what this means.

11. Finding Probability Crude death rate in Canada= 7.5

a. P(died within a year)= 7.5/1000= 0.0075

b. P(two randomly selected who died in year)= P(first died) • P(second died)= 0.0075 • 0.0075= 0.0000563

c. P(two randomly selected who didn't die)= P(first didn't die) • P(second didn't die)= (1 – 0.0075) • (1 – 0.0075)= 0.9925 • 0.9925= 0.9850563

13. Cause-of-Death Ratio.

a. cause-of-death ratio

$$\text{cause-of-death ratio} = \frac{\text{number of deaths due to heart disease}}{\text{total number of deaths}} k = \frac{700{,}142}{2{,}416{,}000} * 100 = 0.2898 * 100 = 28.98$$

b. P(three deaths are not due to heart disease)= $(1 - 0.2898)^3 = 0.7102^3 = 0.358$

15. Number of Deaths It should not be surprising that the number of deaths in the U.S. has been steadily increasing each year since the population would be steadily increasing and getting older. Also, better methods may be available for reporting deaths. This does not mean that anything is happening to the health of the nation.

17. Adjusted Mortality Rate

	0-24	25-64	65 and older	Total
Florida deaths	**4,085**	**35,458**	**128,119**	**167,662**
Florida population	**4,976,942**	**6,436,092**	**2,807,598**	**14,220,632**
U.S. deaths	**75,011**	**464,655**	**1,876,334**	**2,416,000**
U.S. population	**100,464,000**	**149,499,000**	**35,355,000**	**285,318,000**
Proportion in U.S. population	**0.3521**	**0.5240**	**0.1239**	**1.0000**
Adjusted Florida population= Florida total population * U.S. Proportion	**5,007,085**	**7,451,611**	**1,761,936**	**14,220,632**
Weighting factor= Adjusted Florida population/Original Florida population	**1.0061**	**1.1578**	**0.6276**	
Adjusted Florida deaths= Weighing factor * Florida deaths	**4,110**	**41,053**	**80,407**	**125,570**
Age Adjusted Mortality Rate= (Adjusted Florida deaths/Florida population) * 1000				**8.83**

The age adjusted crude mortality rate for Florida is 8.8, down from the non-age adjusted rate of 11.8, is closer to the U.S. crude mortality rate of 8.5.

3 – 8 Counting

In Exercises 1 – 7, evaluate the given expressions and express all results using the usual format for writing numbers (instead of scientific notation).

1. $6! = 6 * 5 * 4 * 3 * 2 * 1 * \quad 0 = 720$

3. $_{25}P_2 = \frac{25!}{(25-2)!} = \frac{25!}{23!} = \frac{25 * 24 * 23!}{23!} = 25 * 24 = 600$

5. $_{25}C_2 = \frac{25!}{(25-2)!\,2!} = \frac{25!}{23!\,2!} = \frac{25 * 24 * 23!}{23!\,2!} = \frac{25 * 24}{2!} = \frac{25 * 24}{2 * 1} = \frac{600}{2} = 300$

7. $_{52}C_5\ _{52}C_5 = \frac{52!}{(52-5)!\,5!} = \frac{52 * 51 * 50 * 49 * 48 * 47!}{47!\,5!} = \frac{52 * 51 * 50 * 49 * 48}{5 * 4 * 3 * 2 * 1} = \frac{311{,}875{,}200}{120} = 2{,}598{,}960$

9. Gender Sequences Two possibilities for each child, number of possible sequences of gender of five babies, number of sequences of five things, each with two possibilities= 2^5= 32

11. Tree Growth Experiment Four treatments: none (N), irrigation only (I), fertilizer only (F), irrigation and fertilization (B). 10 trees selected, how many treatment arrangements possible?
Four possibilities for each of 10 trees

Number of sequences of ten trees, each with one of four possible outcomes= 4^{10}= 1,048,576

13. Age Discrimination

Number of possible selections= ${}_{32}C_4 = \frac{32!}{(32-4)!\,4!} = \frac{32*31*30*29*28!}{28!\,4!} = \frac{863{,}040}{24} = 35{,}960$

Only one way four oldest can be selected

P(four oldest were selected out of 32)= $\frac{1}{35{,}960} = 0.0000278$

This probability is certainly low enough to suspect that the hospital may have discriminated against older physicians since by chance, this outcome of selecting the four oldest physicians would have happened about 3 times in 100,000 times.

15. Selection of Treatment Groups

Six subjects selected from a pool of 15 volunteers

First find number of possible selections in the form of combinations of 6 things selected from 15 things, so we find ${}_{15}C_6$

$${}_{15}C_6 = \frac{15!}{(15-6)!\,6!} = \frac{15*14*13*12*11*10*9!}{9!\,6!} = \frac{15*14*13*14*13*12*11*10}{6*5*4*3*2*1} = \frac{3{,}603{,}600}{720} = 5{,}005$$

There is only one way out of 5,005 samples that the six youngest can be selected

P(six youngest are selected)= $\frac{1}{5005} = 0.000200$

Should this happen, it would appear to be highly likely the sample was selected based on age rather than at random since this would have happened about twice out of 10,000 samples.

17. Social Security Numbers Nine digits in SS number, each can be 0-10

P(of a given SS number)= $\frac{1}{10^9} = \frac{1}{1{,}000{,}000{,}000} = 0.000000001$

19. Elected Board of Directors

a. 12 members of the board of directors, number of slates of candidates of four officers
Order is important since the officer positions are different, so we find ${}_{12}P_4$

$${}_{12}P_4 = \frac{12!}{(12-4)!} = \frac{12*11*10*9*8!}{8!} = 12*11*10*9 = 11{,}880$$

b. Order would not be important, just the number of combination, so we find ${}_{12}C_4$

$${}_{12}C_4 = \frac{12!}{(12-4)!\,4!} = \frac{12*11*10*9*8!}{8!\,4!} = \frac{12*11*10*9}{4*3*2*1} = 495$$

21. Is the Researcher Cheating? 20 newborns, consistently gets 10 girls and 10 boys

a. Number of gender sequences= 2^{20}= 1,048,576

b. Number of gender sequences of 10 boys and 10 girls

$$\text{no. of sequences} = \frac{n!}{n_1!\,n_2!} = \frac{20!}{10!\,10!} = \frac{20*19*....*11*10!}{10!*10!} = \frac{670{,}442{,}572{,}800}{3{,}628{,}800} = 184{,}756$$

c. P(equal number of boys and girls out of 20)= $\frac{\text{number of equal seq.}}{\text{total number of seq.}} = \frac{184{,}756}{1{,}048{,}576} = 0.176$

d. We would tend to disagree with the researcher since the probability of this happening by chance would be less that 2 out of 10 times. This is possible, but not likely.

Review Exercises

In Exercises 1-7, use the data in the accompanying table (based on data from Parke-Davis). The cholesterol leveling drug Lititor consists of atorvastatin calcium.

	10 mg. Atorvastatin	Placebo	Total
Headache	15	65	80
No headache	17	3	20
Total	32	68	100

1. $P(\text{headache})=\dfrac{\text{number of times headache occurs}}{\text{number of subjects in study}}=\dfrac{80}{100}=0.800$

3. $P(\text{headache or treated with Atorvastatin})=\dfrac{\text{number of times happens}}{\text{number of subjects in study}}=\dfrac{97}{100}=0.970$

5. P(two selected that both used placebos)= 68/100 * 67/99= 0.6800 * 0.6768= 0.460

7. P(headache | treated with Atorvastatin)= 15/32= 0.469

9. **Acceptance Sampling** 2500 aspirin tablets, 2% are defective
P(at least one defect)= 1 – P(none of the four samples are defective)=
1 – [P(first is not defective) * P(second is not defective) *...* P(fourth is not defective)]=
$1-0.98^4= 1-0.9224= 0.0776$

11. **Selecting Members** 10 members of board

a. P(three wealthiest members selected)
First find number of possible selections in the form of combinations of 3 things selected from 10 things, order is not important so we find $_{10}C_3$

$$_{10}C_3=\frac{10!}{(10-3)!\,3!}=\frac{10*9*8*7!}{7!\,3!}=\frac{10*9*8}{3*2*1}=\frac{720}{6}=120$$

There is only one way out of 120 samples that the three wealthiest can be selected

$P(\text{three wealthiest out of 10 are selected})=\dfrac{1}{120}=0.00833$

b. Different slates of three officers to be elected from 10 members order is important since the officer positions are different, so we find $_{10}P_3$

$$_{10}P_3=\frac{10!}{(10-3)!}=\frac{10*9*8*7!}{7!}=10*9*8=720$$

13. **Chlamydia Rate** Rate was 278.32 per 100,000 population

a. $P(\text{selected person has Chlamydia})=\dfrac{\text{number out of 100,000}}{100{,}000}=\dfrac{278.32}{100{,}000}=0.0027832$

b. P(two selected have Chlamydia)= P(first has Chlamydia) * P(second has Chlamydia)=
0.002783 * 0.002783= 0.00000775

c. P(two selected don't have Chlamydia)= P(first doesn't) * P(second doesn't)=
0.997217 * 0.997217= 0.9944413

Cumulative Review Exercises

1. Treating Chronic Fatigue Syndrome

a. Mean= $\bar{x} = \frac{x}{n} = \frac{84}{21} = 4.00$

b. Median= middle score since odd number of score= 11th score out of 21= 4.00

c. Standard deviation (to find standard deviation and variance, lets first find the sum of squares SS)

$$\text{Sum of Squares} = \frac{n(\ x^2) - (\ x)^2}{n} = \frac{21(430) - (84)^2}{21} = \frac{9030 - 7056}{21} = \frac{1974}{21} = 94.00$$

$$s = \sqrt{\frac{SS}{n-1}} = \sqrt{\frac{94.00}{21-1}} = \sqrt{\frac{94.00}{20}} = \sqrt{4.700} = 2.168$$

d. Variance= s^2= 2.168^2= 4.700

e. Yes, if it had not been effective, it is likely the mean would be close to zero rather than 4, which was 1.85 standard deviations above zero.

f. P(randomly selected would be positive)= $\frac{\text{number above 0}}{\text{total number of scores}} = \frac{18}{21} = 0.857$

g. P(two selected, both positive)= P(first is positive) * P(second is positive)=
18/21 * 17/20= 0.8571 * 0.8500= 0.729

h. P(18 out of 18 are positive)
number of different sequences= 2^{18}= 262,144
number of ways all can be positive is one
P(18 out of 18 are positive)= 1/262,144= 0.00000381
Yes, the treatment appears to be effective since the probability that this result could have happened by chance when the program was not effective would be about four times out of a million.

Chapter 4. Discrete Probability Distributions

4 – 2 Random Variables

In Exercises 1, identify the given random variable as being discrete or continuous.

1. **a.** Height of giraffe is a continuous random variable, since the observation can be a fractional value.
b. Number of bald eagles is a discrete random variable, since the observation can take on only whole numbers.
c. Exact gestation time is a continuous random variable, since the observation can be a fractional value.
d. Number of blue whales is a discrete random variable, since the observation can take on only whole numbers.
e. Number of manatees killed is a discrete random variable, since the observation can take on only whole numbers.

In Exercises 3 – 7, determine whether a probability distribution is given. In those cases where a probability distribution is not described, identify the requirements that are not satisfied. In those cases where a probability is described, find its mean and standard deviation.

There are two requirements for a probability distribution:
1. Each $P(x)$ must be equal to or greater than 0 and equal to or less than 1.
2. The sum of probabilities, $\Sigma P(x)$, must be equal to 1.

3. **Gender Selection** This is a probability distribution since for each x, $0 \le P(x) \le 1$ and
$\Sigma P(x) = 0.125 + 0.375 + 0.375 + 0.125 = 1.000$
Mean $(\mu) = \Sigma[\, x * P(x)] = (0 * 0.125) + (1 * 0.375) + (2 * 0.375) + (3 * 0.125) = 1.50$
Standard Deviation $\sigma = \sqrt{\ [x^2 * P(x)] - \mu^2} = \sqrt{(0^2 * 0.125) + (1^2 * 0.375) + (2^2 * 0.375) + (3^2 * 0.125) - 1.5^2} =$
$\sqrt{3.00 - 1.5^2} = \sqrt{3.00 - 2.25} = \sqrt{0.75} = 0.866$

5. **Genetics Experiment** This is not a probability distribution since $\Sigma P(x) = 0.94 \ne 1.00$

7. **Genetic Disorder** This is a probability distribution since for each x, $0 \le P(x) \le 1$ and
$\Sigma P(x) = 0.4219 + 0.4219 + 0.1406 + 0.0156 = 1.0000$
Mean, $\mu = \Sigma[\, x * P(x)] = (0 * 0.4219) + (1 * 0.4219) + (2 * 0.1406) + (3 * 0.0156) = 0.75$
Standard Deviation

$$\sigma = \sqrt{\ [x^2 * P(x)] - \mu^2} = \sqrt{(0^2 * 0.4219) + (1^2 * 0.4219) + (2^2 * 0.1406) + (3^2 * 0.0156) - 0.75^2} =$$

$$\sqrt{1.1247 - 0.75^2} = \sqrt{1.1247 - 0.5625} = \sqrt{0.5622} = 0.75$$

9. **Gender Selection Technique Effectiveness** From Table 4-1
a. $P(9) = 0.122$
b. $P(9 \text{ or more}) = 0.122 + 0.061 + 0.022 + 0.006 + 0.001 + 0.000 = 0.212$
c. The probability from part (b) since any outcome 9 or above achieves the same criterion of being unusually high.
d. No, $P(9 \text{ or more})$ is not unusual, $P(9 \text{ or more}) > 0.05$. This would happen by chance about one time out of every five samples. We would conclude there is not sufficient evidence to conclude that the technique is effective.

11. **Gender Selection Technique Effectiveness** From Table 4-1
a. Include probabilities of 11 or more, $P(11 \text{ or more}) = 0.022 + 0.006 + 0.001 = 0.029$
b. Yes, $P(11 \text{ or more})$ is unusual, $P(11 \text{ or more}) < 0.05$. This would happen by chance only about three times out of every 100 samples. We would conclude there is sufficient evidence to conclude that the technique is effective.

13. Finding Mean and Standard Deviation Possible outcomes of gender of four children:

Outcome	# girls (x)	Outcome	# girls (x)
		GGGB	3
BBBB	0	GGBG	3
BBBG	1	GBGG	3
BBGB	1	BGGG	3
BGBB	1	GGGG	4
GBBB	1		
BBGG	2	Number of 0	1
BGGB	2	Number of 1	4
GGBB	2	Number of 2	6
GBGB	2	Number of 3	4
BGBG	2	Number of 4	1
GBBG	2		

There are 16 possible different outcomes (permutations).

Number of Girls (x) out of four	$P(x)$= # girls/16	$x * P(x)$	x^2	$x^2 * P(x)$
0	1/16= 0.0625	0.0000	0	0.0000
1	4/16= 0.2500	0.2500	1	0.2500
2	6/16= 0.3750	0.7500	4	1.5000
3	4/16= 0.2500	0.7500	9	2.2500
4	1/16= 0.0625	0.2500	16	1.0000
Total	1.0000	$\sum[x * P(x)] = 2.0$		$\sum[x^2 * P(x)] = 5.0$

Mean, $\mu = \Sigma[(x) * P(x)] = 2.00$

Standard deviation, $\sigma = \sqrt{\Sigma[x^2 * P(x)] - \mu^2} = \sqrt{5.0 - 2.0^2} = \sqrt{5.0 - 4.0} = \sqrt{1.00} = 1.00$

4 – 3 Binomial Probability Distributions

In Exercises 1 – 7, determine whether the given procedure results in a binomial distribution. For those that are not binomial, identify at least one requirement that is not satisfied.

There are four requirements for a binomial distribution:

1. There are a fixed number of trials or observations
2. The trials are independent
3. Each trial has two possible outcomes
4. The probabilities remain constant for each trial

1. This is not a binomial distribution. The number of trials (people) is not fixed. Possible outcomes are not classified into two categories; they could be any number. There could be any number of different answers to the question.

3. This is not a binomial distribution. The outcomes (answers) are not classified into two categories.

5. This is a binomial distribution, all requirements are met. The outcomes are "male or female".

7. This is not a binomial distribution. The outcomes (answers) are not classified into two categories. The couples could have any number of children.

9. **Finding Probabilities when Guessing Answers** *P*(wrong) = 4/5 = 0.8, *P*(correct) =1/5 = 0.2

a. *P*(WWC) = 0.8 * 0.8 * 0.2 = 0.128

b. Three possible arrangements: WWC, WCW, CWW

P(WWC) = 0.8 * 0.8 * 0.2 = 0.128

P(WCW) = 0.8 * 0.2 * 0.8 = 0.128

P(CWW) = 0.2 * 0.8 * 0.8 = 0.128

c. *P*(exactly one correct answer in 3 guesses) = *P*(WWC) + *P*(WCW) + *P*(CWW)= 0.128 + 0.128 + 0.128= 0.384

In Exercises 11 – 15, assume that a procedure yields a binomial distribution with a trial repeated n times. Use Table A-1 to find the probabilities of x successes given the probability p of success on a given trial.

11. n= 2, x= 0, p=0.01 0.01 Column, 2-0 Row $P(0)$= 0.980

13. n= 4, x= 3, p= 0.95 0.95 Column, 4-3 Row $P(3)$= 0.171

15. n=10, x= 4, p= 0.95 0.95 Column, 10-4 Row $P(4)$= 0.000^+

Note: for answer 15, the numbers are greater than zero, but when rounded to three decimal places, they are 0.000.

In Exercises 17 & 19, assume *that a procedure yields a binomial distribution with a trial repeated n times. Use the binomial probability formula to find the probabilities of x successes given the probability p of success on a given trial.*

17. n=6, x= 4, p= 0.55

$$P(x) = \frac{n!}{(n-x)!x!} * p^x * q^{n-x}$$

$$P(4) = \frac{6!}{(6-4)!4!} * 0.55^4 * 0.45^{6-4} = \frac{6*5*4!}{2!4!} * 0.55^4 * 0.45^2 =$$

$$15 * 0.0915 * 0.2025 = 0.280$$

19. n= 8, x= 3, p= 0.25

$$P(x) = \frac{n!}{(n-x)!x!} * p^x * q^{n-x}$$

$$P(3) = \frac{8!}{(8-3)!3!} * 0.25^3 * 0.75^{8-3} = \frac{8*7*6*5!}{5!3!} * 0.25^3 * 0.75^5 =$$

$$56 * 0.01563 * 0.2373 = 0.208$$

In Exercises 21 & 23, refer to the Minitab display in the margin. The probabilities were obtained by entering the values of n= 6 and p= 0.167. In a clinical test of the drug Lipator (atorvastatin), 16.7% of the subjects treated with 10 mg of atorvastatin experienced headaches (based on data from Parke-Davis). In each case, assume that 6 subjects are randomly selected and treated with 10 mg of atorvastatin, then find the indicated probability.

21. *P*(at least 5 have headache)= *P*(5 or 6)= *P*(5) + *P*(6)= 0.0006 + 0.0000= 0.0006
It would be very unusual for 5 out of 6 subjects to get a headache. This would happen only 6 times out of 1000 samples by chance.

23. *P*(more than 1 have headache)= *P*(2 or 3 or 4 or 5 or 6)= *P*(2) + *P*(3) + *P*(4) + *P*(5) + *P*(6)=
0.2014 + 0.0538 + 0.0081 + 0.0006 + 0.0000= 0.2639
P(not having more than one headache out of 6) = 1 – *P*(2, 3, 4, 5, or 6)= 1.0000 – 0.2639= 0.7361). It would not be unusual to not have more than 1 headache out of 6 patients.

25. Drug Reaction n= 8, p= 0.04

a. x= 3, *P*(3 out of 8)

$$P(x)=\frac{n!}{(n-x)!x!}*p^x*q^{n-x}$$

$$P(8)=\frac{8!}{(8-3)!3!}*0.04^3*0.96^{8-3}=\frac{8*7*6*5!}{5!3!}*0.04^3*0.96^5=$$

$$56*0.000064*0.8154=0.00292$$

b. *x*= 8, *P*(8 out of 8)

$$P(x)=\frac{n!}{(n-x)!x!}*p^x*q^{n-x}$$

$$P(8)=\frac{8!}{(8-8)!8!}*0.04^8*0.96^0=\frac{8!}{0!8!}*0.04^8*0.96^0=$$

$$1*0.00000000000655*1=0.00000000000655=0.000^+$$

c. If all 8 experienced headaches, this would be evidence that this placebo group was certainly different than the 4% group. It is highly unlikely the all 8 in the placebo group would have a headache.

27. Acceptance Sampling *n*= 24, *p*= 0.04, *x*= 1 or 0, *P*(0 or 1)= *P*(0) + *P*(1)

$$P(x)=\frac{n!}{(n-x)!x!}*p^x*q^{n-x}$$

$$P(0)=\frac{24!}{(24-0)!0!}*0.04^0*0.96^{24-0}=\frac{24!}{24!0!}*0.04^0*0.96^{24}=1*1*0.3754=0.3754$$

$$P(1)=\frac{24!}{(24-1)!1!}*0.04^1*0.96^{24-1}=\frac{24*23!}{23!1!}*0.04^1*0.96^{23}=24*0.04*0.3911=0.3754$$

$$P(0\text{ or }1)=P(0)+P(1)=0.3754+0.3754=0.7508$$

The probability of this shipment being accepted is 0.751.

29. Identifying Gender Discrimination *n*= 20, *p*= 0.5, *P*(2 or less)

$$P(x)=\frac{n!}{(n-x)!x!}*p^x*q^{n-x}$$

$$P(0)=\frac{20!}{(20-0)!0!}*0.5^0*0.5^{20-0}=\frac{20!}{20!0!}*0.5^0*0.5^{20}=1*1*0.000000953=0.000000953$$

$$P(1)=\frac{20!}{(20-1)!1!}*0.5^1*0.5^{20-1}=\frac{20*19!}{19!1!}*0.5^1*0.5^{19}=20*0.5*0.00000191=0.0000191$$

$$P(2)=\frac{20!}{(20-2)!2!}*0.5^2*0.5^{20-2}=\frac{20*19*18!}{18!2!}*0.5^2*0.5^{18}=190*0.25*0.00000381=$$

0.0001812

$$P(2\text{ or less})=P(0)+P(1)+P(2)=0.000000953+0.0000191+0.00018120=0.000201$$

Yes, this result would tend to support evidence that discrimination occurred since by chance this would have happened about 2 times out of 1000 samples.

31. Geometric Distribution x= 7, p= 0.2

$$P(x)=p(1-p)^{x-1}=0.2(1\text{-}0.2)^{7-1}=0.2*0.8^6=0.0524$$

33. **Multinomial Distribution**, n= 20, Six categories (genetic genotypes): A, B, C, D, E, and F
Find P(5A's and 4B's and 3C's and 2D's and 3E's)
$P(A)= p_1$, $P(B)= p_2$, $P(C)= p_3$, $P(D)= p_4$, $P(E)= p_5$, $P(F)= p_6$

$$p_1 = p_2 = p_3 = p_4 = p_5 = p_6 = \frac{1}{6} = 0.16667$$

$$\frac{n!}{(x_1!)(x_2!)(x_3!)(x_4!)(x_5!)(x_6!)} p_1^{x_1} p_2^{x_2} p_3^{x_3} p_4^{x_4} p_5^{x_5} p_6^{x_6} =$$

$$\frac{20!}{(5!)(4!)(3!)(2!)(3!)(3!)} * 0.16667^5 * 0.16667^4 * 0.16667^3 * 0.16667^2 * 0.16667^3 * 0.16667^3 =$$

$$1.95546E12 * 0.000129 \quad * 0.000772 * 0.004630 * 0.027779 * 0.004630 * 0.004630 =$$

$$1.95546E12 * 2.73818E-16 = 0.000535$$

4 – 4 Exercises

In Exercises 1 & 3, assume that a procedure yields a binomial distribution with n trials and the probability of success for one trial is p. Use the given values of n and p to find the mean μ and standard deviation σ. Also, use the range rule of thumb to find the minimal usual value $\mu - 2\sigma$ and the maximum usual value $\mu + 2\sigma$.

1. $n= 400$, $p= 0.2$, $q= 1 - p= 1 - 0.2= 0.8$
$\mu = np = 400 * 0.2 = 80.0$

$\sigma = \sqrt{npq} = \sqrt{400 * 0.2 * 0.8} = \sqrt{64.0} = 8.0$

maximum usual value $= \mu + 2\sigma = 80.0 + 2 * 8.0 = 96.0$

minimum usual value $= \mu - 2\sigma = 80.0 - 2 * 8.0 = 64.0$

3. $n= 1984$, $p= 0.75$, $q= 1 - p= 1 - 0.75= 0.25$
$\mu = np = 1984 * 0.75 = 1488$

$\sigma = \sqrt{npq} = \sqrt{1984 * 0.75 * 0.25} = 19.29$

maximum usual value $= \mu + 2\sigma = 1488 + 2 * 19.29 = 1526.6$

minimum usual value $= \mu - 2\sigma = 1488 - 2 * 19.29 = 1449.4$

5. **Guessing Answers**, guessing 7 out of 10 correct when $n= 10$ and $p= 0.5$

a. $n=10$, $p= ½= 0.5$, $q= 1 - p = 1 - 0.5= 0.5$
$\mu = np = 10 * 0.5 = 5$

$\sigma = \sqrt{npq} = \sqrt{10 * 0.5 * 0.5} = \sqrt{2.50} = 1.58$

maximum usual value$= \mu + 2\sigma = 5 + 2 * 1.58 = 8.16$

minimum usual value $= \mu - 2\sigma = 5 - 2 * 1.58 = 1.84$

b. No, the maximum usual value is 8.16, so a value of 7 correct out of 10 would not be considered unusual since it is between 1.84 and 8.16.

7. **Playing Roulette** $P(\text{winning})= 1/38= 0.0263$, number of trials$= n= 100$

a. $n=100$, $p= 0.0263$, $q= 1 - p = 1 - 0.263= 0.974$
$\mu = np = 100 * 0.0263 = 2.63$

$\sigma = \sqrt{npq} = \sqrt{100 * 0.0263 * 0.974} = \sqrt{2.562} = 1.60$

minimum usual value $= ? - 2? = 2.63 - 2 * 1.60 = 2.63 - 3.20 = -0.57$

maximum usual value $= ? + 2? = 2.63 + 2 * 1.60 = 2.63 + 3.20 = 5.83$

b. No, it would not be unusual to not win at least once in the 100 rotations of the wheel since 0 is in the range of minimum usual value to maximum usual values.

9. Amazing Results of Experiments in Gender Selection 15 couples in control group and 15 couples in experimental group, each group has one child

a. Possible outcomes of variable x, number of girls, $P(\text{girl})= 0.50$
Using Table A-1, $n= 15$, $p= 0.50$

x (number of girls)	$P(x)$
0	0.000
1	0.000
2	0.003
3	0.014
4	0.042
5	0.092
6	0.153
7	0.196
8	0.196
9	0.153
10	0.092
11	0.042
12	0.014
13	0.003
14	0.000
15	0.000

b. $n= 15$, $p= 0.50$, $q= 0.50$
$\mu = np = 15 * 0.50 = 7.50$
$\sigma = \sqrt{npq} = \sqrt{15 * 0.50 * 0.50} = \sqrt{3.75} = 1.94$

c. Is 10 girls and 5 boys unusual?
maximum usual value $= \mu + 2\sigma = 7.5 + 2 * 1.94 = 11.38$
mimmum usual value $= \mu - 2\sigma = 7.5 - 2 * 1.94 = 3.62$
No, 10 girls out of 15 would not be unusual since 10 is in the range of minimum usual value to maximum usual values.

11. Cholesterol-Reducing Drug 863 patients given Lipitor, 19 experienced flu symptoms, $P(\text{flu for not treated})= 0.019$

a. $n= 863$, $p= 0.019$, $q= 0.981$
$\mu = np = 863 * 0.019 = 16.40$
$\sigma = \sqrt{npq} = \sqrt{863 * 0.019 * 0.981} = \sqrt{16.085} = 4.01$

b. Is it unusual to find 19 out of 863 with flu symptoms?
maximum usual value $= \mu + 2\sigma = 16.40 + 2 * 4.01 = 24.42$
minimum usual value $= \mu - 2\sigma = 16.40 - 2 * 4.01 = 8.38$
No, it is not unusual to have 19 patients with flu symptoms out of 863 patients since 19 is in the range of the minimum usual value to maximum usual values.

c. Flu symptoms do not appear to be an adverse reaction that should be of concern to users of Lipitor.

13. Car Crashes 34% of those in age 20-24 have car crashes in one year, in a sample of 500 randomly selected New Your City drivers age 20-24, 42% had accidents

a. Number in sample having accidents= 42/100 * 500= 0.42 * 500= 210

b. $n= 500$, $p= 0.34$, $q= 0.66$
$\mu = np = 500 * 0.34 = 170$
$\sigma = \sqrt{npq} = \sqrt{500 * 0.34 * 0.66} = \sqrt{112.20} = 10.59$

c. Is 42% unusually high for the NYC drivers in this age group?
maximum usual value $= \mu + 2\sigma = 170 + 2 * 10.59 = 191.18$
minimum usual value $= \mu - 2\sigma = 170 - 2 * 10.59 = 148.82$

Yes, the result of 42% or 210 drivers is unusually high compared with 34% rate in the general population of 20-24 year olds since 210 is not in the range of usual minimum and maximum values. 210 accidents is higher than the usual maximum value.

4 – 5 Exercises

In Exercises 1 & 3, assume the Poisson Distribution applies and Proceed to use the given mean to find the indicated probability.

1. $\mu = 2,\ x = 3,\ P(3) = \dfrac{\mu^x * e^{-\mu}}{x!} = \dfrac{2^3 * e^{-2}}{3!} = \dfrac{8 * 2.71828^{-2}}{6} = \dfrac{1.0827}{6} = 0.1804$

3. $\mu = 100,\ x = 99,\ P(x) = \dfrac{100^{99} * e^{-100}}{99!} = \dfrac{3.72E^{154}}{9.333E^{155}} = 0.0399$ (Excel used for this computatio

5. **Radioactive Delay** for Cessium 137, over 365 days decay was 1,000,000,000 down to 977,287

a. find mean decay per day

$$\mu = \frac{\text{number of atoms}}{\text{number of days}} = \frac{1{,}000{,}000 - 977{,}287}{365} = \frac{22{,}713}{365} = 62.227$$

b. P(on a given day, 50 atoms lost)

$$P(50 \text{ on a given day}) = \frac{62.227^{50} * e^{-62.227}}{50!} = \frac{4.9997E89 * 9.4444E-28}{3.0414E64}$$

$$= \frac{4.7219E62}{3.0414E64} = 0.0155$$

7. **Deaths from Horse Kicks** 196 horse kick deaths during 280 corps years

$$\mu = \frac{\text{number of deaths}}{\text{number of corpsyears}} = \frac{196}{280} = 0.7$$

In the following: $e^{-0.7} = 2.71828^{-0.7} = 0.4966$

a. $P(0) = \dfrac{0.7^0 * e^{-0.7}}{0!} = \dfrac{0.4966}{1} = 0.497$

b. $P(1) = \dfrac{0.7^1 * e^{-0.7}}{1!} = \dfrac{0.7 * 0.4966}{1} = \dfrac{0.3476}{1} = 0.348$

c. $P(2) = \dfrac{0.7^2 * e^{-0.7}}{2!} = \dfrac{0.49 * 0.4966}{2} = \dfrac{0.2433}{2} = 0.122$

d. $P(3) = \dfrac{0.7^3 * e^{-0.7}}{3!} = \dfrac{0.343 * 0.4966}{6} = \dfrac{0.1703}{6} = 0.0284$

e. $P(4) = \dfrac{0.7^4 * e^{-0.7}}{4!} = \dfrac{0.2401 * 0.4966}{24} = \dfrac{0.1192}{24} = 0.00497$

Comparison of actual results with Poisson Distribution for 280 Corps-years

Deaths	Expected number of corps-years from Poisson distribution	Actual number of horse kick deaths in corps-years
0	0.4973*280= 139.24	144
1	0.3481*280= 97.47	91
2	0.1218*280= 34.10	32
3	0.0284*280= 7.95	11
4	0.0050*280= 1.40	2

Yes, the Poisson distribution serves as a good device for predicting the actual results. There are some deviations that would be expected by chance. In general the shape and frequency of the values are very similar.

9. **Dandelions** mean number of dandelions in a given area is 7 per square meter
μ= 7 per square meter
In the following: $e^{-7} = 2.71828^{-7} = 0.0009119$

a. $P(0) = \frac{7^0 * e^{-7}}{0!} = \frac{1 * 0.0009119}{1} = \frac{0.0009119}{1} = 0.000912$

b. $P(\text{at least } 1) = 1 - P(0) = 1 - 0.000912 = .999088$

c. $P(1) = \frac{7^1 * e^{-7}}{1!} = \frac{7 * 0.0009119}{1} = \frac{0.0009119}{1} = 0.00638$

$P(2) = \frac{7^2 * e^{-7}}{2!} = \frac{49 * 0.0009119}{2} = \frac{0.04468}{2} = 0.0223$

$P(2 \text{ at most}) = P(0) + P(1) + P(2) = 0.000912 + 0.00638 + 0.0223 = 0.0296$

Review Exercises

1. **a.** A random variable is a variable that has a single value, usually numeric, determined by chance, for each outcome of a procedure or experiment.

b. A probability distribution is a description of the possible values a random variable can assume with the associated probability for each possible outcome. A valid probability distribution has values of the probabilities of $0 \le P(x) \le 1$ for each value of x and the sum of the probabilities must equal 1.

c. The accompanying table describes a probability distribution because the sum of all the probabilities = 1, and each probability value is $0 \le P(x) \le 1$.

d. Mean of this distribution, $\mu = \Sigma[x * P(x)] =$

$(0 * 0.528) + (1 * 0.360) + (2 * 0.098) + (3 * 0.013) + (4 * 0.001) + (5 * 0.000) =$

$0 + 0.360 + 0.196 + 0.039 + 0.004 + 0.000 = 0.599$

e. Standard deviation of this distribution, $? = \sqrt{\Sigma[x^2 * P(x)] - \mu^2} =$

$\sqrt{(0^2 * 0.528) + (1^2 * 0.360) + (2^2 * 0.098) + (3^2 * 0.013) + (4^2 * 0.001) + (5^2 * 0.000) - 0.599^2} =$

$\sqrt{0.885 - 0.359} = \sqrt{0.526} = 0.725$

f. Yes, it would be very unusual to select 5 condoms at random that all failed the test. That would happen very infrequently, less than 5 times out of 100 repetitions, a standard we often use to judge unusual events.

3. **Reasons for Being Fired** 17% indicate "inability to get along with others" as reason for firing at the Kansas Agriculture Company
n= 5, p= 0.17, q= 0.83, x= 4

a. P(at least 4 out of 5 cite this reason)

$P(x) = \frac{n!}{x!\,(n-x)!} * p^x * q^{n-x}$

$P(4) = \frac{5!}{4!(5-4)!} 0.17^4 * 0.83^{5-4} = \frac{5 * 4!}{4!\,1!} * 0.000835 * 0.83^1 = \frac{5}{1} * 0.000835 * 0.83 =$

0.00347

$P(5) = \frac{5!}{5!(5-5)!} 0.17^5 * 0.83^{5-5} = \frac{5!}{5!} * 0.000142 * 0.83^0 = 1 * 0.000142 * 1 = 0.000142$

$P(4 \text{ or more}) = P(4) + P(5) = 0.00347 + 0.000142 = 0.00361$

b. If she finds four out of 5 firings due to "inability to get along with others" as the reason, then this company would differ from other companies. The probability of this happening by chance would be very low, about 4 times out of 1000 and this would be considered to be very unusual compared with the other companies where the percentage giving this as the reason is 17%.

Cumulative Review Exercises

1. **Weights: Analysis of Last Digits** last digits in data, 0 – 9, distribution of frequency each digit is observed.

a. Mean and standard deviation

x	f	$f*x$	x^2	$f*x^2$
0	7	0	0	0
1	14	14	1	14
2	5	10	4	20
3	11	33	9	99
4	8	32	16	128
5	4	20	25	100
6	5	30	36	180
7	6	42	49	294
8	12	96	64	768
9	6	54	81	486
	n= 78	Σx= 331		Σx^2= 2089

$$\mu = \frac{x}{n} = \frac{331}{78} = 4.244$$

$$\sigma = \sqrt{\frac{x^2 - \frac{(\ x)^2}{n}}{n}} = \sqrt{\frac{2089 - \frac{331^2}{78}}{78}} = \sqrt{\frac{2089 - \frac{109561}{78}}{78}} = \sqrt{\frac{2089 - 1404.63}{78}} =$$

$$\sqrt{\frac{684.37}{78}} = \sqrt{8.774} = 2.962$$

b. Relative Frequency Table

x	f	$P(x)$	$x*P(x)$	x^2	$x^2*P(x)$
0	7	0.0875	0.0000	0	0.0000
1	14	0.1750	0.1750	1	0.1750
2	5	0.0625	0.1250	4	0.2500
3	11	0.1375	0.4125	9	1.2375
4	8	0.1000	0.4000	16	1.6000
5	4	0.0500	0.2500	25	1.2500
6	5	0.0625	0.3750	36	2.2500
7	6	0.0750	0.5250	49	3.6750
8	12	0.1500	1.2000	64	9.6000
9	6	0.1000	0.9000	81	8.1000
	n= 78	$\sum P(x)$= 1.00	4.3625		28.1375

$$\text{Mean} = \sum[(x * P(x)] = 4.363$$

$$\sigma^2 = \sum\left[\mathrm{x}^2 * P(x)\right] - \mu^2 = 28.138 - 4.363^2 = 28.138 - 19.036 = 9.102$$

$$\sigma = \sqrt{9.102} = 3.017$$

c. Probability Distribution where each digit equally likely

x	$P(x)$	$x * P(x)$	x^2	$x^2 * P(x)$
0	0.1	0.0	0	0.0
1	0.1	0.1	1	0.1
2	0.1	0.2	4	0.4
3	0.1	0.3	9	0.9
4	0.1	0.4	16	1.6
5	0.1	0.5	25	2.5
6	0.1	0.6	36	3.6
7	0.1	0.7	49	4.9
8	0.1	0.8	64	6.4
9	0.1	0.9	81	8.1
Total	$\Sigma(x)=1$	$\sum[x*P(x)] = 4.5$		$\sum[x^2 * P(x)] = 28.5$

Mean, $\mu = \Sigma[x * P(x)] = 4.5$

Standard deviation, $\sigma = \sqrt{\Sigma[x^2 * P(x)] - \mu^2} = \sqrt{28.5 - 20.25} = \sqrt{8.25} = 2.872$

d. The above table (**c**) describes what we would expect from a random selection of last digits-an equal likelihood of each occurring. However, in Tables a and b, which describe our sample, we see some deviation from this expected distribution. We would have expected the relative frequency to be as similar to our probability distribution in Table c as possible. While the means and standard deviations do not vary by a great deal, there are some digits that seem to occur more often than expected (1, 3, and 8) and a few that seem to occur less often than expected (2, 5, and 6). This could be assessed more accurately with another method that will be discussed later.

Chapter 5. Normal Probability Distributions

5-2 The Standard Normal Distribution

In Exercises 1 & 3, refer to the continuous uniform distribution depicted in figure 5-2, assume that a class length between 50.0 min and 52.0 min is randomly selected, and find the probability that the given time is selected.

1. P(class less than 50.3 minutes)= $0.5 * (50.3-50) = 0.5 * 0.3 = 0.15$

3. P(class between 50.5 minutes and 50.8 minutes)= $0.5 * (50.8 - 50.5) = 0.5 * 0.3 = 0.15$

In Exercises 5 & 7, assume that voltages in a circuit vary between 6 volts and 12 volts, and voltages are spread evenly over the range of possibilities, so that there is a uniform distribution. Find the probability of the given range of voltage levels.

5. For a discrete probability distribution, $\Sigma P(x) = 1$. Since the values on the x axis range from 6 to 12, this is a range of 6.0. To get the closed area within the rectangle to be equal to 1, the height of the rectangle has to be 1/6 = 0.167 and these are placed adjacent to each other to cover all values in the full range of 6 to 12

P(voltage greater than 10 volts) = $\frac{1}{6} * (12-10) = \frac{1}{6} * 2 = 2/6 = 1/3 = 0.333$

7. P (voltage between 7 and 10 volts) = $\frac{1}{6} * (10-7) = \frac{1}{6} * 3 = 3/6 = 1/2 = 0.500$

In Exercises 9 - 27, assume that the readings on scientific thermometers are normally distributed with a mean of 0°C and a standard deviation of 1.00°C. A thermometer is randomly selected and tested. In each case, draw a sketch, and find the probability of each reading in degrees Celsius.

9. Less than −0.25. The probability distribution of readings is a standard normal distribution because the readings are normally distributed with a mean of 0 and standard deviation of 1. We need to find the area below $z = -0.25$. From Table A-2, this is 0.4013.
So, $P(x < -0.25) = 0.4013$.

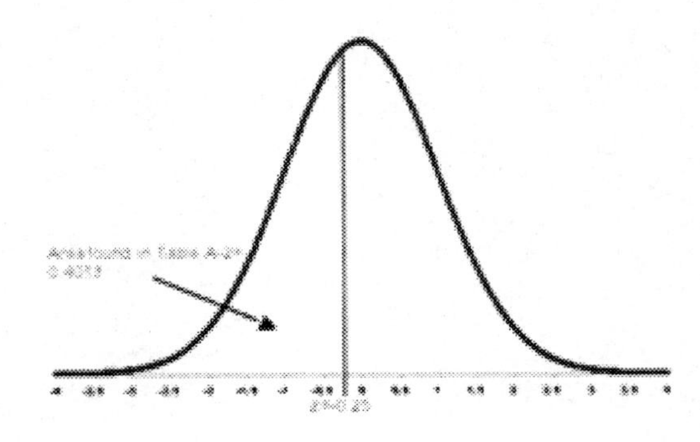

11. Probability of a thermometer reading less than 0.25°C, $z = +0.25$
Area below z of +0.25= 0.5987,
$P(x < +0.25) = 0.5987$

13. Probability of a thermometer reading greater than 2.33°C,
$z = +2.33$
Area below z of $+2.33 = 0.9901$,
$P(x > +2.33) = 1 - 0.9901 = 0.0099$

15. Probability of a thermometer reading greater than −2.33°C,
$z = -2.33$
Area below z of $-2.33 = 0.0099$,
$P(x > -2.33) = 1 - 0.0099 = 0.9901$

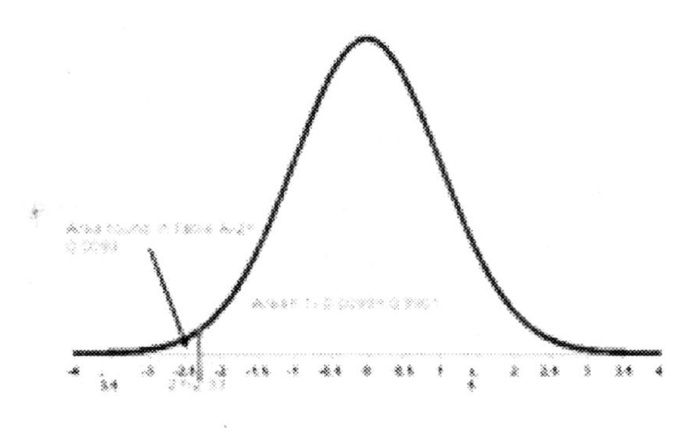

17. Probability of a thermometer reading between 0.5°C and 1.5°C, between $z = +0.50$ and $z = +1.50$, Area below z of $+1.50 = 0.9332$ and area below z of $+0.50 = 0.6915$
$P(+0.50 < x < +1.50) = 0.9332 - 0.6915$
$= 0.2417$

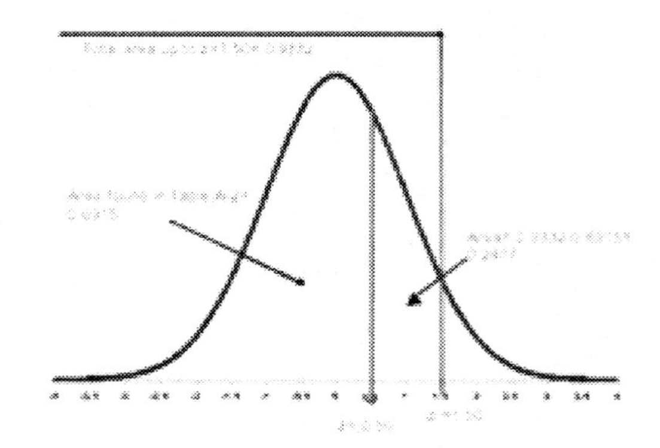

19. Probability of a thermometer reading between −2.00°C and −1.0°C, $z = -2.00$ and $z = -1.00$
Area below z of -1.00 is 0.1587 and area below z of -2.00 is 0.0228
$P(-2.00 < x < -1.00) = 0.1587 - 0.0228 = 0.1359$

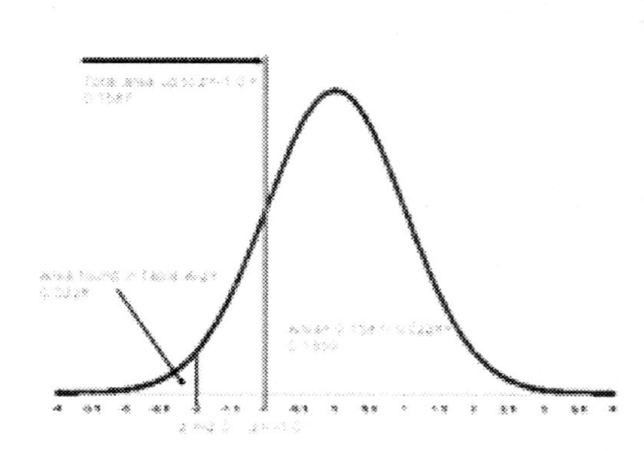

21. Probability of a thermometer reading between −2.67°C and 1.28°C, $z = -2.67$ and $z = +1.28$
Area below z of +1.28 is 0.8997 and area below z of −2.67 is 0.0038
$P(-2.67 < x < +1.28) = 0.8997 - 0.0038$
$= 0.8959$

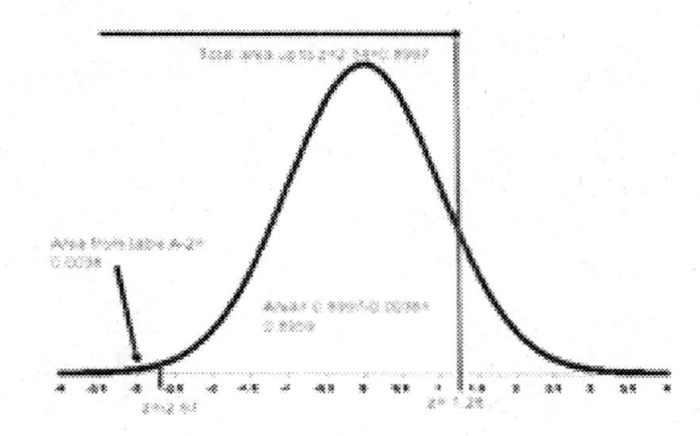

23. Probability of a thermometer reading between −0.52°C and 3.75°C, $z = -0.52$ and $z = +3.75$
Area below z of +3.75 is 0.9999 and area below z of −0.52 is 0.3015
$P(-0.52 < x < +3.75) = 0.9999 - 0.3015 = 0.6984$

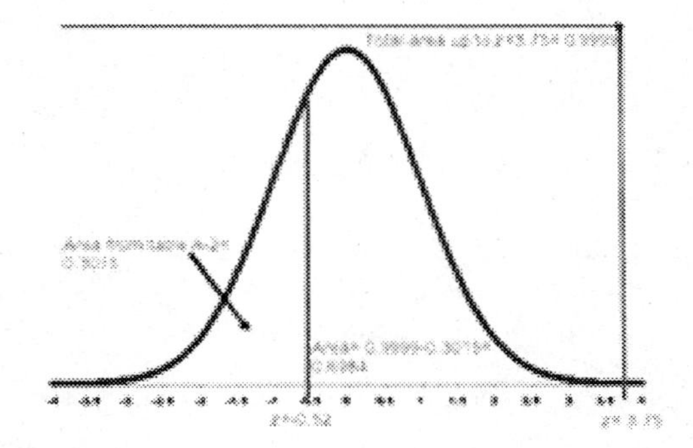

25. Probability of a thermometer reading greater than 3.57°C, $z = +3.57$
Area below z of +3.57=0.9999
$P(x > +3.57) = 1 - 0.9999 = 0.0001$

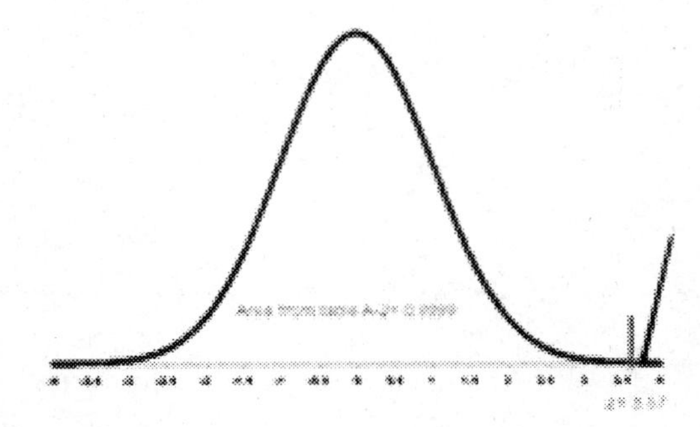

27. Probability of a thermometer reading greater than 0°C, $z = 0.00$
Area below z of 0.00= 0.5000
$P(x > 0.00) = 1 - 0.5000 = 0.5000$

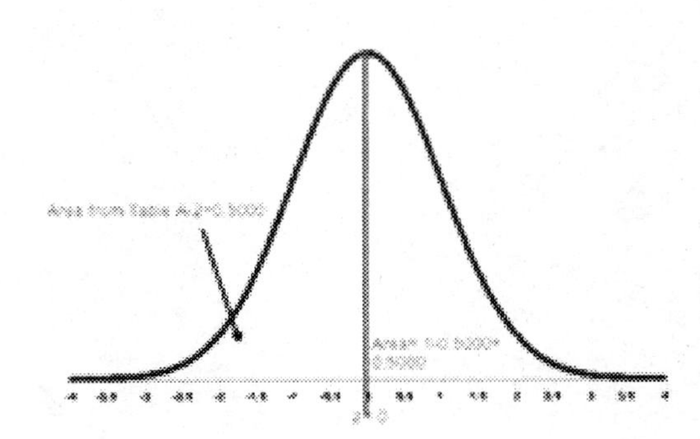

In Exercises 29 & 31, find the indicated area under the curve of the standard normal distribution, then convert it to a percentage and fill in the blank. The results form the basis for the empirical rule introduced in Section 2-5.

29. About 68.26% of the area is between $z = -1$ and $z = +1$ (or within one standard deviation of the mean). Since the area below z= −1.00 is 0.1587 the area between the mean and z= −1.00 is 0.5000 – 0.1587 = 0.3413, then the total area between z= −1.00 and z= +1.00 is
$2 * 0.3413 = 0.6826$, converted to a percentage is $0.6826 * 100\%$ 68.26%

31. About 99.74% of the area is between $z = -3$ and $z = +3$ (or within three standard deviation of the mean). Since the area below z= -3.00 is 0.0013 the area between the mean and z= −3.00 is 0.5000 – 0.0013 = 0.4987, then the total area between z= −3.00 and z= +3.00 is
$2 * 0.4987 = 0.9974$, converted to a percentage is $0.9974 * 100\% = 99.74\%$

Finding Probability. *In Exercises 33-36, assume that the readings on the thermometers are normally distributed with a mean of 0°C and a standard deviation of 1.00°C. Find the indicated probability, where z is the reading in degrees.*

33. $P(-1.96 < z < 1.96)$ = (Area below z= +1.960) – (Area below z= −1.960) = 0.9750 – 0.0250 = 0.9500

35. $P(z > -2.575)$ = 1 – (Area below z= −2.575) = 1 – 0.0050 = 0.9950

In Exercises 37 & 39, assume that the readings on the thermometers are normally distributed with a mean of 0°C and a standard deviation of 1.00°C. A thermometer is randomly selected and tested. In each case, draw a sketch, and find the temperature reading corresponding to the given information.

37. 0.90 in the body of the table corresponds to a z score of +1.28. So, the 90th percentile is the temperature reading of
$\mu + (1.28 * \sigma) = 0 + (1.28 * 1.00) = 1.28°C$.

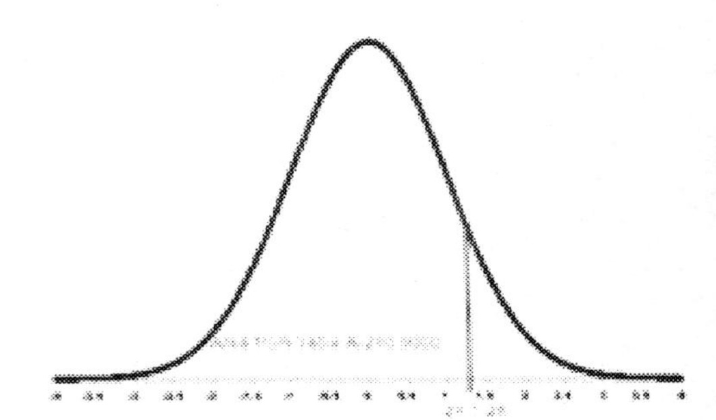

39. 0.05 in the body of the table corresponds to a z score of −1.645. So, the 5th percentile is the temperature reading of
$\mu + (-1.645 * \sigma) = 0 + (-1.645 * 1.00)$
$= -1.645°C$.

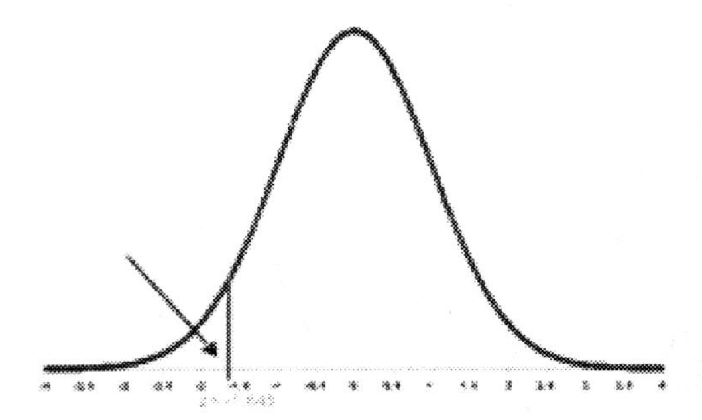

41. **a.** The percentage of data that are between one standard deviation from the mean corresponds to the area between $-1.00z$ and $+1.00z$ scores. This area is 68.26%.

b. The percentage of data that are between 1.96 standard deviations from the mean corresponds to the area between $-1.96z$ and $+1.96z$ scores. This area is 95.00%.

c. The percentage of data that are between three standard deviations from the mean corresponds to the area between $-3.00z$ and $+3.00z$ scores. This area is 99.74%.

d. The percentage of data that are between one standard deviation below the mean and two standard deviations above the mean one corresponds to the area between $-1.00z$ and $+2.00z$ scores. This is $0.9772 - 0.1587 = 0.8185$. This area is 81.85%

e. The percentage of data that are more than two standard deviations away from the mean corresponds to $1 -$ Area between $-2.00z$ and $+2.00z$ scores $= 1 - 0.9544 = 0.0456$ or 4.56%.

5-3 Applications of Normal Distributions

In Exercises 1-7, assume that adults have IQ scores that are normally distributed with a mean of 100 and a standard deviation of 15 (as on the Wechsler test). (Hint: Draw a graph in each case.)

1. The IQ of 115 is converted to a z score as follows:

$$z = \frac{x-\mu}{\sigma} = \frac{115-100}{15} = \frac{15}{15} = +1.00$$

Referring to Table A-2, $z = +1.00$ corresponds to an area of 0.8413, so $P(\text{IQ} < 115) = 0.8413$

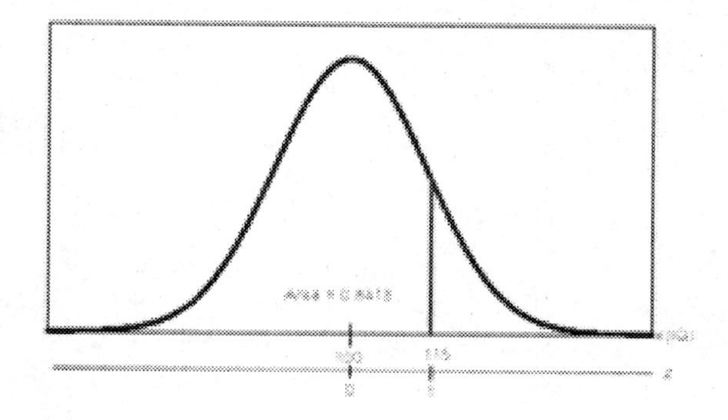

3. The IQs of 90 and 110 are converted to a z scores as follows:

$$z = \frac{x-\mu}{\sigma} = \frac{90-100}{15} = \frac{-10}{15} = -0.67$$

$$z = \frac{x-\mu}{\sigma} = \frac{110-100}{15} = \frac{10}{15} = +0.67$$

Referring to Table A-2, $z = -0.67$ corresponds to an area of 0.2514 and $z = +0.67$ corresponds to an area of 0.7486, so $P(90 < \text{IQ} < 110) = 0.7486 - 0.2514 = 0.4972$

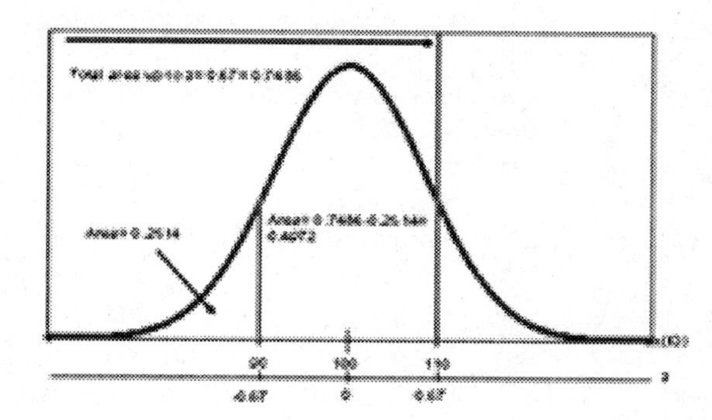

5. We find 0.2 in the body of the table and find the corresponding z score The z score for a cumulative area of 0.20 $= -0.84$

$$x = \mu + (z * \sigma) = 100 + (-0.84 * 15) = 100 + (-12.6) = 87.4$$

The P_{20} for IQ score = 87.4

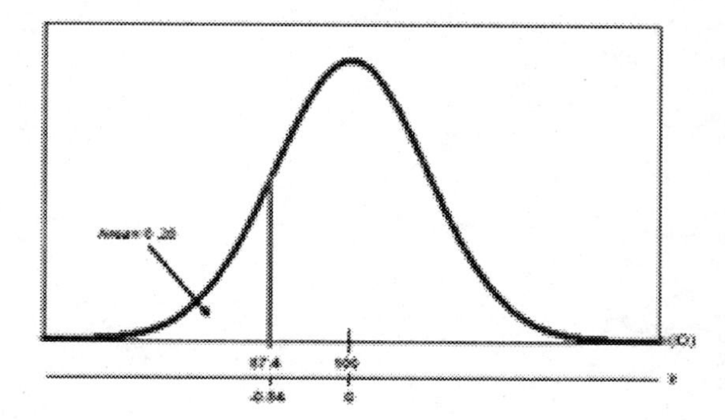

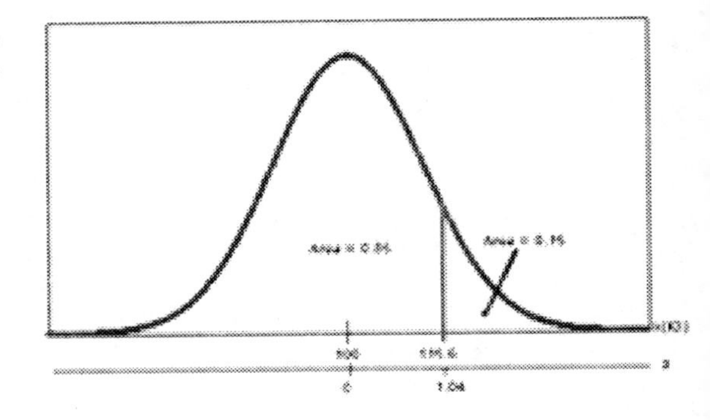

7. The IQ score separating the top 15% from the others is the same score that separates the bottom (100 – 15) % from the others 100 – 15 = 85. We find 0.85 in the body of the table and find the corresponding z score. The z score for a cumulative area of 0.85 = 1.04

$x = \mu + (z * \sigma) = 100 + (1.04 * 15) = 100 + 15.6 = 115.6$

The IQ score separating the top 15% from the others = 115.6

9. Body Temperature

a. $\mu = 98.20, \quad \sigma = 0.62, \; x = 100.6$

$$z = \frac{x - \mu}{\sigma} = \frac{100.6 - 98.2}{0.62} = \frac{2.4}{0.62} = +3.87$$

From Table A-2, P(Temperature < 100.6) = P (z< +3.87) = 0.9999.

P(Temperature> +3.87) = P(z > +3.87) = 1 – 0.9999 = 0.0001

This corresponds to 0.01%. Yes, this percentage suggests that the cutoff of 100.6°C is appropriate.

b. Since we want 5% of the people to exceed the required temperature, we use (100 – 5)%to find the area to the left of the cutoff line first. This corresponds to an area 0.95.

From Table A-2, this corresponds to a z score of +1.645.

$x = \mu + (z * \sigma) = 98.2 + (1.645 * 0.62) = 98.2 + 1.02 = 99.22$

Thus, 5% of the people will exceed 99.2°C

11. Designing Helmets, $\mu = 6, \quad \sigma = 1$

To find the cutoff points for the smallest 2.5% and the largest 2.5%, we find the z scores for the areas 0.025 and (1 – 0.025) or 0.975. From the table, these are –1.96 and +1.96 respectively.

$x = \mu + (z * \sigma) = 6 + (-1.96 * 1) = 6 - 1.96 = 4.04 \approx 4$

$x = \mu + (z * \sigma) = 6 + (+1.96 * 1) = 6 + 1.96 = 7.96 \approx 8$

The minimum and maximum head breadths are 4 inches and 8 inches respectively.

In Exercises 13 & 15, assume that heights of women are normally distributed with a mean given by $\mu = 63.6$ in. and a standard deviation given by $\sigma = 2.5$ in. (based on data from the National Health Survey). In each case, draw a graph.

13. Beanstalk Club Height Requirement

$\mu = 63.6, \quad \sigma = 2.5,$

$$z = \frac{x - \mu}{\sigma} = \frac{70 - 63.6}{2.5} = \frac{6.4}{2.5} = +2.56$$

This corresponds to a probability of 0.9948. So, 99.48% of the women have height < 70 in. Therefore (100 – 99.48) or 0.52% of the women meet the requirement of being at least 70in. in height.

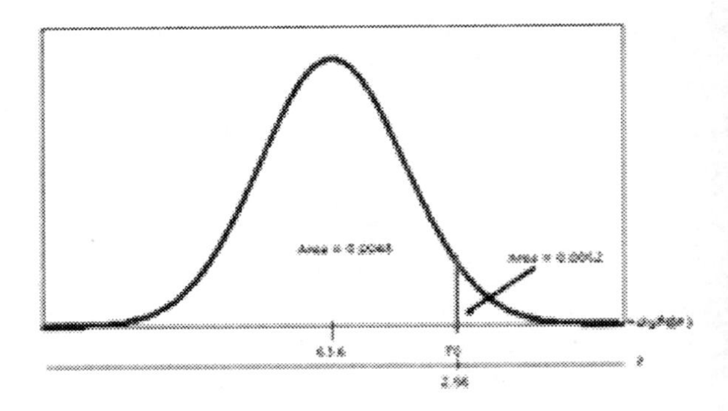

15. Height Requirement for Rockettes

We need to find the z scores and areas for 66.5 in. and 71.5 in.

$$z = \frac{x-\mu}{\sigma} = \frac{66.5-63.6}{2.5} = \frac{2.9}{2.5} = +1.16$$

$$z = \frac{x-\mu}{\sigma} = \frac{71.5-63.6}{2.5} = \frac{7.9}{2.5} = +3.16$$

The areas for these z scores are 0.8770 and 0.9992 respectively. The probability of being between these heights is 0.9992 – 0.8770 = 0.1222. The probability of meeting this new height is 0.1222. Only 12.22% of women meet this requirement. Yes, it seems that the height of the Rockettes is well above the mean.

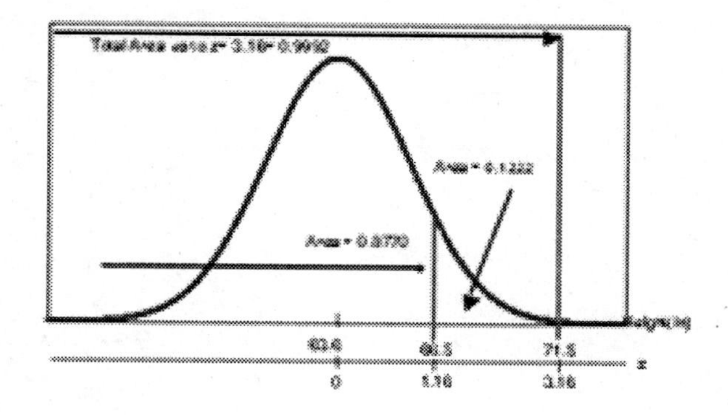

17. Birth Weights, $\mu = 3420, \quad \sigma = 495$

To find the cutoff weights for the lightest 2% we need to find to find the z score corresponding to the area 0.02. From the Table, the z score is -2.05. We then use the formula:

$x = \mu + (z * \sigma) = 3420 + (-2.05 * 495)$
$= 3420 - 1014.75 = 2405.25$
. Therefore, the weight of 2405g separates the lightest 2% of American babies from the others.

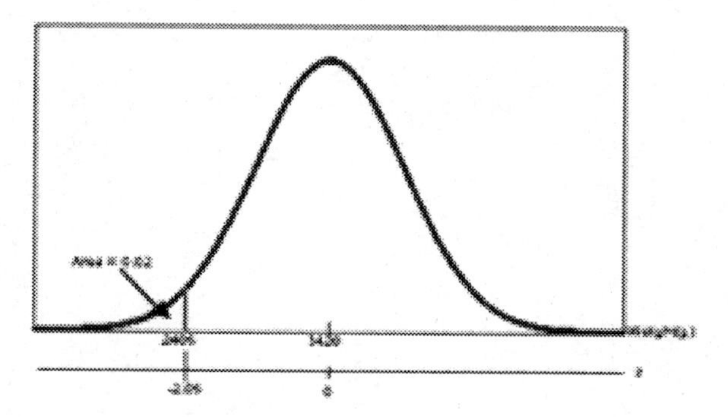

19. Units of Measurement, $\mu = 143, \quad \sigma = 29$

a. z scores are measured in units of number of standard deviations from the mean, but they do not possess the units of the original variable

b. The mean will be 0, the standard deviation will be 1, and the distribution will be normal since the original distribution is normal. z scores have the same shape of distribution as does the original variable distribution; converting to z scores does not result in a normal distribution of z scores if the original distribution was not normally distributed

c. After converting to kg., the distribution will be normal since the original distribution is normal, 1 lb= 0.4536 kg

143 lb = (0.4536 * 143) kg = 64.86 kg = mean

29 lb = (0.4536 * 29) kg = 13.15 kg = standard deviation

5-4 Sampling Distributions and Estimators

1. Survey of Voters

No, we cannot assume that the survey was done incorrectly because the value of a statistic varies from sample to sample due to sampling variability. In this example, the values for the sample proportion are different because of sampling variability. A variation of 49% and 51% would seem to happen by chance relatively often.

3. Sampling Distribution of Body Temperatures

No, the histogram will not show the shape of a sampling distribution of sampling means. It will show the distribution of individual values within one sample. A sampling distribution will show a distribution of all possible means of similar samples with the same sample size.

5. Phone Center

Selecting samples with replacement, there will be $3^2 = 9$ equally likely samples.

Sample Number	**a.** Sample	Sample Mean, $\bar{x}$	**b.** Probability
1	10,10	10.0	1/9
2	10, 6	8.0	1/9
3	10, 5	7.5	1/9
4	6, 10	8.0	1/9
5	6, 6	6.0	1/9
6	6, 5	5.5	1/9
7	5, 10	7.5	1/9
8	5, 6	5.5	1/9
9	5, 5	5.0	1/9
Sum of Sample Means	$\bar{x} =$	63.0	
Mean of statistic values	$\mu = \frac{\bar{x}}{9} =$	7.0	
Population parameter	$\mu = \frac{10+6+5}{3} = \frac{21}{3} =$	7.0	

Sampling Distribution

Sample Mean, $\bar{x}$	Probability
10.0	1/9
8.0	2/9
7.5	2/9
6.0	1/9
5.5	2/9
5.0	1/9

b. The probability of each sample is 1/9. The distribution of sample means is bi-modal and somewhat flat.

c. Mean of sample statistics= $\mu = \frac{\bar{x}}{9} = \frac{63}{9} = 7.0$

d. Yes, the mean of the sampling distribution is equal to the mean of the population of the three values. Yes, these means are always equal, but only if every possible sample is included.

7. Heights of L.A. Lakers

Selecting samples with replacement, there will be $5^2 = 25$ equally likely samples.

Sample Number	**a.** Sample	Sample Mean, $\bar{x}$	Probability
1	85, 85	85.0	1/25
2	85, 79	82.0	1/25
3	85, 82	83.5	1/25
4	85, 73	79.0	1/25
5	85, 78	81.5	1/25
6	79, 79	79.0	1/25
7	79, 85	82.0	1/25
8	79, 82	80.5	1/25
9	79, 73	76.0	1/25
10	79, 78	78.5	1/25
11	82, 82	82.0	1/25
12	82, 85	83.5	1/25
13	82, 79	80.5	1/25
14	82, 73	77.5	1/25
15	82, 78	80.0	1/25
16	73, 73	73.0	1/25
17	73, 85	79.0	1/25
18	73, 79	76.0	1/25
19	73, 82	77.5	1/25
20	73, 78	75.5	1/25
21	78, 78	78.0	1/25
22	78, 85	81.5	1/25
23	78, 79	78.5	1/25
24	78, 82	80.0	1/25
25	78, 73	75.5	1/25
Sum of Sample Means	$\bar{x} =$	1985	
Mean of statistic values	$\mu = \frac{\bar{x}}{25} =$	79.4	
Population parameter	$\mu = \frac{85+79+82+73+78}{5} = \frac{397}{5} =$	79.4	

Sampling Distribution

Sample Mean, $\bar{x}$	Probability
85.0	1/25
83.5	2/25
82.0	3/25
81.5	2/25
80.5	2/25
80.0	2/25
79.0	3/25
78.5	2/25
78.0	1/25
77.5	2/25
76.0	2/25
75.5	2/25
73.0	1/25

b. The probability of each sample occurring is 1/25. The sampling distribution of means consists of the 25 sample means with their corresponding probabilities. It has more than one mode and it is not symmetrical.

c. The means of the sampling distribution is $\mu = \frac{\Sigma \bar{x}}{n} = \frac{1985}{25} = 79.4$

d. Yes, the mean of the sampling distribution is equal to the mean of the population of the five heights listed above. Yes, these means are always equal as long as every possible sample is included.

9. Quality Control

Selecting samples with replacement, there will be $5^2 = 25$ equally likely samples.
D1= 1, D2= 1, A1= 0, A2=0, A3=0

Sample Number	**a.** Sample	Sample Mean $\bar{x}$	Probability
1	D1, D1= 1, 1	1.0	1/25
2	D1, D2= 1, 1	1.0	1/25
3	D1, A1= 1, 0	0.5	1/25
4	D1, A2= 1, 0	0.5	1/25
5	D1, A3= 1, 0	0.5	1/25
6	D2, D2= 1, 1	1.0	1/25
7	D2, D1= 1, 1	1.0	1/25
8	D2, A1= 1, 0	0.5	1/25
9	D2, A2= 1, 0	0.5	1/25
10	D2, A3= 1, 0	0.5	1/25
11	A1, A1= 0, 0	0.0	1/25
12	A1, A2= 0, 0	0.0	1/25
13	A1, A3= 0, 0	0.0	1/25
14	A1, D1= 0, 1	0.5	1/25
15	A1, D2= 0, 1	0.5	1/25
16	A2, A2= 0, 0	0.0	1/25
17	A2, A3= 0, 0	0.0	1/25
18	A2, D1= 0, 1	0.5	1/25
19	A2, D2= 0, 1	0.5	1/25
20	A2, A1= 0, 0	0.0	1/25
21	A3, A3= 0, 0	0.0	1/25
22	A3, D1= 0, 1	0.5	1/25
23	A3, D2= 0, 1	0.5	1/25
24	A3, A1= 0, 0	0.0	1/25
25	A3, A2= 0, 0	0.0	1/25
Sum of Sample Means	$\bar{x} =$	10.0	
Mean of statistic values	$\mu = \frac{\bar{x}}{25} = \frac{10}{25} =$	0.40	
Population parameter	$\mu = \frac{1+1+0+0+0}{5} = \frac{2}{5} =$	0.40	

Sampling Distribution

Sample Mean, $\bar{x}$	Probability
0.0	9/25
0.5	12/25
1.0	4/25

b. The sampling distribution consists of the 25 proportions and their corresponding probabilities of 1/25 each. The sampling distribution has one mode, but it is not symmetrical.

c. The mean of the sampling distribution is $\mu = \frac{\Sigma\bar{x}}{n} = \frac{10}{25} = 0.40$

d. Yes, the mean of the sampling distribution is equal to the population proportion of defects. Yes, the mean of the sampling distribution of proportions always equals the population proportion as long as every possible sample is included.

11. Mean Absolute Deviation

From Table 5-2, $x = 1, 2, 5$, $\mu = 2.67$

Population Mean Absolute Deviation, see this formula in Section 2-5.

$$\frac{|x-\bar{x}|}{n} = \frac{|1-2.67|+|(2-2.67)|+|(5-2.67)|}{3} = \frac{1.67+0.67+2.33}{3} = \frac{4.67}{3} = 1.56$$

Sample Number	Sample	Sample Mean $\bar{x}$	Absolute Deviation $\lvert d\rvert = \left\lvert\frac{(x_1 - x_2)}{2}\right\rvert$
1	1, 1	1.0	0.0
2	1, 2	1.5	0.5
3	1, 5	3.0	2.0
4	2, 1	1.5	0.5
5	2, 2	2.0	0.0
6	2, 5	3.5	1.5
7	5, 1	3.0	2.0
8	5, 2	3.5	1.5
9	5, 5	5.0	0.0

$$MAD = |\bar{d}| = \frac{|d|}{n} = \frac{8}{9} = 0.89$$

Since MAD = 0.89 ≠ 1.56 (the population absolute mean deviation) the mean absolute deviation is not a good estimate of the population mean absolute deviation.

5-5 The Central Limit Theorem

In Exercises 1-5, assume that men's weights are normally distributed with a mean given by μ = 172 lb and a standard deviation given by σ = 29 lb (based on data from the National Health Survey).

1. **a.** $P(x < 167)$

$z = \frac{x-\mu}{\sigma} = \frac{167-172}{29} = \frac{-5}{29} - 0.17$. From Table A-2, $P(z < -0.17) = 0.4325$.

There is a 0.4325 probability that an individual man will weigh less than 167 lb.

b. $P(\bar{x} < 167)$

$z = \frac{x-\mu}{\sigma/\sqrt{n}} = \frac{167-172}{29/\sqrt{36}} = \frac{-5}{29/6} = \frac{-5}{4.833} = -1.03$. From Table A-2, $P(z < -1.03) = 0.1515$. There is a 0.1515 probability that a group of 36 men will have a mean weight less than 167 lb.

3. **a.** $P(170 < x < 175)$

$z = \frac{x-\mu}{\sigma} = \frac{170-172}{29} = \frac{-2}{29} = -0.07$, $z = \frac{x-\mu}{\sigma} = \frac{175-172}{29} = \frac{3}{29} = +0.10$.

From Table A-2, $P(z < -0.07) = 0.4721$ and $P(z < +0.10) = 0.5398$.

The difference is 0.5398 – 0.4721 = 0.0677. There is a 0.0677 probability that an individual man will weigh between 170 lb and 175 lb

b. $P(170 < \bar{x} < 175)$

$$z = \frac{x-\mu}{\sigma/\sqrt{n}} = \frac{170-172}{29/\sqrt{64}} = \frac{-2}{29/8} = \frac{-2}{3.625} = -0.55$$

$$z = \frac{x-\mu}{\sigma/\sqrt{n}} = \frac{175-172}{29/\sqrt{64}} = \frac{3}{29/8} = \frac{3}{3.625} = +0.83$$

From Table A-2, $P(z < -0.55) = 0.2912$, and $P(z < +0.83) = 0.7967$.
The difference is $0.7967 - 0.2912 = 0.5055$. There is a 0.5055 probability that a group of 64 men will have a mean weight between 170 lb and 175 lb

5. **a.** $P(\bar{x} > 160)$

$$z = \frac{x-\mu}{\sigma/\sqrt{n}} = \frac{160-172}{29/\sqrt{25}} = \frac{-12}{29/5} = \frac{-12}{5.80} = -2.07$$

From Table A-2, $P(z < -2.07) = 0.0192$
Therefore $P(z > -2.07) = 1 - 0.0192 = 0.9808$. There is a 0.9808 probability that a group of 25 men will weigh more than 160 lb.

b. The central limit theorem can be used in part (a) because the original distribution is a normal distribution and we assume the sampling distribution would be normal even though the sample size is less than 30.

7. **Redesign of Ejection Seats,** $\mu = 143$, $\sigma = 29$

a. $P(140 < x < 211)$

$$z = \frac{x-\mu}{\sigma} = \frac{140-143}{29} = \frac{-3}{29} = -0.10, \quad z = \frac{x-\mu}{\sigma} = \frac{211-143}{29} = \frac{68}{29} = +2.34$$

From Table A-2, $P(z < -0.10) = 0.4602$ and $P(z < +2.34) = 0.9904$.
The difference is $0.9904 - 0.4602 = 0.5302$. There is a 0.5302 probability that an individual woman will weigh between 140 lb and 211 lb.

b. $P(140 < \bar{x} < 211)$

$$z = \frac{x-\mu}{\sigma/\sqrt{n}} = \frac{140-143}{29/\sqrt{36}} = \frac{-3}{29/6} = \frac{-3}{4.833} = -0.62$$

$$z = \frac{x-\mu}{\sigma/\sqrt{n}} = \frac{211-143}{29/\sqrt{36}} = \frac{68}{29/6} = \frac{68}{4.833} = +14.07$$

From Table A-2, $P(z < -0.62) = 0.2676$ and $P(z < +14.07) \sim 0.9999$. The difference is $0.9999 - 0.2676 = 0.7323$. There is a 0.7323 probability that a group of 36 women will have a mean weight between 140 lb and 211 lb

c. The results from part (a) are more important because the seats will be occupied by individual women, and not by groups of women.

9. **Designing a Roller Coaster,** $\mu = 14.4$, $\sigma = 1$

a. $P(\bar{x} > 16.0)$

$$z = \frac{x-\mu}{\sigma/\sqrt{n}} = \frac{16-14.4}{1/\sqrt{2}} = \frac{1.6}{1/1.414} = \frac{1.6}{0.707} = +2.26.$$

From Table A-2, $P(z < +2.26) = 0.9881$. Therefore, $P(z > 2.26) = 1 - 0.9881 = 0.0119$. The probability that the mean of the 2 men is greater than 16 in. is 0.0119.

b. No, most riders will be able to fit since the probability of both riders having a mean hip breadth of greater than 16in. is very low.(0.0119). Yes, this design appears to be acceptable.

11. Blood Pressure, $\mu = 114.8$, $\sigma = 13.1$

a. $z = \frac{x-\mu}{\sigma} = \frac{140-114.8}{13.1} = \frac{25.2}{13.1} = +1.92.$

From Table A-2, $P(z < +1.92) = 0.9726$.Therefore, $P(z > +1.92) = 1 - 0.9726 = 0.0274$. There is a 0.0274 probability that an individual woman will have a systolic blood pressure greater than 140.

b. $z = \frac{x-\mu}{\sigma/\sqrt{n}} = \frac{140-114.8}{13.1/\sqrt{4}} = \frac{25.2}{13.1/2} = \frac{25.2}{6.55} = +3.85.$

From Table A-2, $P(z < +3.85) = 0.9999$. Therefore, $P(z > 3.85) = 1 - 0.9999 = 0.0001$. There is a 0.0001 probability that a group of 4 women will have a mean systolic blood pressure greater than 140.

c. The central limit theorem can be used in part (b) because the original distribution is a normal distribution, even though the sample size is less than 30.

d. No. Although the mean result for the 4 women is less than 140, the individual values could be above or below 140 due to sampling variability.

13. Elevator Design, $\mu = 172$, $\sigma = 29$, $n = 16$, $P = 0.975$

We first find the z score for the area $P = 0.975$ from the body of table A-2.This corresponds to $z = +1.96$. We then use the formula:

$$x = \mu + z * \frac{\sigma}{\sqrt{n}} = 172 + 1.96 * \frac{29}{\sqrt{16}} = 172 + 1.96 * \frac{29}{4.0} = 172 + (1.96 * 7.25) = 172 + 14.21 = 186.21$$

To get the total value for 16 men, $186.21 * 16 = 2979.4$. This is the maximum total allowable weight if we want a 0.975 probability of this weight not being exceeded with 16 men.

15. Correcting for a Finite Population, $\mu = 143$, $\sigma = 29$, $N=120$, $n = 8$

a. If we do not want to exceed this limit, we need to find the probability of the 8 of them having a total weight less than 1300 lb. A total capacity of 1300 lb for the 8 women means 1300/8 = 162.5 lb per woman on average.

$$z = \frac{x-\mu}{\frac{\sigma}{\sqrt{n}} * \sqrt{\frac{N-n}{N-1}}} = \frac{162.5-143}{\frac{29}{\sqrt{8}} * \sqrt{\frac{120-8}{120-1}}} = \frac{19.5}{\frac{29}{2.828} * \sqrt{\frac{112}{119}}} =$$

$$\frac{19.5}{10.25 * \sqrt{0.941}} = \frac{19.5}{10.25 * 0.970} = \frac{19.5}{9.94} = 1.96$$

From Table A-2, $P(z < +1.96) = 0.975$.

The probability of their total weight not exceeding 1300lb = 0.9750.

b. We first find the z score for the area $P = 0.9900$ from the body of Table A-2. This corresponds to $z = +2.33$. We then use the formula:

$$x = \mu + z * \frac{\sigma}{\sqrt{n}} * \sqrt{\frac{N-n}{N-1}} = 143 + 2.33 * \frac{29}{\sqrt{8}} * \sqrt{\frac{120-8}{120-1}} = 143 + 2.34 * \frac{29}{2.828} * \sqrt{0.941} =$$

$$143 + 2.34 * 10.25 * 0.970 = 143 + 23.18 = 166.18$$

To get the total value for 8 women, = $166.18 * 8 = 1329$ lb. This is the maximum allowable weight of passengers in the elevator if we want a 0.99 probability that the elevator will not be overloaded.

5-6 Normal as Approximation to Binomial

In Exercises 1-7, the given values are discrete. Use the continuity correction and describe the region of the normal distribution that corresponds to the indicated probability. For example, the probability of "more than 20 girls" corresponds to the area of the normal curve described with this answer: "the area to the right of 20.5."

1. The probability of "more than 15 males with blue eyes" corresponds to the area of the normal curve to the right of 15.5, $P(x > 15) = P_c(x > 15.5)$

3. The probability of "fewer than 100 bald eagles sighted in a week" corresponds to the area of the normal curve to the left of 99.5, $P(x < 100) = P_c(x < 99.5)$

5. The probability of "no more than 4 students absent in a biostatistics class" corresponds to the area of the normal curve to the left of 4.5, $P(x \le 4) = P_c(x < 4.5)$

7. The probability that "the number of rabbit offspring is between 8 and 10 inclusive" corresponds to the area of the normal curve between 7.5 and 10.5,
$P(8 \le x \le 10) = P_c(7.5 < x < 10.5)$

In Exercises 9 & 11, do the following. (a) Find the indicated binomial probability by using Table A-1. (b) If np ≥ 5 and nq ≥ 5, also estimate the indicated probability by using the normal distribution as an approximation to the binomial distribution; if np < 5 or nq < 5, then state that the normal approximation is not suitable.

9. **a.** $n = 14$, $p = 0.5$, From Table A-1, $P(9) = 0.122$.
b. Normal approximation
$np = nq = 14 * 0.5 = 7$ (both ≥ 5, normal approximation is justified)
$\mu = np = 14 * 0.5 = 7$

$$\sigma = \sqrt{npq} = \sqrt{14 * 0.5 * 0.5} = \sqrt{3.5} = 1.871$$

$$z = \frac{x - \mu}{\sigma} = \frac{8.5 - 7}{1.871} = \frac{1.5}{1.871} = 0.80$$

$$z = \frac{x - \mu}{\sigma} = \frac{9.5 - 7}{1.871} = \frac{2.5}{1.871} = 1.34$$

z = 0.80 corresponds to a probability of 0.7881
z = 1.34 corresponds to a probability of 0.9099
$P(9)$ from Normal Approximation= 0.9099 – 0.7881 = 0.1218 (very good approximation to 0.122)

11. **a.** $n = 15$, $p = 0.9$, From Table A-1, $P(14) = 0.343$, $P(15) = 0.206$
P (at least 14) = P (14 or more)= $P(14) + P(15) = 0.343 + 0.206 = 0.549$
b. Normal approximation
$np = 15 * 0.9 = 13.5$, $nq = 15 * 0.1 = 1.5$
$nq = 1.5$ which is < 5. Therefore the normal approximation is not justified

13. **Probability of More Than 55 Girls,** $np = 50$, $nq = 50$ (both ≥ 5, normal distribution justified)
$n = 100$, $p = 0.5$

$$\mu = np = 100 * 0.5 = 50$$

$$\sigma = \sqrt{npq} = \sqrt{100 * 0.5 * 0.5} = \sqrt{25} = 5$$

$$z = \frac{x - \mu}{\sigma} = \frac{55.5 - 50}{5} = \frac{5.5}{5} = +1.1$$

$x = 55$, finding $P_c(x > 55.5)$
P (girls > 55) = P (z > +1.1) = 1 – 0.8643 = 0.1357
No, since P(girls > 55) is greater than 0.05, it is not unusual to get more than 55 girls out of 100 births.

15. **Probability of at Least Passing,** $np = 50$, $nq = 50$ (both ≥ 5, normal distribution justified)
$n = 100$, $p = 0.5$ (true or false answer)

$$\mu = np = 100 * 0.5 = 50$$

$$\sigma = \sqrt{npq} = \sqrt{100 * 0.5 * 0.5} = \sqrt{25} = 5$$

$$z = \frac{x - \mu}{\sigma} = \frac{59.5 - 50}{5} = \frac{9.5}{5} = +1.9$$

$P(x \ge 60)$, finding $P_c(x > 59.5)$
P (score ≥ 60) = P (z > +1.9) = 1 – 0.9713 = 0.0287
No, since P(score ≥ 60) is less than 0.05, it is unusual to get a score of at least 60 by guessing

17. Mendel's Hybridization Experiment, $np = 145$, $nq = 435$ (both ≥ 5, normal distribution justified)

$n = 580,\ p = 0.25$

$\mu = np = 580 * 0.25 = 145$

$\sigma = \sqrt{npq} = \sqrt{580 * 0.25 * 0.75} = \sqrt{108.75} = 10.43$

$z = \dfrac{x - \mu}{\sigma} = \dfrac{151.5 - 145}{10.43} = \dfrac{6.5}{10.43} = +0.62$

$P(x \geq 152)$, finding $P_c(x > 151.5)$

P (z at least 0.62) $= P(z > +0.62) = 1 - 0.7324 = 0.2676$

No, there is no evidence that the Mendelian rate of 25% is wrong because it is not unusual to get 152 yellow pods out of 580 seedlings, $p = 0.2676$

19. Probability of at Least 50 Color-Blind Men, $np = 54$, $nq = 546$ (both ≥ 5, normal distribution justified)

$\text{n} = 600,\ \text{p} = 0.09,$

$\mu = \text{np} = 600 * 0.09 = 54$

$\sigma = \sqrt{\text{npq}} = \sqrt{600 * 0.09 * 0.91} = \sqrt{49.14} = 7.01$

$\text{z} = \dfrac{x - \mu}{\sigma} = \dfrac{49.5 - 54}{7.01} = \dfrac{-4.5}{7.01} = -0.64$

$P(x \geq 50)$, finding $P_c(x > 49.5)$

P(z at least -0.64) $= P(z > -0.64) = 1 - 0.2611 = 0.7389$

It is quite likely to have 50 color blind men among this group of 600 men ($P = 0.7389$). However, the researchers cannot be very confident since there is still quite some chance of not getting up to 50 men.

21. Identifying Gender Discrimination, $np = 31$, $nq = 31$ (both ≥ 5, normal distribution justified)

$n = 62,\ p = 0.5$

$\mu = np = 62 * 0.5 = 31$

$\sigma = \sqrt{npq} = \sqrt{62 * 0.5 * 0.5} = \sqrt{15.5} = 3.937$

$z = \dfrac{x - \mu}{\sigma} = \dfrac{21.5 - 31}{3.937} = \dfrac{-9.5}{3.937} = -2.41$

$P(x \leq 21)$, finding $P_c(x < 21.5)$

$P\ (z < -2.41) = 0.0080$

It is unusual to have 21 female employees out of 62 new employees being hired assuming no gender discrimination. ($P = 0.0080$) These results support the charge of gender discrimination taking place.

23. Acceptance Sampling, $np = 5$, $nq = 45$ (both ≥ 5, normal distribution justified)

$n = 50,\ p = 0.1$

$\mu = np = 50 * 0.1 = 5$

$\sigma = \sqrt{npq} = \sqrt{50 * 0.1 * 0.9} = \sqrt{4.5} = 2.12$

$z = \dfrac{x - \mu}{\sigma} = \dfrac{1.5 - 5}{2.12} = \dfrac{-3.5}{2.12} = -1.65$

$P(x \geq 2)$, finding $P(x > 1.5)$

$P(z < -1.65) = 0.0495$, $P(z > -1.65) = 1 - 0.0495 = 0.9505$

Yes, this plan would detect defects at the 10% level about 95% of the time.

25. **Cloning Survey,** $np = 506$, $nq = 506$ (both ≥ 5, normal distribution justified)
$n = 1012, \ p = 0.5$

$\mu = np = 1012 * 0.5 = 506$

$\sigma = \sqrt{npq} = \sqrt{1012 * 0.5 * 0.5} = \sqrt{253} = 15.91$

$z = \dfrac{x - \mu}{\sigma} = \dfrac{900.5 - 506}{15.91} = \dfrac{394.5}{15.91} = +24.80$

$P(x \geq 900)$, finding $P(x > 900.5)$

$P\ (z < +24.80) \approx 0.9999$, $P(z > +24.80) = 1 - 0.9999 = 0.0001$

The probability of having 89% (900) of 1012 people in a sample assuming a general probability of 0.5 is very low. Yes, this evidence supports the claim that a majority of people are opposed to cloning

5-7 Assessing Normality

(Note: In Section 5-7 all graphics were generated using SPSS)

In Exercises 1 & 3, examine the normal quantile plot and determine whether it depicts data that have a normal distribution.

1. The data are not normally distributed since the data plot dots depart from being a straight line that follows the normal quantile plot that is expected if the data are normally distributed.

3. The data are normally distributed since the data plot dots are very close to a straight line that follows the normal quantile plot that is expected if the data are normally distributed.

Determining Normality. *In Exercises 5-8, refer to the indicated data set and determine whether the requirement of a normal distribution is satisfied. Assume that this requirement is loose in the sense that the population distribution need not be exactly normal, but it must be a distribution that is basically symmetric with only one mode.*

5. **BMI,** Data Set 1 in Appendix B

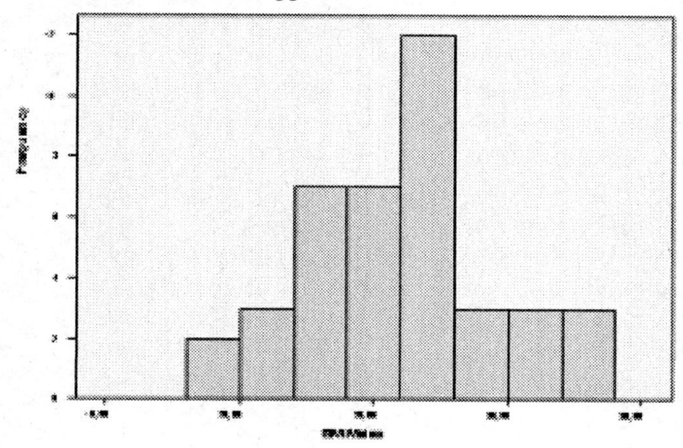

The histogram above shows a distribution with one mode, relatively symmetrical, and bell-shaped. It can be said to approximate a normal distribution.

7. **Water Conductivity**

The histogram above shows a distribution with one mode. However, the distribution is not symmetrical and bell-shaped so it would not be considered to be approximately normal.

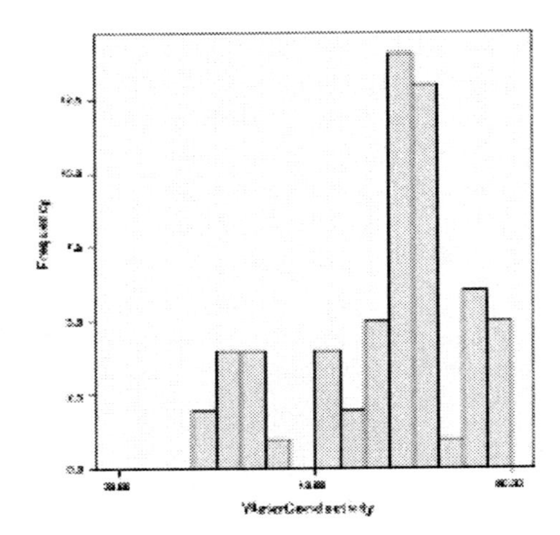

In Exercises 9 & 112, use the data from the indicated exercise in this section. Use a TI-83/84Plus calculator or software (such as SPSS, SAS, STATDISK, Minitab. or Excel) capable of generating normal quantile plots (or normal probability plots). Generate the graph, then determine whether the data appear to come from a normally distributed population.

NOTE: The following Normal Quantile Plots, except that in Exercises 15, were generated by JMP. When using the SPSS option for standardized or *z* scores, both axes are put into *z* score units, not just the Y-axis.

9. From Exercise 5

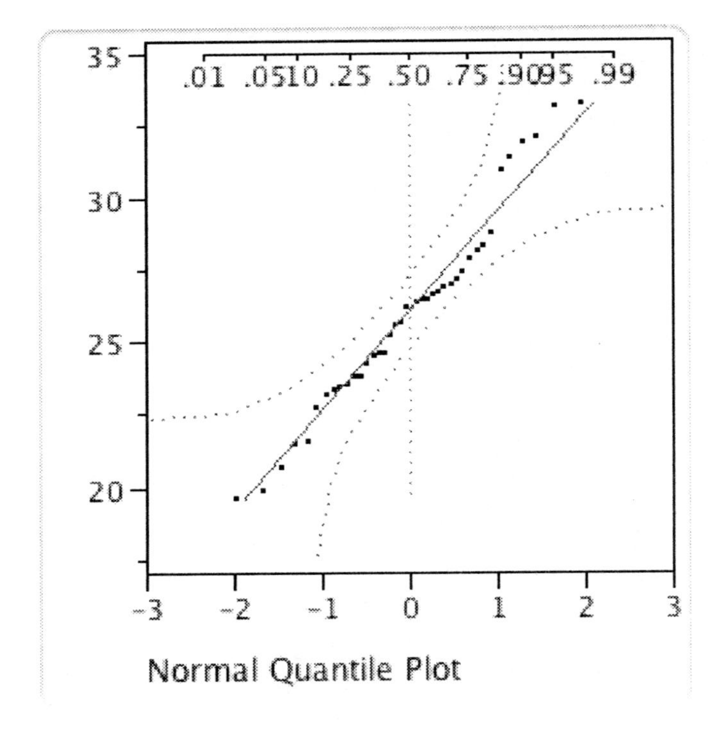

The BMI data from Exercise 5 seems to come from a normal distribution. Most of the points are very close to the straight line.

11. From Exercise 7

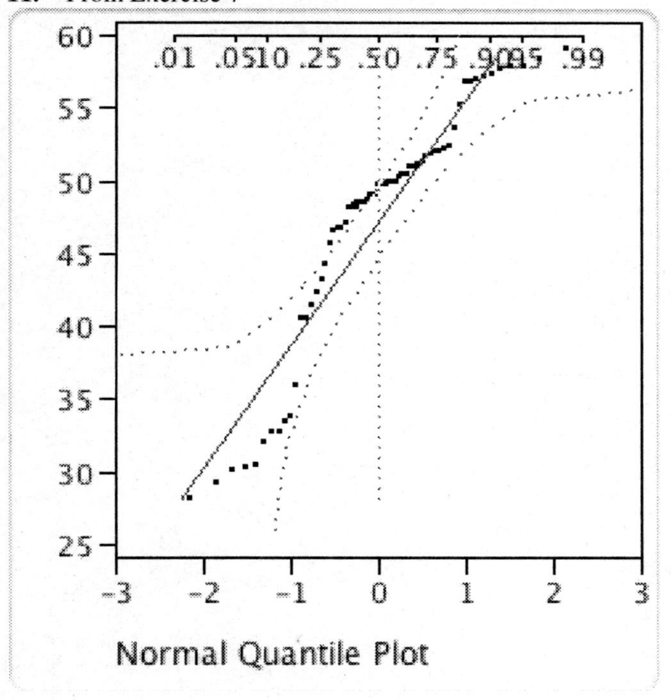

The data on the conductivity variable are not normally distributed. The points depart quite a bit from the straight line.

13. Comparing Data Sets

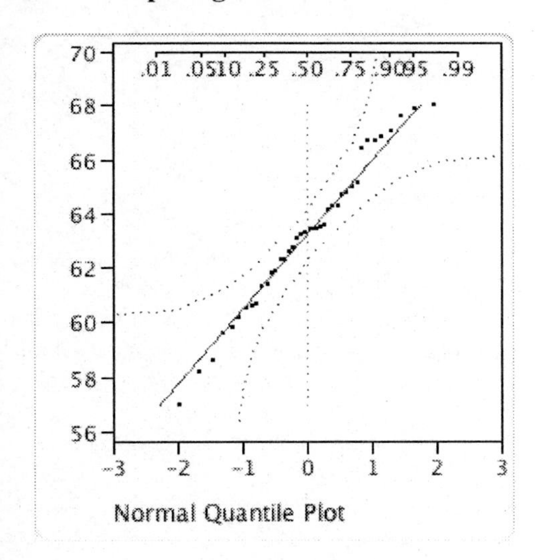

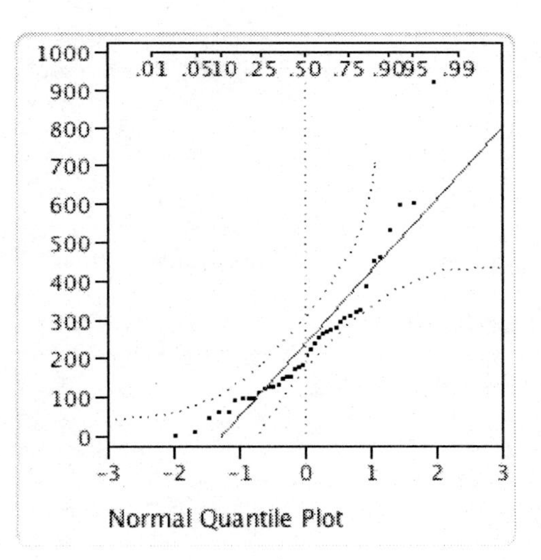

The distribution for height appears to be normal, but the distribution for cholesterol does not appear to be normal. This could be because cholesterol levels depend on diet and many other human behaviors in different ways that do not yield normally distributed results while height is a more natural variable less influenced by human behaviors.

In Exercise 15, use the given data values and identify the corresponding z scores that are used for a normal quantile plot, then construct the normal quantile plot and determine whether the data appear to be from a population with a normal distribution.

15. Heights of L.A. Lakers

Sorting the data by order gives us 73, 78, 79, 82, 85
n = 5, 1/2n, 3/2n, 5/2n, 7/2n, 9/2n = 0.1, 0.3, 0.5, 0.7, 0.9
Corresponding z scores, using Table A-2 for these areas are:
−1.28, −0.52, 0.00, +0.52, and +1.28
We now pair the sorted heights with their corresponding z scores:
(73, −1.28) (78, −0.52) (79, 0) (82, +0.52) (85, +1.28)
We plot these (x,y) coordinates to get the normal quantile plot.

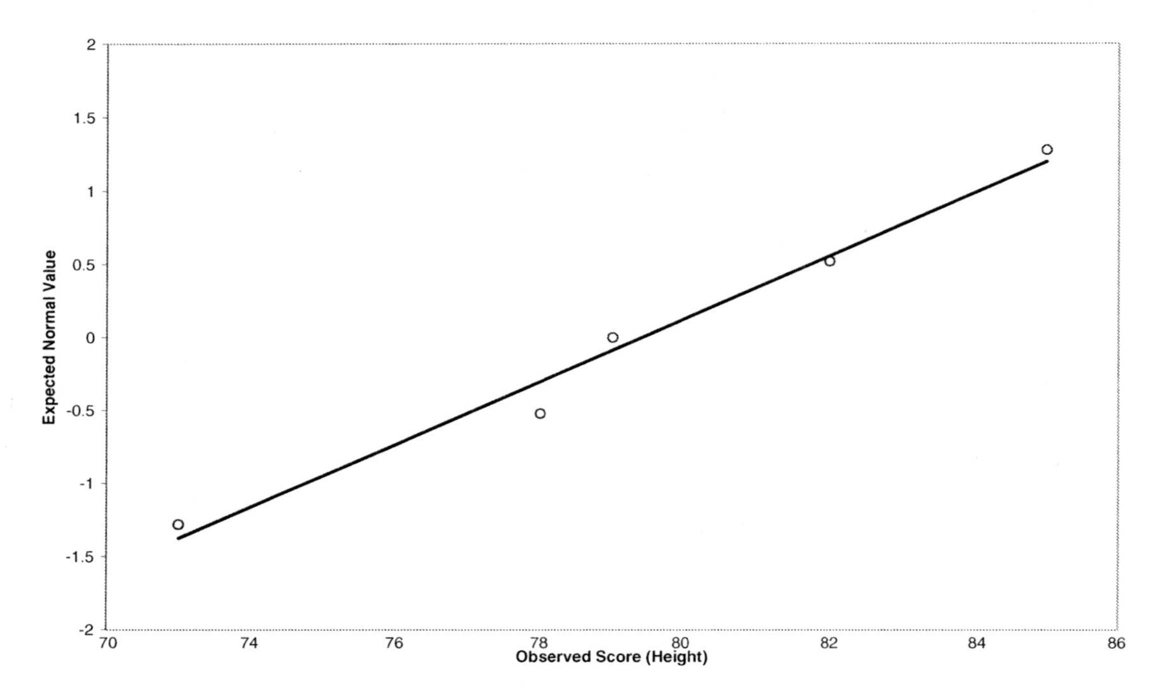

This distribution looks like it approximates a normal distribution.

17. Using Standard Scores

No, the transformation to z scores involves subtracting a constant and dividing by a constant, so the plot of the (x,z,) points will always be a straight line, regardless of the nature of the distribution.

Review Exercises

1. High Cholesterol Levels, $\mu = 178.1$, $\sigma = 40.7$

a. $P(x > 260)$

$$z = \frac{x-\mu}{\sigma} = \frac{260-178.1}{40.7} = \frac{81.9}{40.7} = +2.01$$

$P(x > 260) = P(z > +2.01)$, Using Table A-2, $P(z < +2.01) = 0.9778$

$P(z > +2.01) = 1 - P(z \le +2.01) = 1 - 0.9778 = 0.0222$

b. $P(170 < x < 200)$

$$z = \frac{x-\mu}{\sigma} = \frac{170-178.1}{40.7} = \frac{-8.1}{40.7} = -0.20 \qquad z = \frac{x-\mu}{\sigma} = \frac{200-178.1}{40.7} = \frac{21.9}{40.7} = +0.54$$

$P(z < +0.54) = 0.7054$, $P(z < -0.20) = 0.4207$

$P(170 < x < 200) = P(-0.20 < z < +0.54) = 0.7054 - 0.4207 = 0.2847$

c. $P(170 < \bar{x} < 200)$, with $n = 9$

$$z = \frac{x-\mu}{\sigma/\sqrt{n}} = \frac{170-178.1}{40.7/\sqrt{9}} = \frac{-8.1}{40.7/3} = \frac{-8.1}{13.57} = -0.60$$

$$z = \frac{x-\mu}{\sigma/\sqrt{n}} = \frac{200-178.1}{40.7/\sqrt{9}} = \frac{21.9}{40.7/3} = \frac{21.9}{13.57} = +1.61$$

From Table A-2, $P(z < -0.60) = 0.2743$ and $P(z < +1.61) = 0.9463$. The difference is $0.9463 - 0.2743 = 0.6720$. There is a 0.6720 probability that a group of 9 men will have a mean cholesterol level between 170 mg/dL and 200 mg/dL

d. The top 3% is equivalent to bottom 97%. From Table A-2, the area 0.97 corresponds to a z score of +1.88

$x = \mu + (z * \sigma) = 178.1 + (+1.88 * 40.7) = 254.6$

Therefore, the cutoff for men should be a cholesterol level of 254.6

3. Blue Genes, since $np = 25$ and $nq = 75$, both > 5, use of normal approximation to a binomial distribution, with continuity correction, is justified

$P(x \le 19)$, find $P_c(x < 19.5)$

$n = 100$, $p = 0.25$

$\mu = np = 100 * 0.25 = 25$

$\sigma = \sqrt{npq} = \sqrt{100 * 0.25 * 0.75} = \sqrt{18.75} = 4.33$

$$z = \frac{x-\mu}{\sigma} = \frac{19.5-25}{4.33} = \frac{-5.5}{4.33} = -1.27$$

From Table A-2, the area below a z score of −1.27 is 0.1020. Since $P = 0.1020 > 0.05$, it would not be considered to be unusual to have 19 or fewer offspring with blue eyes out of 100 births.

5. Sampling Distributions

a. With a sample size of 100, which is considered a large sample size, we would expect the distribution of sample means to be normally distributed regardless of the shape of distribution from which the samples are drawn. The basis for making this claim is the Central Limit Theorem.

b. The standard deviation of the sample means is referred to as the standard error of the mean. If $\sigma = 512$ and samples are of size, $n = 100$, it is found as:

$$\sigma_{\bar{x}} = \frac{\sigma}{\sqrt{n}} = \frac{512}{\sqrt{100}} = \frac{512}{10} = 51.2$$

c. With a sample size of 1200, which is considered a very large sample size, we would expect the distribution of sample proportions from x/n to be normally distributed even though the original distribution is a binomial distribution. The basis for making this claim is the Central Limit Theorem.

7. Testing for Normality, From Data Set 6 in Appendix B, Bear Neck Size

From the graphs below, the distribution is approximately normal. The histogram, with a normal distribution superimposed on it, has one mode and is roughly bell-shaped and the normal quantile plot has most of the points on the straight line.

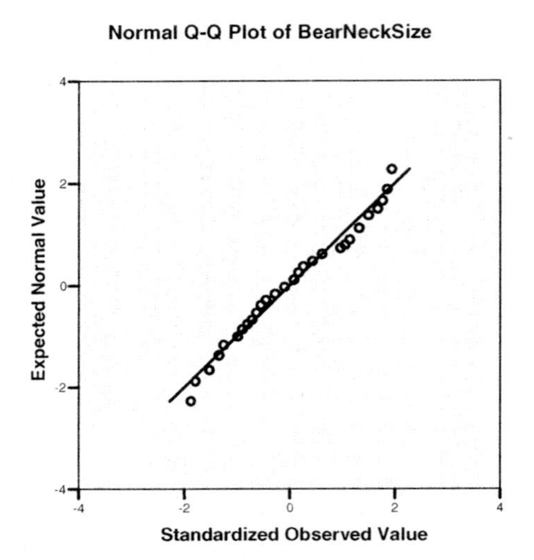

Cumulative Review Exercises

1. Eye Measurement Statistics

Ordered scores: 55 59 62 63 66 66 66 67 in mm

a. Sample Mean

$$\bar{x} = \frac{x}{n} = \frac{67+66+59+62+63+66+66+55}{8} = \frac{504}{8} = 63.0$$

b. Since there are a even number of scores, the median is the middle point between the two middle, Median, $\tilde{x} = (63+66)/2 = 64.5$

c. The mode is the number that occurs the most frequent = 66 (occurs 3 times)

d. Standard deviation

$x = 504 \qquad x^2 = 31876$

$$s^2 = \frac{n\ x^2 - (\ x)^2}{n(n-1)} = \frac{8(31876) - (504)^2}{8(8-1)} = \frac{255008 - 254016}{8 * 7} = \frac{992}{56} = 17.714$$

$$s = \sqrt{s^2} = \sqrt{17.714} = 4.21$$

e. $z = \frac{x - \bar{x}}{s} = \frac{59 - 63}{4.21} = \frac{-4}{4.21} = -0.95$

f. 6 of the 8 numbers are greater than 59, 6/8= 0.75 or 75%

g. Assuming a normal distribution, the area below a z score of −0.95, $P(z < -0.95) = 0.1711$

$P(z > -0.95) = 1 - 0.1711 = 0.8289$. This corresponds to 82.89%

h. This data set is ratio level of measurement since there are equal intervals of measurement and there is a natural staring point at zero.

i. The exact un-rounded distances are continuous data that can be any value on the continuum.

Chapter 6. Estimates and Samples with One Sample

6-2 Estimating a Population Proportion

In exercises 1 & 3, find the critical value $z_{\alpha/2}$, that corresponds to the given confidence level.

1. By inspecting the table on p. 261, we see that the critical value is 2.575.

3. For a 98% confidence interval, α=.02, and so α/2=.01. The area to the left is then 1–.01=.99. Referring to Table A-2 we find that the area .99 corresponds to the z-score 2.33, which is the critical value.

5. The upper limit of the interval is 0.280 and the lower limit of the interval is 0.220. Using the formulas above, $\hat{p}$ is 0.250 and the margin of error is 0.030. So we express the confidence interval as 0.250 ±0.030.

7. The upper limit of the interval is 0.704 and the lower limit of the interval is 0.604. Using the formulas above, $\hat{p}$ is 0.654 and the margin of error is 0.050. So we express the confidence interval as 0.654 ±0.050.

In Exercises 9 & 11, use the given confidence interval limits to find the point estimate $\hat{p}$ and the margin of error E.

The formulas used to compute $\hat{p}$ and E in exercises 9-12 are

$\hat{p}$ = (upper confidence limit) + (lower confidence limit)/2, and
E = (upper confidence limit) – (lower confidence limit)/2
(see page 268)

9. The upper limit of the interval is 0.484 and the lower limit of the interval is 0.444. Using the formulas above, $\hat{p} = 0.464$ and the margin of error $E = 0.020$.

11. The upper limit of the interval is 0.678 and the lower limit of the interval is 0.632. Using the formulas above, $\hat{p} = 0.655$ and the margin of error $E = 0.023$.

In Exercises 13 & 15, assume that a sample is used to estimate a population proportion p. Find the margin of error E that corresponds to the given statistics and confidence level.

We use the following formula, provided on page 261, to calculate the margin of error, E. The formula for $\hat{p}$ is $\hat{p}$ = x/n. The formula for $\hat{q}$ is 1- $\hat{p}$.

$$E = z_{\alpha/2}\sqrt{\frac{\hat{p}\hat{q}}{n}}$$

13. On page 261, the table of confidence levels and critical values indicates that for a confidence level of 95%, the critical value is 1.96. Also, n = 800, x = 200, and so $\hat{p}$=200/800=0.250 and $\hat{q}$=1–0.250=0.750. So the margin of error is

$$E = z_{\alpha/2}\sqrt{\frac{\hat{p}\hat{q}}{n}} = (1.96)\sqrt{\frac{(0.250)(0.750)}{800}} = (1.96)(0.015) = 0.030$$

15. On page 261, the table of confidence levels and critical values indicates that for a confidence level of 99%, the critical value is 2.575. The sample size n=1000. The percent of successes is 45%, and so $\hat{p}$=0.450 and $\hat{q}$=0.550. So the margin of error is

$$E = z_{\alpha/2}\sqrt{\frac{\hat{p}\hat{q}}{n}} = (2.575)\sqrt{\frac{(0.450)(0.550)}{1000}} = (2.575)(0.016) = 0.041$$

In Exercises 17 & 19, use the sample data and confidence level to construct the confidence interval estimate of the population proportion p.

We use the formula for confidence intervals for the population proportion p, found on page 262, to calculate the confidence intervals.

$$\hat{p} - E < p < \hat{p} + E$$

We also use the following formula, provided on page 261, to calculate the margin of error, E. The formula for $\hat{p}$ is $\hat{p}$ = x/n. The formula for $\hat{q}$ is 1- $\hat{p}$.

$$E = z_{\alpha/2}\sqrt{\frac{\hat{p}\hat{q}}{n}}$$

17. On page 261, the table of confidence levels and critical values indicates that for a confidence level of 95%, the critical value is 1.96. The sample size n=400 and number of successes x=300. So $\hat{p}$=300/400=0.750 and $\hat{q}$=1–0.750=0.250. The margin of error is

$$E = z_{\alpha/2}\sqrt{\frac{\hat{p}\hat{q}}{n}} = (1.96)\sqrt{\frac{(0.750)(0.250)}{400}} = (1.96)(0.022) = 0.042$$

This means the confidence interval is

$$\hat{p} - E < p < \hat{p} + E$$

$$0.750 - 0.042 < p < 0.750 + 0.042$$

$$0.708 < p < 0.792$$

19. For a 98% confidence interval, α=.02, and so α/2=.01. The area to the left is then 1–.01=.99. Referring to Table A-2 we find that the area .99 corresponds to the z-score 2.33, which is the critical value. The sample size is n=1655 and the number of successes is 176, so $\hat{p} = x/n = 176/1655 = 0.106$ and $\hat{q} = 1 - p = 0.894$. So the margin of error is

$$E = z_{\alpha/2}\sqrt{\frac{\hat{p}\hat{q}}{n}} = (2.33)\sqrt{\frac{(0.106)(0.894)}{1655}} = (2.33)(0.008) = 0.018$$

In Exercises 21 & 23, use the given data to find the minimum sample size required to estimate a population proportion or percentage.

We use the formula for sample size calculation for population proportion confidence intervals, found on page 266, to calculate the minimum sample size required.

When an estimate for $\hat{p}$ is known: $n = \frac{[z_{\alpha/2}]^2\,\hat{p}\hat{q}}{E^2}$

When an estimate for $\hat{p}$ is unknown: $n = \frac{[z_{\alpha/2}]^2 \cdot 0.25}{E^2}$

21. On page 261, the table of confidence levels and critical values indicates that for a confidence level of 99%, the critical value is 2.575. No estimate is known for $\hat{p}$. The margin of error is 0.060. Using the formula, we find that

$$n = \frac{[z_{\alpha/2}]^2 \cdot 0.25}{E^2} = \frac{[2.575]^2 \cdot 0.25}{(0.060)^2} = 460.46$$

We round up, so the minimum required sample size is n=461.

23. On page 261, the table of confidence levels and critical values indicates that for a confidence level of 95%, the critical value is 1.96. The estimate for $\hat{p}$ is 0.185, which makes $\hat{q}$=0.815. The margin of error is 0.050. Using the formula, we find that

$$n = \frac{[z_{\alpha/2}]^2 \cdot \hat{p}\hat{q}}{E^2} = \frac{[1.96]^2 \cdot 0.185 \cdot 0.815}{(0.050)^2} = 231.69$$

We round up, so the minimum required sample size is n=232.

25. Interpreting Calculator Display

a. We are 95% confident that the interval 0.04891 to 0.05308 actually does contain the true value of p, the proportion of males aged 18=20 who drove while impaired in the last month.

b. Alcohol-impaired driving does appear to be a problem for males aged 18-20, as the data suggests that approximately 1 in 20 such males drive while impaired each month.

c. It would be conservative to use the upper limit of the confidence interval, and so it would be prudent to use the value 5.3%.

27. Mendelian Genetics

a. The sample size n=705+224=929, and there were x=705 with red flowers. The requirements for this procedure are satisfied, as $n\hat{p} = x = 705$ and $n\hat{q} = n - x = 224$. $\hat{p}$=705/929=.759 and $\hat{q}$=1−$\hat{p}$=.241. On page 261, the table of confidence levels and critical values indicates that for a confidence level of 95%, the critical value is 1.96. The margin of error is given by

$$E = z_{\alpha/2}\sqrt{\frac{\hat{p}\hat{q}}{n}} = (1.96)\sqrt{\frac{(0.759)(0.241)}{929}} = (1.96)(0.014) = 0.028.$$

So the confidence interval is $(\hat{p} - E, \hat{p} + E) = (0.759 - 0.028, 0.759 + 0.028) = (0.731, 0.787)$

b. Though the results are not precisely 75%, the results are well within the margin of error, and so do not contradict the theory of Mendel.

29. Smoking and College Education

a. The sample size is n=785, $\hat{p}$=0.183 and $\hat{q}$=1−$\hat{p}$=0.817. The requirements for this procedure are satisfied, as $n\hat{p} = 785 * 0.183 = 143.655$ and $n\hat{q} = 785 * 0.817 = 641.345$. For a 98% confidence interval, α=.02, and so α/2=.01. The area to the left is then 1–.01=.99. Referring to Table A-2 we find that the area .99 corresponds to the z-score 2.33, which is the critical value. The margin of error is given by

$$E = z_{\alpha/2}\sqrt{\frac{\hat{p}\hat{q}}{n}} = (2.33)\sqrt{\frac{(0.183)(0.817)}{785}} = (2.33)(0.014) = 0.032.$$

So the confidence interval is

$(\hat{p} - E, \hat{p} + E) = (0.183 - 0.032, 0.183 + 0.032) = (0.151, 0.215)$

b. We are 98% confident that the true percentage of college students who smoke is between 15.1% and 21.5%, and so we are 98% confident that that percentage is lower than the 27% rate for the general public. This would imply that the smoking rate for college students is substantially different than the rate for the general public.

31. Sample Size for Plant Growers

a. For a 94% confidence interval, α=.06, and so α/2=.03. The area to the left is then 1–.03=.97. Referring to Table A-2 we find that the area .97 corresponds to the z-score 1.88, which is the critical value. The estimate for $\hat{p}$ is 0.860, which makes $\hat{q}$=0.140. The margin of error is 0.030. Using the formula for sample size from page 266, we find that

$$n = \frac{[z_{\alpha/2}]^2 \cdot \hat{p}\hat{q}}{E^2} = \frac{[1.88]^2 \cdot 0.860 \cdot 0.140}{(0.030)^2} = 472.824$$

We round up, so the minimum required sample size is n=473.

b. Again, the critical value is 1.88. No estimate is known for $\hat{p}$. The margin of error is 0.030. Using the formula for sample size from p. 266, we find that

$$n = \frac{[z_{\alpha/2}]^2 \cdot 0.25}{E^2} = \frac{[1.88]^2 \cdot 0.25}{(0.030)^2} = 981.778$$

We round up, so the minimum required sample size is n=982.

c. This would make the sample a self-selected sample, and people with a strong interest would be inclined to participate. It may occur that those who like to grow plants in their homes would then be overrepresented, inflating the value of $\hat{p}$.

33. Cell Phones and Cancer

a. The sample size n=420,095, and there were x=135 cell phone users who developed brain or nervous system cancer. The requirements for this procedure are satisfied, as $n\hat{p} = x = 135$ and $n\hat{q} = n - x = 419{,}960$. $\hat{p}$=135/420,095=0.000321 and $\hat{q}$=1−$\hat{p}$=0.999679. On page 261, the table of confidence levels and critical values indicates that for a confidence level of 95%, the critical value is 1.96. The margin of error is given by

$$E = z_{\alpha/2}\sqrt{\frac{\hat{p}\hat{q}}{n}} = (1.96)\sqrt{\frac{(0.000321)(0.999679)}{420{,}095}} = (1.96)(0.0000276) = 0.0000542.$$

So the confidence interval is

$(\hat{p} - E, \hat{p} + E) = (0.000321 - 0.000054, 0.000321 + 0.000054) = (0.000267, 0.000375)$

b. We are 95% confident that the true percentage of cell phone users who develop brain or nervous system cancer is between 0.0267% and 0.0375%. The rate of such cancer in those who do not use cell phones is 0.034%. This percentage is within the confidence interval, so it does not appear that the rate of such cancer in cell phone users is different than the rate in those who do not use cell phones.

35. Gender Selection

a. The sample size n=51, and there were x=39 of the births using YSORT were boys. The requirements for this procedure are satisfied, as $n\hat{p} = x = 39$ and $n\hat{q} = n - x = 12$. $\hat{p}$=39/51=0.765 and $\hat{q}$=1−$\hat{p}$=0.235. On page 261, the table of confidence levels and critical values indicates that for a confidence level of 99%, the critical value is 2.575. The margin of error is given by

$$E = z_{\alpha/2}\sqrt{\frac{\hat{p}\hat{q}}{n}} = (2.575)\sqrt{\frac{(0.765)(0.235)}{51}} = (2.575)(0.059) = 0.153.$$

So the confidence interval is

$(\hat{p} - E, \hat{p} + E) = (0.765 - 0.153, 0.765 + 0.153) = (0.612, 0.918)$

b. We are 99% confident that the true percentage of female births following the use of YSORT is between 61.2% and 91.8%. Assuming that boys comprise 50% of births, we are 99% confident that the birthrate of boys following the use of the YSORT method is higher than 50%, and so it would appear that the YSORT method is effective.

37. Wearing Hunter Orange

a. The sample size n=123, and there were x=6 hunters who were wearing orange at the time of the injury. The requirements for this procedure are satisfied, as $n\hat{p} = x = 6$ and $n\hat{q} = n - x = 117$. $\hat{p}$=6/123=0.049 and $\hat{q}$=1−$\hat{p}$=0.951. On page 261, the table of confidence levels and critical values indicates that for a confidence level of 95%, the critical value is 1.96. The margin of error is given by

$$E = z_{\alpha/2}\sqrt{\frac{\hat{p}\hat{q}}{n}} = (1.96)\sqrt{\frac{(0.049)(0.951)}{123}} = (1.96)(0.019) = 0.038.$$

So the confidence interval is

$(\hat{p} - E, \hat{p} + E) = (0.049 - 0.038, 0.049 + 0.038) = (0.011, 0.087)$

b. The sample size n=1115, and there were x=811 hunters who routinely wear orange. The requirements for this procedure are satisfied, as $n\hat{p}=x=811$ and $n\hat{q}=n-x=304$. $\hat{p}$=811/1115=0.727 and $\hat{q}$=1−$\hat{p}$=0.273. On page 261, the table of confidence levels and critical values indicates that for a confidence level of 95%, the critical value is 1.96. The margin of error is given by

$$E=z_{\alpha/2}\sqrt{\frac{\hat{p}\hat{q}}{n}}=(1.96)\sqrt{\frac{(0.727)(0.273)}{1115}}=(1.96)(0.013)=0.026.$$

So the confidence interval is

$(\hat{p}-E,\hat{p}+E)=(0.727-0.026,0.727+0.026)=(0.701,0.753)$

c. If hunters who wear orange were as likely to get injured as those who did not wear orange, then the percentage of injuries sustained by hunters who wore orange should be approximately equal to the percentage of hunters who wear orange. The confidence interval for the percentage of hunters who were wearing orange when they were injured is significantly lower than the confidence interval for the percentage of hunters who routinely wear orange, with no overlap. The results do then seem to indicate that hunters who wear orange are les likely to be injured because of being mistaken for game.

39. Using Finite Population Correction Factor
The population size is N=10,000 households. As in Exercise 31, the critical value is 1.88. No estimate is known for $\hat{p}$, so we allow $\hat{p}\hat{q}=0.25$ in the formula. The margin of error is 0.030. Using the formula from Exercise 39 on page 272, we find that

$$n=\frac{N\hat{p}\hat{q}[z_{\alpha/2}]^2}{\hat{p}\hat{q}[z_{\alpha/2}]^2+(N-1)E^2}=\frac{10{,}000\cdot 0.25[1.88]^2}{0.25[1.88]^2+(10{,}000-1)(0.030)^2}=\frac{8836}{9.8827}=894.088$$

We round up, so the minimum required sample size is n=895.

6-3 Estimating a Population Mean: σ Known

In Exercises 1 & 3, find the critical value $z_{\alpha/2}$, that corresponds to the given confidence level.

1. For a 98% confidence interval, α=.02, and so α/2=.01. The area to the left is then 1–.01=.99. Referring to Table A-2 we find that the area .99 corresponds to the z-score 2.33, which is the critical value.

3. For a 96% confidence interval, α=.04, and so α/2=.02. The area to the left is then 1–.02=.98. Referring to Table A-2 we find that the area .98 corresponds to the z-score 2.05, which is the critical value.

In Exercises 5 & 7, determine whether the given conditions justify using the margin of error $E=z_{\alpha/2}\sigma/\sqrt{n}$ when finding a confidence interval estimate of the population mean μ.

5. Since the sample size is greater than 30, using $E=z_{\alpha/2}\sigma/\sqrt{n}$ is justified.

7. Since the original population is normal, and σ is known, using $E=z_{\alpha/2}\sigma/\sqrt{n}$ is justified.

In Exercises 9 & 11, use the given confidence level and sample data to find (a) the margin of error E and (b) a confidence interval estimate for the population mean μ.

The formulas to compute margin of error and the confidence interval are

$E=z_{\alpha/2}\sigma/\sqrt{n}$ and $\bar{x}-E<\mu<\bar{x}+E$

9. On page 261, the table of confidence levels and critical values indicates that for a confidence level of 95%, the critical value is 1.96. The sample size is n=100, the sample mean is $\bar{x} = 95{,}000$ and the population standard deviation is $\sigma = 12{,}345$. The margin of error is

$E = z_{\alpha/2}\sigma / \sqrt{n} = (1.96) \cdot (12{,}345) / \sqrt{100} = 2419.62$

This means the confidence interval is

$\bar{x} - E < \mu < \bar{x} + E$

$95{,}000 - 2419.62 < \mu < 95{,}000 + 2419.62$

$92{,}580.38 < \mu < 97{,}419.62$

11. On page 261, the table of confidence levels and critical values indicates that for a confidence level of 90%, the critical value is 1.645. The sample size is n=25, the sample mean is $\bar{x} = 5.24$ and the population standard deviation is $\sigma = 2.50$. The margin of error is

$E = z_{\alpha/2}\sigma / \sqrt{n} = (1.645) \cdot (2.50) / \sqrt{25} = .823$

This means the confidence interval is

$\bar{x} - E < \mu < \bar{x} + E$

$5.24 - .823 < \mu < 5.24 + .823$

$4.417 < \mu < 6.063$

In Exercises 13 & 15, use the given margin of error, confidence level, and population standard deviation σ to find the minimum sample size required to estimate the population mean μ.

We use the formula for sample size calculation for population mean confidence intervals, found on page 279, to calculate the minimum sample size required.

$$n = \left[\frac{z_{\alpha/2}\sigma}{E}\right]^2$$

13. The margin of error, *E*, is 125 and the population standard deviation σ = 500. On page 261, the table of confidence levels and critical values indicates that for a confidence level of 95%, the critical value is 1.96. Using the formula,

$$n = \left[\frac{z_{\alpha/2}\sigma}{E}\right]^2 = \left[\frac{1.96 \cdot 500}{125}\right]^2 = [7.84]^2 = 61.466$$

We round up, so the minimum required sample size is n=62.

15. The margin of error, *E*, is 5 and the population standard deviation σ = 48. On page 261, the table of confidence levels and critical values indicates that for a confidence level of 90%, the critical value is 1.645. Using the formula,

$$n = \left[\frac{z_{\alpha/2}\sigma}{E}\right]^2 = \left[\frac{1.645 \cdot 48}{5}\right]^2 = [15.792]^2 = 249.387$$

We round up, so the minimum required sample size is n=250.

In Exercises 17 & 19 refer to the accompanying TI-83/84 Plus calculator display of a 95% confidence interval generated by using the methods of this section. The sample display results from using a sample of 80 measured cholesterol levels of randomly selected adults.

17. The point estimate for the population mean μ is $\bar{x} = 318.1$.

19. Since the lower confidence interval limit is $\bar{x} - E$, we find that $E = 318.1 - 262.09 = 56.01$. So the confidence interval may be expressed as 318.1 ± 56.01

21. Everglades Temperatures

The requirements are met, as the 61 days are randomly selected, it is assumed that $\sigma = 1.7$, and the sample size is n = 61 which is >30. On page 261, the table of confidence levels and critical values indicates that for a confidence level of 95%, the critical value is 1.96. The sample mean is $\bar{x} = 30.4$. The margin of error is

$$E = z_{\alpha/2}\sigma/\sqrt{n} = (1.96)\cdot(1.7)/\sqrt{61} = 0.427$$

This means the confidence interval is

$$\bar{x} - E < \mu < \bar{x} + E$$

$$30.4 - 0.427 < \mu < 30.4 + 0.427$$

$$30.0 < \mu < 30.8$$

It was not realistic to assume that σ is known.

23. Cotinine Levels of Smokers

The requirements are met, as the 40 smokers are randomly selected, it is assumed that $\sigma = 119.5$, and the sample size is n = 40 which is greater than 30. On page 261, the table of confidence levels and critical values indicates that for a confidence level of 90%, the critical value is 1.645. The sample mean is $\bar{x} = 172.5$. The margin of error is

$$E = z_{\alpha/2}\sigma/\sqrt{n} = (1.645)\cdot(119.5)/\sqrt{40} = 31.082$$

This means the confidence interval is

$$\bar{x} - E < \mu < \bar{x} + E$$

$$172.5 - 31.082 < \mu < 172.5 + 31.082$$

$$141.4 < \mu < 203.6$$

It was not realistic to assume that σ is known.

25. Sample Size for Mean IQ of Biology Majors

The margin of error, *E*, is 2 and the population standard deviation is conservatively estimated to be $\sigma = 15$. On page 261, the table of confidence levels and critical values indicates that for a confidence level of 95%, the critical value is 1.96. Using the formula,

$$n = \left[\frac{z_{\alpha/2}\sigma}{E}\right]^2 = \left[\frac{1.96\cdot 15}{2}\right]^2 = [14.7]^2 = 216.09$$

We round up, so the minimum required sample size is n=217.

27. Sample Size using Range Rule of Thumb

Estimates for the minimum and maximum ages for typical textbooks currently used in colleges will vary. Suppose that the minimum age is 0 and the maximum age is 10. The range is then 10. The range rule of thumb estimates σ = range/4. In this example, the estimated value of $\sigma = 2.5$.

The margin of error, *E*, is 0.25 and the population standard deviation is estimated, by the range rule of thumb, to be $\sigma = 2.5$. On page 261, the table of confidence levels and critical values indicates that for a confidence level of 90%, the critical value is 1.645. Using the formula,

$$n = \left[\frac{z_{\alpha/2}\sigma}{E}\right]^2 = \left[\frac{1.645\cdot 2.5}{0.25}\right]^2 = [16.45]^2 = 270.603$$

We round up, so the minimum required sample size is n=271.

29. Sample Size Using Sample Data

The minimum and maximum diastolic blood pressure for females, as found in Appendix B, Data Set 1, are 41 and 102. The range is then 61. The range rule of thumb estimates σ = range/4. In this example, the estimated value of $\sigma = 15.25$.

The margin of error, E, is 3 and the population standard deviation is estimated, by the range rule of thumb, to be $\sigma = 10$. On page 261, the table of confidence levels and critical values indicates that for a confidence level of 95%, the critical value is 1.96. Using the formula,

$$n = \left[\frac{z_{\alpha/2}\sigma}{E}\right]^2 = \left[\frac{1.96 \cdot 15.25}{3}\right]^2 = [9.963]^2 = 99.268$$

We round up, so the minimum required sample size is n=100.

If instead the sample standard deviation, s = 11.626 is used in place of the known population standard deviation, σ, we instead get the following:

$$n = \left[\frac{z_{\alpha/2}\sigma}{E}\right]^2 = \left[\frac{1.96 \cdot 11.626}{3}\right]^2 = [7.596]^2 = 57.694$$

We round up, so the minimum required sample size is n=58.

The sample sizes found are fairly different (actual difference = 43). It is more likely that the sample size n=123 is closer to the correct sample size.

31. Sample Size with Finite Population Correction Factor

The margin of error, E, is 2 and the population standard deviation is conservatively estimated to be $\sigma = 15$. On page 261, the table of confidence levels and critical values indicates that for a confidence level of 95%, the critical value is 1.96. The population from which the sample is to be taken is finite, $N = 200$, and so the formula which incorporates the finite population correction factor should be used.

$$n = \frac{N\sigma^2(z_{\alpha/2})^2}{(N-1)E^2 + \sigma^2(z_{\alpha/2})^2} = \frac{200 \cdot (15)^2(1.96)^2}{199(2)^2 + (15)^2(1.96)^2} = \frac{172872}{1660.36} = 104.117$$

We round up, so the minimum required sample size is n=105.

6-4 Estimating a Population Mean: σ Not Known

In Exercises 1–7, do one of the following, as appropriate: (a) Find the critical value $z_{\alpha/2}$, (b) find the critical value $t_{\alpha/2}$, (c) state that neither the normal nor the t distribution applies.

1. Since the population appears to be normal, and σ is unknown, it is appropriate to use the t distribution. For a 95% confidence interval, α=.05, and so $\alpha/2$=.025. Since n=5, the degrees of freedom df=n–1=4. Using Table A-3, we see that $t_{.025} = 2.776$.

3. Since the population appears to be very skewed, and the sample size is less than 30 (n=15), neither the normal nor the t distribution applies.

5. Since the population appears to be normal, and σ is unknown, it is appropriate to use the t distribution. For a 90% confidence interval, α=.10, and so $\alpha/2$=.05. Since n=92, the degrees of freedom df=n–1=91. Table A-3 does not include 91 degrees of freedom, so we use the closest number of degrees of freedom, 90. From Table A-3, we see that $t_{.05} = 1.662$.

7. Since the population appears to be normal and σ is known, it is appropriate to use the normal distribution. For a 98% confidence interval, α=.02, and so $\alpha/2$=.01. The area to the left is then 1–.01=.99. Referring to Table A-2 we find that the area .99 corresponds to the z-score 2.33, which is the critical value, $z_{\alpha/2}$.

In Exercise 9, use the given confidence level and sample data to find (a) the margin of error and (b) the confidence interval for the population mean μ. Assume that the population has a normal distribution.

The formulas to compute margin of error and the confidence interval are
$E = t_{\alpha/2}\dfrac{s}{\sqrt{n}}$ and $\bar{x} - E < \mu < \bar{x} + E$

9. For a 95% confidence interval, α=.05, and so α/2=.025. Since n=15, the degrees of freedom df=n–1=14. Using Table A-3, we see that $t_{.025} = 2.145$.

a The margin of error is $E = t_{.025}\dfrac{s}{\sqrt{n}} = 2.145\dfrac{108}{\sqrt{15}} = 59.81$

b The confidence interval is
$\bar{x} - E < \mu < \bar{x} + E$
$496 - 59.81 < \mu < 496 + 59.81$
$436.19 < \mu < 555.81$

In Exercise 11, use the given data and the corresponding TI-83/84 Plus calculator display to express the confidence interval in the format of $\bar{x} - E < \mu < \bar{x} + E$. Also write a statement that interprets the confidence interval.

The formula used to compute E in exercises 11 and 12is
E = [(upper confidence limit) – (lower confidence limit)]/2
(see page 293)

11. The lower confidence limit is 112.84 and the upper confidence limit is 121.56. This makes the margin of error E = [121.56–112.84]/2 = 4.36. The sample mean $\bar{x}$ =117.2. The confidence interval is
$\bar{x} - E < \mu < \bar{x} + E$
$117.2 - 4.36 < \mu < 117.2 + 4.36$
$112.84 < \mu < 121.56$
On the basis of the sample data, we are 95% confident that the limits of 112.84 and 121.56 do actually contain the population mean IQ score of biology students.

In Exercises 13-23, construct the indicated confidence intervals.

13. Historical Corn Data

a. We assume that the data is normally distributed. For a 95% confidence interval, α=.05, and so α/2=.025. Since n=11, the degrees of freedom df=n–1=10. Using Table A-3, we see that $t_{.025} = 2.228$. Calculating $\bar{x}$ from the sample data gives
$\bar{x} = \Sigma x / n = 20627/11 = 1875.182$
Calculating the sample standard deviation from the data gives
$s = \sqrt{\dfrac{n\Sigma(x^2) - (\Sigma x)^2}{n(n-1)}} = \sqrt{\dfrac{11 \cdot 39787267 - (20627)^2}{11 \cdot 10}} = 332.850$
The margin of error is $E = t_{.025}\dfrac{s}{\sqrt{n}} = 2.228\dfrac{332.850}{\sqrt{11}} = 223.598$
The confidence interval is
$\bar{x} - E < \mu < \bar{x} + E$
$1875.182 - 223.598 < \mu < 1875.182 + 223.598$
$1651.584 < \mu < 2098.780$

b. The intervals overlap, and so it is possible that the mean yields of kiln-dried and non-kiln dried corn seeds are equal. There is no evidence that the method of kiln-drying the seeds changes the yield of the seeds.

15. Heights of Parents

a. It appears that the data is not far from normal. For a 99% confidence interval, α=.01, and so $\alpha/2$=.005. Since n=20, the degrees of freedom df=n–1=19. Using Table A-3, we see that $t_{.005} = 2.861$. The sample mean $\bar{x}$=4.4 with a sample standard deviation s=4.2.

The margin of error is $E = t_{.005}\dfrac{s}{\sqrt{n}} = 2.861\dfrac{4.2}{\sqrt{20}} = 2.687$

The confidence interval is

$\bar{x} - E < \mu < \bar{x} + E$

$4.4 - 2.687 < \mu < 4.4 + 2.687$

$1.713 < \mu < 7.087$

b. It is 99% likely that the true mean difference in mother's and father's height is between 1.713 inches and 7.087 inches. The confidence interval does not contain 0, so the confidence interval supports the sociologist's claim that women tend to marry men who are taller than themselves.

17. Shoveling Heart Rates

a. We will assume that the data is not far from normal. For a 95% confidence interval, α=.05, and so $\alpha/2$=.025. Since n=10, the degrees of freedom df=n–1=9. Using Table A-3, we see that $t_{.005} = 2.262$. The sample mean $\bar{x}$=175 with a sample standard deviation s=15.

The margin of error is $E = t_{.025}\dfrac{s}{\sqrt{n}} = 2.262\dfrac{15}{\sqrt{10}} = 10.730$

The confidence interval is

$\bar{x} - E < \mu < \bar{x} + E$

$175 - 10.730 < \mu < 175 + 10.730$

$164.270 < \mu < 185.730$

b. We will assume that the data is not far from normal. For a 95% confidence interval, α=.05, and so $\alpha/2$=.025. Since n=10, the degrees of freedom df=n–1=9. Using Table A-3, we see that $t_{.005} = 2.262$. The sample mean $\bar{x}$=124 with a sample standard deviation s=18.

The margin of error is $E = t_{.025}\dfrac{s}{\sqrt{n}} = 2.262\dfrac{18}{\sqrt{10}} = 12.876$

The confidence interval is

$\bar{x} - E < \mu < \bar{x} + E$

$124 - 12.876 < \mu < 124 + 12.876$

$111.124 < \mu < 136.876$

c. The value 185 would be of great concern, as it is a value from the interval, and is very high. It may be that, on average, those who shovel manually achieve this high heart rate.

d. The confidence interval for heart rates for those shoveling manually and those using a snow thrower do not overlap and are significantly far apart. This is strong evidence that the mean heart rates for the two groups are significantly different.

19. Skull Breadths

First, we find a 95% confidence interval for the skulls from 4000B.C. and then find a 95% confidence interval for the skulls from 150 A.D. We will assume both samples come from populations that appear normally distributed, which indicates the use of t intervals is appropriate. For a 95% confidence interval, α=.05, and so α/2=.025. Since n=12 for both samples, each interval will be based on the t distribution with degrees of freedom df=n−1=11. Using Table A-3, we see that $t_{.025} = 2.201$. Calculating $\bar{x}$ for the sample of skulls from 4000 B.C. gives

$\bar{x} = \Sigma x / n = 1544/12 = 128.7$

Calculating the sample standard deviation for the sample of skulls from 4000 B.C. gives

$$s = \sqrt{\frac{n\Sigma(x^2) - (\Sigma x)^2}{n(n-1)}} = \sqrt{\frac{12 \cdot 198898 - (1544)^2}{12 \cdot 11}} = 4.6$$

The margin of error is $E = t_{.025}\frac{s}{\sqrt{n}} = 2.201\frac{4.6}{\sqrt{12}} = 2.9$

The confidence interval is

$\bar{x} - E < \mu < \bar{x} + E$

$128.7 - 2.9 < \mu < 128.7 + 2.9$

$125.8 < \mu < 131.6$

We are 95% confident that the true mean skull breadth for male Egyptian skulls from 4000B.C. is between 125.8 and 131.6.

For the skulls form 150 A.D, we do the same. Calculating $\bar{x}$ for the sample of skulls from 150 A.D gives

$\bar{x} = \Sigma x / n = 1600/12 = 133.3$

Calculating the sample standard deviation for the sample of skulls from 150 A.D gives

$$s = \sqrt{\frac{n\Sigma(x^2) - (\Sigma x)^2}{n(n-1)}} = \sqrt{\frac{12 \cdot 213610 - (1600)^2}{12 \cdot 11}} = 5.0$$

The margin of error is $E = t_{.025}\frac{s}{\sqrt{n}} = 2.201\frac{5.0}{\sqrt{12}} = 3.2$

The confidence interval is

$\bar{x} - E < \mu < \bar{x} + E$

$133.3 - 3.2 < \mu < 133.3 + 3.2$

$130.1 < \mu < 136.5$

We are 95% confident that the true mean skull breadth for male Egyptian skulls from 150 A.D is between 130.1 and 136.5.

There is some overlap in the two confidence intervals, and so it can not be concluded that there is a difference in the mean skull breadths. So there is not enough evidence to say that the head sizes does not seem to have changed.

21. Body Mass Index

a. Since the sample size is n=40 and σ is unknown, the t interval is indicated. For a 99% confidence interval, α=.01, and so α/2=.005. Since n=40, the degrees of freedom df=n–1=39. The degrees of freedom 39 is not included in Table A-3, so we use the degrees of freedom closest to 39, which is 38. Using Table A-3, we see that, for df=38, $t_{.005} = 2.712$. Calculating $\bar{x}$ for the BMI for males gives

$\bar{x} = \Sigma x / n = 1039.9/40 = 26.00$

Calculating the sample standard deviation for the sample of BMI for males

$$s = \sqrt{\frac{n\Sigma(x^2) - (\Sigma x)^2}{n(n-1)}} = \sqrt{\frac{40 \cdot 27493.8 - (1039.9)^2}{40 \cdot 39}} = 3.431$$

The margin of error is $E = t_{.005}\frac{s}{\sqrt{n}} = 2.712\frac{3.431}{\sqrt{40}} = 1.47$

The confidence interval is

$\bar{x} - E < \mu < \bar{x} + E$

$26.00 - 1.47 < \mu < 26.00 + 1.47$

$24.53 < \mu < 27.47$

We are 99% confident that the true mean BMI for males is between 24.53 and 27.47.

b. Since the sample size is n=40 and σ is unknown, the t interval is indicated. For a 99% confidence interval, α=.01, and so α/2=.005. Since n=40, the degrees of freedom df=n–1=39. The degrees of freedom 39 is not included in Table A-3, so we use the degrees of freedom closest to 39, which is 38. Using Table A-3, we see that, for df=38, $t_{.005} = 2.712$. Calculating $\bar{x}$ for the BMI for females gives

$\bar{x} = \Sigma x / n = 1029.6/40 = 25.74$

Calculating the sample standard deviation for the sample of BMI for females

$$s = \sqrt{\frac{n\Sigma(x^2) - (\Sigma x)^2}{n(n-1)}} = \sqrt{\frac{40 \cdot 27984.5 - (1029.6)^2}{40 \cdot 39}} = 6.166$$

The margin of error is $E = t_{.005}\frac{s}{\sqrt{n}} = 2.712\frac{6.166}{\sqrt{40}} = 2.64$

The confidence interval is

$\bar{x} - E < \mu < \bar{x} + E$

$25.74 - 2.64 < \mu < 25.74 + 2.64$

$23.10 < \mu < 28.38$

We are 99% confident that the true mean BMI for females is between 23.10 and 28.38.

c. We are 99% confident that the true mean BMI for males is between 24.53 and 27.47, and we are 99% confident that the true mean BMI for females is between 23.10 and 28.38. Since there is much overlap between the two intervals, it is possible that the mean BMI for males is not larger than that for females.

23. Sepal Widths of Irises

We find a 95% confidence interval for the sepal widths of each class. Both samples have sample size n=50, which indicates the use of t intervals is appropriate. For a 95% confidence interval, α=.05, and so α/2=.025. Since n=50 for both samples, each interval will be based on the t distribution with degrees of freedom df=n–1=49. The degrees of freedom 49 is not included in Table A-3, so we use the degrees of freedom closest to 49, which is 50. Using Table A-3, we see that $t_{.025} = 2.009$. Calculating $\bar{x}$ for the sepal widths of class setosa gives

$\bar{x} = \Sigma x / n = 170.9/50 = 3.42$

Calculating the sample standard deviation for the sepal widths of class setosa

$$s = \sqrt{\frac{n\Sigma(x^2) - (\Sigma x)^2}{n(n-1)}} = \sqrt{\frac{50 \cdot 591.25 - (170.9)^2}{50 \cdot 49}} = 0.381$$

The margin of error is $E = t_{.025}\frac{s}{\sqrt{n}} = 2.009\frac{0.381}{\sqrt{50}} = 0.11$

The confidence interval is

$\bar{x} - E < \mu < \bar{x} + E$

$3.42 - 0.11 < \mu < 3.42 + 0.11$

$3.31 < \mu < 3.53$

We are 95% confident that the true mean sepal widths of class setosa is between 3.31 mm and 3.53 mm.

For the sepal widths of class versicolor, we do the same. Calculating $\bar{x}$ for the sample of the sepal widths of class versicolor gives

$\bar{x} = \Sigma x / n = 138.5/50 = 2.77$

Calculating the sample standard deviation for the sample of the sepal widths of class versicolor gives

$$s = \sqrt{\frac{n\Sigma(x^2) - (\Sigma x)^2}{n(n-1)}} = \sqrt{\frac{50 \cdot 388.47 - (138.5)^2}{50 \cdot 49}} = 0.314$$

The margin of error is $E = t_{.025}\frac{s}{\sqrt{n}} = 2.009\frac{0.314}{\sqrt{50}} = 0.09$

The confidence interval is

$\bar{x} - E < \mu < \bar{x} + E$

$2.77 - 0.09 < \mu < 2.77 + 0.09$

$2.68 < \mu < 2.86$

We are 95% confident that the true mean sepal widths of class versicolor is between 2.68 mm and 2.86 mm.

For the sepal widths of class virginica, we do the same. Calculating $\bar{x}$ for the sample of the sepal widths of class virginica gives

$\bar{x} = \Sigma x / n = 148.7/50 = 2.97$

Calculating the sample standard deviation for the sample of the sepal widths of class virginica gives

$$s = \sqrt{\frac{n\Sigma(x^2) - (\Sigma x)^2}{n(n-1)}} = \sqrt{\frac{50 \cdot 447.33 - (148.7)^2}{50 \cdot 49}} = 0.322$$

The margin of error is $E = t_{.025}\frac{s}{\sqrt{n}} = 2.009\frac{0.322}{\sqrt{50}} = 0.09$

The confidence interval is

$\bar{x} - E < \mu < \bar{x} + E$

$2.97 - 0.09 < \mu < 2.97 + 0.09$

$2.88 < \mu < 3.06$

We are 95% confident that the true mean sepal widths of class virginica is between 2.88 mm and 3.06 mm.

No pair of the intervals overlap, but the confidence intervals are rather close to one another. This would indicate differences, but dramatic may be too strong a word.

25. Yeast Cell Counts

The sample size for the yeast cell counts is 400, so a t interval is indicated. For a 95% confidence interval, α=.05, and so α/2=.025. Since n=400, the degrees of freedom df=n–1=399. The degrees of freedom 399 is not included in Table A-3, so we use the degrees of freedom closest to 399, which is 400. Using Table A-3, we see that $t_{.025} = 1.966$. Calculating $\bar{x}$ for the yeast cell counts gives

$\bar{x} = \Sigma x / n = 1872/400 = 4.68$

Calculating the sample standard deviation for the sample of hemoglobin counts for males

$$s = \sqrt{\frac{n\Sigma(x^2)-(\Sigma x)^2}{n(n-1)}} = \sqrt{\frac{400 \cdot 10544 - (1872)^2}{400 \cdot 399}} = 2.114$$

The margin of error is $E = t_{.025}\frac{s}{\sqrt{n}} = 1.966\frac{2.114}{\sqrt{400}} = 0.208$

The confidence interval is

$\bar{x} - E < \mu < \bar{x} + E$

$4.68 - 0.208 < \mu < 4.68 + 0.208$

$4.472 < \mu < 4.888$

We are 95% confident that the true mean yeast cell count is between 4.472 and 4.888 cells, which is entirely contained within the required cell count range. The sample data then appear to be acceptable.

27. Using the Wrong Distribution

The normal distribution's critical values are smaller, so the resulting confidence interval would be narrower than it should be.

6-5 Estimating a Population Variance

In Exercises 1 & 3, find the critical values χ_L^2 and χ_R^2 that correspond to the given confidence level and sample size.

1. For a 95% confidence interval, α=0.05, and so α/2=0.025 and 1-α/2=0.975. The degrees of freedom is df=n-1=15. Referring to Table A-4 we find that the critical values are $\chi_L^2 = \chi_{.975}^2 = 6.262$ *and* $\chi_R^2 = \chi_{.025}^2 = 27.488$.

3. For a 99% confidence interval, α=0.01, and so α/2=0.005 and 1-α/2=0.995. The degrees of freedom is df=n-1=79. Since df=79 is not on the table, we use the closest value, df=80. Referring to Table A-4 we find that the critical values are $\chi_L^2 = \chi_{.995}^2 = 51.172$ and $\chi_R^2 = \chi_{.005}^2 = 116.321$.

In Exercises 5 & 7, use the given confidence level and sample data to find a confidence interval for the population standard deviation σ. In each case, assume that a simple random sample has been selected from a population that has a normal distribution.

5. For a 95% confidence interval, α=0.05, and so α/2=0.025 and 1-α/2=0.975. The degrees of freedom is df=n-1=19. Referring to Table A-4 we find that the critical values are $\chi_L^2 = \chi_{.975}^2 = 8.907$ and $\chi_R^2 = \chi_{.025}^2 = 32.852$.

The confidence interval is

$$\sqrt{\frac{(n-1)s^2}{\chi_R^2}} < \sigma < \sqrt{\frac{(n-1)s^2}{\chi_L^2}}$$

$$\sqrt{\frac{19 \cdot 12{,}345^2}{32.852}} < \sigma < \sqrt{\frac{19 \cdot 12{,}345^2}{8.907}}$$

$9{,}388.3 < \sigma < 18{,}030.3$

7. For a 90% confidence interval, α=0.10, and so α/2=0.05 and 1-α/2=0.95. The degrees of freedom is df=n-1=29. Referring to Table A-4 we find that the critical values are $\chi_L^2 = \chi_{.95}^2 = 17.708$ and $\chi_R^2 = \chi_{.05}^2 = 42.557$. The confidence interval is

$$\sqrt{\frac{(n-1)s^2}{\chi_R^2}} < \sigma < \sqrt{\frac{(n-1)s^2}{\chi_L^2}}$$

$$\sqrt{\frac{29 \cdot 2.50^2}{42.557}} < \sigma < \sqrt{\frac{29 \cdot 2.50^2}{17.708}}$$

$$2.064 < \sigma < 3.199$$

In Exercises 9 & 11, assume that each sample is a simple random sample obtained from a normally distributed population.

9. We are to be 95% confident we are within 10% of σ. We use Table 6-2 from page 305 to see that a sample size of 191 is necessary.

11. We are to be 99% confident we are within 1% of the population variance, σ^2. We use Table 6-2 from page 305 to see that a sample size of 133,448 is necessary. This sample size is not practical in most cases.

In Exercises 13-21, assume that each sample is a simple random sample from a population with a normal distribution.

We use the following formula, found on page 302, in each exercise to find the confidence interval for the population standard deviation.

$$\sqrt{\frac{(n-1)s^2}{\chi_R^2}} < \sigma < \sqrt{\frac{(n-1)s^2}{\chi_L^2}}$$

13. Historical Corn Data

We assume that the data is normally distributed and the sample is a simple random sample. For a 95% confidence interval, α=0.05, and so α/2=0.025 and 1-α/2=0.975. The degrees of freedom is df=n-1=10. Referring to Table A-4 we find that the critical values are $\chi_L^2 = \chi_{.975}^2 = 3.247$ and $\chi_R^2 = \chi_{.025}^2 = 20.483$. Calculating the sample standard variance from the data gives

$$s^2 = \frac{n\Sigma(x^2) - (\Sigma x)^2}{n(n-1)} = \frac{11 \cdot 38475192 - (20256)^2}{11 \cdot 10} = 117{,}468.873$$

The confidence interval is

$$\sqrt{\frac{(n-1)s^2}{\chi_R^2}} < \sigma < \sqrt{\frac{(n-1)s^2}{\chi_L^2}}$$

$$\sqrt{\frac{10 \cdot 117{,}468.873}{20.483}} < \sigma < \sqrt{\frac{10 \cdot 117{,}468.873}{3.247}}$$

$$239.5 < \sigma < 601.5$$

15. Monitoring Lead in Air

We assume that the data is normally distributed and the sample is a simple random sample. For a 95% confidence interval, α=0.05, and so α/2=0.025 and 1-α/2=0.975. The degrees of freedom is df=n-1=5. Referring to Table A-4 we find that the critical values are $\chi_L^2 = \chi_{.975}^2 = 0.831$ and $\chi_R^2 = \chi_{.025}^2 = 12.833$. Calculating the sample variance, we have

$$s^2 = \frac{n\Sigma(x^2)-(\Sigma x)^2}{n(n-1)} = \frac{6\cdot 32.520-(9.23)^2}{6\cdot 5} = 3.664$$

The confidence interval is

$$\sqrt{\frac{(n-1)s^2}{\chi_R^2}} < \sigma < \sqrt{\frac{(n-1)s^2}{\chi_L^2}}$$

$$\sqrt{\frac{5\cdot 3.664}{12.833}} < \sigma < \sqrt{\frac{5\cdot 3.664}{0.831}}$$

$1.195 < \sigma < 4.695$

The data is skewed to the right, and so may not be from a normally distributed population.

17. Pulse Rates

a. We assume that the data is normally distributed and the sample is a simple random sample. For a 95% confidence interval, α=0.05, and so α/2=0.025 and 1-α/2=0.975. The degrees of freedom is df=n-1=39. Since df=39 is not on the table, we use the closest value, df=40. Referring to Table A-4 we find that the critical values are $\chi_L^2 = \chi_{.975}^2 = 24.433$ and $\chi_R^2 = \chi_{.025}^2 = 59.342$. The sample standard deviation for male pulse rates is s=11.3.

The confidence interval is

$$\sqrt{\frac{(n-1)s^2}{\chi_R^2}} < \sigma < \sqrt{\frac{(n-1)s^2}{\chi_L^2}}$$

$$\sqrt{\frac{39\cdot 11.3^2}{59.342}} < \sigma < \sqrt{\frac{39\cdot 11.3^2}{24.433}}$$

$9.16 < \sigma < 14.28$

b. The critical values are as above. The sample standard deviation for female pulse rates is s=12.5.

The confidence interval is

$$\sqrt{\frac{(n-1)s^2}{\chi_R^2}} < \sigma < \sqrt{\frac{(n-1)s^2}{\chi_L^2}}$$

$$\sqrt{\frac{39\cdot 12.5^2}{59.342}} < \sigma < \sqrt{\frac{39\cdot 12.5^2}{24.433}}$$

$10.13 < \sigma < 15.79$

c. The confidence intervals overlap significantly, so it does not appear that the variances for pulse rates in men and women are different.

19. Body Mass Index

a. We assume that the data is normally distributed and the sample is a simple random sample. For a 99% confidence interval, α=0.01, and so α/2=0.005 and 1-α/2=0.975. The degrees of freedom is df=n-1=9. Since df=39 is not on the table, we use the closest value, df=40. Referring to Table A-4 we find that the critical values are $\chi_L^2 = \chi_{.995}^2 = 20.707$ and $\chi_R^2 = \chi_{.005}^2 = 66.766$. Calculating the sample standard deviation for the sample of BMI for males

$$s = \sqrt{\frac{n\Sigma(x^2)-(\Sigma x)^2}{n(n-1)}} = \sqrt{\frac{40\cdot 27493.8-(1039.9)^2}{40\cdot 39}} = 3.431$$

The confidence interval is

$$\sqrt{\frac{(n-1)s^2}{\chi_R^2}} < \sigma < \sqrt{\frac{(n-1)s^2}{\chi_L^2}}$$

$$\sqrt{\frac{39 \cdot 3.431^2}{66.766}} < \sigma < \sqrt{\frac{39 \cdot 3.431^2}{20.707}}$$

$2.6 < \sigma < 4.7$

b. The critical values are as above. Calculating the sample standard deviation for the sample of BMI for females

$$s = \sqrt{\frac{n\Sigma(x^2) - (\Sigma x)^2}{n(n-1)}} = \sqrt{\frac{40 \cdot 27984.5 - (1029.6)^2}{40 \cdot 39}} = 6.166$$

The confidence interval is

$$\sqrt{\frac{(n-1)s^2}{\chi_R^2}} < \sigma < \sqrt{\frac{(n-1)s^2}{\chi_L^2}}$$

$$\sqrt{\frac{39 \cdot 6.166^2}{66.766}} < \sigma < \sqrt{\frac{39 \cdot 6.166^2}{20.707}}$$

$4.7 < \sigma < 8.5$

c. The confidence intervals are adjacent, but do not overlap. So it appears that the population variances for BMI for males and females are different.

Review Exercises

1. Mendel's Genetics Experiments

a. We are finding the point estimate, $\hat{p}$, for a population proportion, and will convert to percentages when we find the estimate. The sample size, n=787+277=1064. The number of short stem offspring is x=277. Calculating $\hat{p} = x/n = 277/1064 = 0.260$. A point estimate for the percentage is then 26.0%.

b. We are finding the confidence interval for a population proportion, and will convert to percentages when we find the interval. The sample size, n=787+277=1064. The number of short stem offspring is x=277. Calculating $\hat{p} = x/n = 277/1064 = 0.260$. The requirements for this procedure are satisfied, as $n\hat{p} = x = 277$ and $n\hat{q} = n - x = 787$. $\hat{p}$=0.260 and $\hat{q}$=1−$\hat{p}$=0.740. On page 261, the table of confidence levels and critical values indicates that for a confidence level of 95%, the critical value is 1.96. The margin of error is given by

$$E = z_{\alpha/2}\sqrt{\frac{\hat{p}\hat{q}}{n}} = (1.96)\sqrt{\frac{(0.260)(0.740)}{1064}} = (1.96)(0.013) = 0.026.$$

So the confidence interval is

$(\hat{p} - E, \hat{p} + E) = (0.240 - 0.026, 0.240 + 0.026) = (0.214, 0.266)$.

A confidence interval for the percentage of offspring that are short stemmed would then be (21.4%,26.6%).

c. Here, we are asked for a sample size. On page 261, the table of confidence levels and critical values indicates that for a confidence level of 99%, the critical value is 2.575. No estimate is known for $\hat{p}$. The margin of error is 2.5 percentage points. Converted to decimal, the margin of error is to be 0.025. Using the formula from page 266, we find that

$$n = \frac{[z_{\alpha/2}]^2 \cdot 0.25}{E^2} = \frac{[2.575]^2 \cdot 0.25}{(0.025)^2} = 2652.25$$

We round up, so the minimum required sample size is n=2653.

3. Estimating Variation

a. We are finding a confidence interval for the standard deviation of the heights of trees in the control group. For a 95% confidence interval, α=0.05, and so α/2=0.025 and 1-α/2=0.975. The degrees of freedom is df=n-1=19. Referring to Table A-4 we find that the critical values are $\chi_L^2 = \chi_{.975}^2 = 8.907$ and $\chi_R^2 = \chi_{.025}^2 = 32.852$. Calculating the sample standard deviation for the sample of heights of trees in the control group

$$s = \sqrt{\frac{n\Sigma(x^2) - (\Sigma x)^2}{n(n-1)}} = \sqrt{\frac{20 \cdot 528.42 - (99)^2}{20 \cdot 19}} = 1.421$$

The confidence interval is

$$\sqrt{\frac{(n-1)s^2}{\chi_R^2}} < \sigma < \sqrt{\frac{(n-1)s^2}{\chi_L^2}}$$

$$\sqrt{\frac{19 \cdot 1.421^2}{32.852}} < \sigma < \sqrt{\frac{19 \cdot 1.421^2}{8.907}}$$

$$1.08 < \sigma < 2.08$$

b. We are finding a confidence interval for the standard deviation of the heights of trees in the irrigation treatment group. For a 95% confidence interval, α=0.05, and so α/2=0.025 and 1-α/2=0.975. The degrees of freedom is df=n-1=19. Referring to Table A-4 we find that the critical values are $\chi_L^2 = \chi_{.975}^2 = 8.907$ and $\chi_R^2 = \chi_{.025}^2 = 32.852$. Calculating the sample standard deviation for the heights of trees in the irrigation treatment group

$$s = \sqrt{\frac{n\Sigma(x^2) - (\Sigma x)^2}{n(n-1)}} = \sqrt{\frac{20 \cdot 461.84 - (89.6)^2}{20 \cdot 19}} = 1.783$$

The confidence interval is

$$\sqrt{\frac{(n-1)s^2}{\chi_R^2}} < \sigma < \sqrt{\frac{(n-1)s^2}{\chi_L^2}}$$

$$\sqrt{\frac{19 \cdot 1.783^2}{32.852}} < \sigma < \sqrt{\frac{19 \cdot 1.783^2}{8.907}}$$

$$1.36 < \sigma < 2.60$$

c. The two intervals overlap, and so there does not seem to be a significant difference in the standard deviations of the heights of trees in the control group and the irrigation treatment group.

5. Determining Sample Size

We need to find the sample size required for a confidence interval for a population proportion. For a 97% confidence interval, α=.003, and so α/2=0.015. The area to the left is then 1–0.015=.985. Referring to Table A-2 we find that the area .985 corresponds to the z-score 2.17, which is the critical value. No estimate is known for $\hat{p}$. The margin of error is 0.020. Using the formula found on page 266, we find that

$$n = \frac{[z_{\alpha/2}]^2 \cdot 0.25}{E^2} = \frac{[2.17]^2 \cdot 0.25}{(0.020)^2} = 2{,}943.063$$

We round up, so the minimum required sample size of biology majors to insure a maximum of a 2% error in estimating the percentage of US biology majors who get grades of B or higher is n=2,944.

7. Investigating Effect of Exercise

First, we find a 95% confidence interval for the systolic blood pressure pre-exercise with stress from math test and then find a 95% confidence interval for the systolic blood pressure post-exercise with stress from the math test. We assume that each population is normal, and the population standard deviations for the two populations are unknown, which indicates the use of t intervals is appropriate. For a 95% confidence interval, α=.05, and so α/2=.025. Since n=24 for both samples, each interval will be based on the t distribution with degrees of freedom df=n–1=23. Using Table A-3, we see that $t_{.025} = 2.069$. Calculating $\bar{x}$ for the systolic blood pressure pre-exercise with stress from math test gives

$$\bar{x} = \Sigma x / n = 3{,}194.33/24 = 133.10$$

Calculating the sample standard deviation for the systolic blood pressure pre-exercise with stress from math test

$$s=\sqrt{\frac{n\Sigma(x^2)-(\Sigma x)^2}{n(n-1)}}=\sqrt{\frac{24\cdot 430{,}963.44-(3{,}194.33)^2}{24\cdot 23}}=15.889$$

The margin of error is $E=t_{.025}\frac{s}{\sqrt{n}}=2.069\frac{15.889}{\sqrt{24}}=6.710$

The confidence interval is

$\bar{x}-E<\mu<\bar{x}+E$

$133.10-6.710<\mu<133.10+6.710$

$126.39<\mu<139.81$

We are 95% confident that the true mean systolic blood pressure pre-exercise with stress from math testis between 126.39 and 139.81.

For the systolic blood pressure post-exercise with stress from math test, we do the same. Calculating $\bar{x}$ for the sample of the systolic blood pressure post-exercise with stress from math test gives

$\bar{x}=\Sigma x/n=3045.67/24=126.90$

Calculating the sample standard deviation for the sample of the systolic blood pressure post-exercise with stress from math test gives

$$s=\sqrt{\frac{n\Sigma(x^2)-(\Sigma x)^2}{n(n-1)}}=\sqrt{\frac{24\cdot 390{,}825.89-(3045.67)^2}{24\cdot 23}}=13.709$$

The margin of error is $E=t_{.025}\frac{s}{\sqrt{n}}=2.069\frac{13.709}{\sqrt{24}}=5.790$

The confidence interval is

$\bar{x}-E<\mu<\bar{x}+E$

$126.90-5.790<\mu<126.90+5.790$

$121.110<\mu<132.690$

We are 95% confident that the true mean for the systolic blood pressure post-exercise with stress from math test is between 121.110 and 132.690.

There is overlap in the two confidence intervals, and so it can not be concluded that there is a difference in the mean systolic blood pressure pre-exercise with stress from math test and post-exercise with stress from math test.

Cumulative Review Exercises

1. Analyzing Weights of Supermodels

Here is the data listed in order, and the necessary sums.

105 115 119 119 123 125 127 128 128

$\Sigma x = 1089 \quad \Sigma x^2 = 132223$

a. The mean is $\bar{x} = \Sigma x / n = 1089/9 = 121$

b. The median is 123

c. The modes are 119 & 128

d. The midrange is (105+128)/2 = 116.5

e. The range is (128-105) = 23

f. The variance is $s^2 = \dfrac{n\Sigma(x^2) - (\Sigma x)^2}{n(n-1)} = \dfrac{9 \cdot 132223 - (1089)^2}{9 \cdot 8} = 56.75$

g. The standard deviation is $s = \sqrt{\dfrac{n\Sigma(x^2) - (\Sigma x)^2}{n(n-1)}} = \sqrt{\dfrac{9 \cdot 132223 - (1089)^2}{9 \cdot 8}} = 7.533$

h. $Q_1 = 119$

i. $Q_2 = 123$

j. $Q_3 = 127$

k. The data is ratio.

l. The boxplot is below.

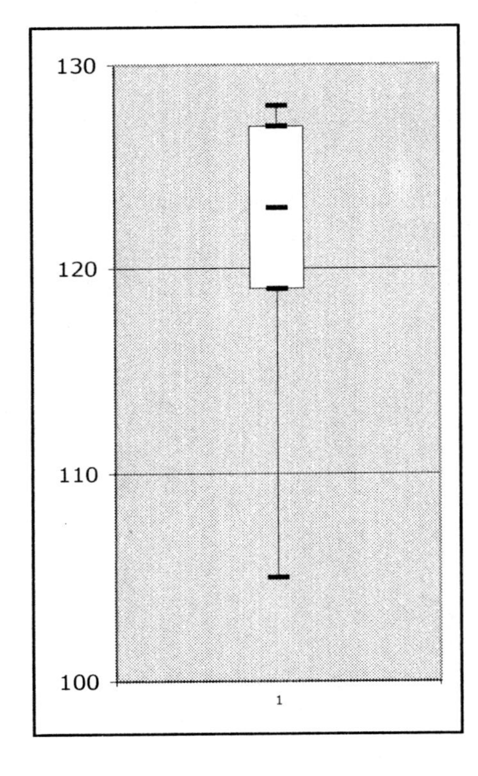

m. We assume that the weights of supermodels are normally distributed. Since σ is unknown, the t interval is indicated. For a 99% confidence interval, α=.01, and so α/2=.005. Since n=9, the degrees of freedom df=n–1=8. Using Table A-3, we see that, for df=9, $t_{.005} = 3.355$. Earlier in the exercise, we found the sample mean and standard deviation, $\bar{x} = 121$ and $s = 7.533$

The margin of error is $E = t_{.005}\dfrac{s}{\sqrt{n}} = 3.355\dfrac{7.533}{\sqrt{9}} = 8.424$

The confidence interval is

$\bar{x} - E < \mu < \bar{x} + E$

$121 - 8.424 < \mu < 121 + 8.424$

$112.6 < \mu < 129.4$

We are 99%confident that the true mean weight for supermodels is between 112.6 and 129.4 pounds.

n. We are finding a confidence interval for the standard deviation of the weights of supermodels. For a 99% confidence interval, α=0.01, and so α/2=0.005 and 1-α/2=0.995. The degrees of freedom is df=n-1=8. Referring to Table A-4 we find that the critical values are $\chi^2_L = \chi^2_{.995} = 1.344$ and $\chi^2_R = \chi^2_{.005} = 21.955$. Earlier in the exercise we found the sample standard deviation, $s = 7.533$.

The confidence interval is

$$\sqrt{\frac{(n-1)s^2}{\chi^2_R}} < \sigma < \sqrt{\frac{(n-1)s^2}{\chi^2_L}}$$

$$\sqrt{\frac{8 \cdot 7.533^2}{21.955}} < \sigma < \sqrt{\frac{8 \cdot 7.533^2}{1.344}}$$

$4.5 < \sigma < 18.4$

o. The margin of error, *E*, is 2 and the population standard deviation is approximated by the sample standard deviation, $s = 7.533$. On page 261, the table of confidence levels and critical values indicates that for a confidence level of 99%, the critical value is 2.575. Using the formula,

$$n = \left[\frac{z_{\alpha/2}\sigma}{E}\right]^2 = \left[\frac{2.575 \cdot 7.533}{2}\right]^2 = [9.699]^2 = 94.07$$

We round up, so the minimum required sample size is n=95.

p. The upper limit of the 99% confidence interval for supermodels is 129.4, which is less than one standard deviation below the mean weight of women in general. So we cannot say that the weights of supermodels is substantially less than the weights of randomly selected women.

3. Analyzing Survey Results

a. The sample size was 1059 with 413 responding yes, so the percentage is 39.0%

b. The sample size n=413+646=1059, and there were x=413 who answered yes. The requirements for this procedure are satisfied, as $n\hat{p} = x = 413$ and $n\hat{q} = n - x = 646$. $\hat{p}$=413/1059=0.390 and $\hat{q}$=1–$\hat{p}$=0.610. On page 261, the table of confidence levels and critical values indicates that for a confidence level of 95%, the critical value is 1.96. The margin of error is given by

$$E = z_{\alpha/2}\sqrt{\frac{\hat{p}\hat{q}}{n}} = (1.96)\sqrt{\frac{(0.390)(0.610)}{1059}} = (1.96)(0.015) = 0.029.$$

So the confidence interval is

$(\hat{p} - E, \hat{p} + E) = (0.390 - 0.029, 0.390 + 0.029) = (0.361, 0.419)$

c. Yes. The value 0.50 is not within the 95% confidence interval, so we are at least 95% confident that the percentage who would answer yes is different from 50%.

d. A sensible response could be that the procedure takes that into account. At the end, there is a 5% chance that the procedure produced an incorrect estimate, that is, that the true percentage that would answer yes to the question lies outside the confidence interval.

Chapter 7. Hypothesis Testing with One Sample

7-2 Basics of Hypothesis Testing

In exercises 1–4, what do you conclude? (Don't use formal procedures and exact calculations. Use only the rare event rule described in Section 7.1, and make subjective estimates to determine whether events are likely.)

1. Supposing that the proportion of births is evenly split between male and female babies, 26 female births out of 50 is not a rare event. There is not sufficient evidence to conclude that the gender selection method is effective.

3. Supposing that the proportion of adult Americans is about 0.5, the result, that 89% were opposed to cloning humans, would be a rare event, and could not be easily attributed to chance. There is sufficient evidence to conclude that a majority of adult Americans are opposed to cloning humans.

In exercises 5-12, examine the given statement, then express the null hypothesis H_0 and alternative hypothesis H_1 in symbolic form. Be sure to use the correct symbol (μ, p, σ) for the indicated parameter.

5. Let μ be the population mean annual income of workers who have studied statistics. The claim is that this value is greater than \$50,000, or μ>50,000. If this is true then $\mu\leq$50,000 must be false. Since μ>50,000 does not include equality, we let H_1 be μ>50,000 and we let H_0 be μ=50,000.

7. Let p be the population proportion of internet users who make online purchases. The claim is that more than half, or p>0.5. If this is true then $p\leq$0.5 must be false. Since $p\leq$0.5 includes equality, we let H_0 be p=0.5 and we let H_1 be p>0.5.

9. Let σ be the population standard deviation of women's heights. The claim is that this standard deviation is less than 2.8, or σ<2.8. If this is true then $\sigma\geq$2.8 must be false. Since $\sigma\geq$2.8 includes equality, we let H_0 be σ =2.8 and we let H_1 be σ<2.8.

11. Let μ be the population mean amount of rubbing alcohol in containers. The claim is that this value is at least 12, or $\mu\geq$12. If this is true then μ<12 must be false. Since $\mu\geq$12 includes equality, we let H_0 be μ=12 and we let H_1 be μ<12.

In Exercises 13-20,find the critical z values. In each case, assume that the normal distribution applies.

13. In a two-tailed test, the critical values are $\pm z_{\alpha/2}$. Since α=0.05, α/2 = 0.025. The critical values are then $\pm z_{\alpha/2} = \pm z_{.025} = \pm 1.96$.

15. In a right-tailed test, the critical value is z_α. Since α=0.01, the critical value is $z_\alpha = z_{.01} = 2.33$.

17. Since H_1 is $p\neq$0.17, this is a two-tailed test. In a two-tailed test, the critical values are $\pm z_{\alpha/2}$. Since α=0.10, α/2 = 0.05. The critical values are then $\pm z_{\alpha/2} = \pm z_{.05} = \pm 1.645$.

19. Since H_1 is p<0.19, this is a left-tailed test. In a left-tailed test, the critical value is $-z_\alpha$. Since α=0.02, the critical value is $z_\alpha = z_{.02} = 2.05$.

In Exercises 21-24, find the value of the test statistic z using

$$z = \frac{\hat{p} - p}{\sqrt{\frac{pq}{n}}}$$

Recall that $\hat{p} = \frac{x}{n}$ and $q = 1 - p$

21. $H_0 : p = 0.5$, which makes q = 0.5, n = 1025, and $\hat{p} = 0.29$. The test statistic is then

$$z = \frac{\hat{p} - p}{\sqrt{\frac{pq}{n}}} = \frac{0.29 - 0.5}{\sqrt{\frac{0.5 \cdot 0.5}{1025}}} = \frac{-0.21}{0.0156} = -13.447$$

23. $H_0 : p = 0.25$, which makes q = 0.75, n = 400, and $\hat{p} = 0.290$. The test statistic is then

$$z = \frac{\hat{p} - p}{\sqrt{\frac{pq}{n}}} = \frac{0.290 - 0.25}{\sqrt{\frac{0.25 \cdot 0.75}{400}}} = \frac{0.040}{0.02165} = 1.8475$$

In Exercises 25-32, use the given information to find the P-value.* (Hint: *See Figure 7-6.)

25. It is a right tailed test, so the P-value is the area to the right of the test statistic, $z = 0.55$. Using the methods of 5-2, the P-value is 1-0.7088 = 0.2912.

27. It is a two-tailed test, and the test statistic $z = 1.95$, is to the right of center so the p-value is the twice the area to the right of the test statistic. Using the methods of 5-2, the P-value is 2×(1–0.9744) = 0.0512.

29. Since $H_1 : p > 0.29$, it is a right tailed test, so the P-value is the area to the right of the test statistic, $z = 1.97$. Using the methods of 5-2, the P-value is 1-0.9756 = 0.0244.

31. Since $H_1 : p \neq 0.31$, it is a two-tailed test, and the test statistic $z = 0.77$, is to the right of center so the p-value is the twice the area to the right of the test statistic. Using the methods of 5-2, the P-value is 2×(1–0.7794) = 0.4412.

In Exercises 33-36, state the final conclusion in simple, non-technical terms. Be sure to address the original claim.* (Hint: *See Figure 7-7)

33. The original claim did not include equality, H_0 was rejected, so the conclusion is:
The sample data support the claim that the proportion of men who are married is greater than 0.5.

35. The original claim did not include equality, H_0 was not rejected, so the conclusion is:
There is no sufficient sample evidence to support the claim that the proportion of fatal commercial aviation crashes is different from 0.038.

In exercises 37-40, identify the type I error and the type II error that correspond to the given hypothesis.

37. Type I: Conclude there is sufficient evidence to support that the proportion of married women is greater than 0.5 when in actuality $p = 0.5$
Type II: Fail to reject that the proportion of married women is 0.5 when in actuality $p > 0.5$.

39. Type I: Conclude there is sufficient evidence to support that the proportion of fatal commercial aviation crashes is different than 0.038 when in actuality $p = 0.038$
Type II: Fail to reject that the proportion of fatal commercial aviation crashes is 0.038 when in actuality $p \neq 0.038$.

41. The alternate hypothesis is H_1 $p > 0.50$, which makes this a right tailed test. The sample proportion is $\hat{p} = 0.27$, which is less than 0.50. The test statistic is then negative. The critical region is to the right of z = 0, and so the test statistic cannot fall in the critical region, and so there is no chance to reject H_0.

43. We follow the steps outlined in the problem:

a. Step 1

The significance level is 5%, this is a left tailed test, so the critical value is $-z_{0.05} = -1.645$.

Step 2

Solving for $\hat{p}$, and substituting the values p=0.5, n = 1998, and q = 0.5, we find

$$\hat{p} = p - z_{.05}\sqrt{\frac{pq}{n}} = 0.5 - 1.645\sqrt{\frac{0.5 \cdot 0.5}{1998}} = 0.48$$

Step 3

We now find the area to the left of $\hat{p}$, using the z-score based on a population proportion $p = 0.45$.

$$z = \frac{\hat{p} - p}{\sqrt{\frac{pq}{n}}} = \frac{0.48 - 0.45}{\sqrt{\frac{0.45 \cdot 0.55}{1998}}} = 2.70$$

The area to the left of 2.70 is 0.9965, which is the power of the test. In this hypothesis test, it would be very likely that we reject H_0 when the true population proportion is $p = 0.45$.

b. $\beta = (1 - 0.9965) = 0.0035$

7-3 Testing a Claim About a Proportion

1. Mendel's Hybridization Experiments

a. The sample proportion is $\hat{p} = 0.2494$, the claim is that the proportion is 0.25, so $p = 0.25$. The sample size is n = 8023. The requirements are met, since $np = 8023 \times 0.25 = 2005.75$. and $nq = 8023 \times 0.75 = 6017.25$. The test statistic is

$$z = \frac{\hat{p} - p}{\sqrt{\frac{pq}{n}}} = \frac{0.2494 - 0.25}{\sqrt{\frac{0.25 \times 0.75}{8023}}} = -0.124$$

b. The significance level is 0.05, and the claim is that the proportion is 0.25, making this a two tailed test. The critical values are $\pm z_{\alpha/2} = \pm z_{.025} = \pm 1.96$.

c. Since this is a two tailed test with test statistic to the left of center, the P-value is twice the area to the left of the test statistic. This area is found using Table A-2, and is 0.4522. So, $P-value = 2 \times 0.4522 = 0.9044$.

d. Since the P-value is larger than the significance level, we would fail to reject H_0. The conclusion is: There is not sufficient evidence to warrant the rejection of the claim that the proportion of green-flowered peas is 0.25.

e. No. A hypothesis test can only demonstrate sufficient evidence that a parameter is not a specified value, not that the parameter is some value.

In Exercises 3-6, assume that a method of gender selection is being tested with couples wanting to have baby girls. Identify the null hypothesis, alternate hypothesis, test statistic, P-value or critical value(s), conclusion about the null hypothesis, and final conclusion that addresses the original claim. Use the P-value method unless your instructor specifies otherwise.

3. The claim is that the proportion of girls born using the method is greater than 0.50, so this is a right tailed test. The sample size is n = 100 and the number of births that were girls was x = 65, so $\hat{p} = 65/100 = 0.65$. The significance level is 0.01.

H_0: $p = 0.50$

H_1: $p > 0.50$

The test statistic is $z = \frac{\hat{p} - p}{\sqrt{\frac{pq}{n}}} = \frac{0.65 - 0.50}{\sqrt{\frac{0.50 \times 0.50}{100}}} = 3.00$

In a right-tailed test at the 0.01 significance level, the critical value is $z_\alpha = z_{.01} = 2.33$.
Since this is a right tailed test, the P-value is the area to the right of the test statistic. Using Table A-2, we find that the $P-value = 1-0.9987 = 0.0013$.
We reject the null hypothesis.
The sample data support the claim that the proportion of girls born when using this method of gender selection is greater than 0.5.

5. The claim is that the proportion of boys born using the method is less than 0.50, so this is a left-tailed test. The sample size is n = 50 and the number of births that were boys was x = 15, so $\hat{p} = 15/50 = 0.30$. The significance level is 0.05.
H_0: $p = 0.50$
H_1: $p < 0.50$
The test statistic is $z = \frac{\hat{p}-p}{\sqrt{\frac{pq}{n}}} = \frac{0.30-0.50}{\sqrt{\frac{0.50\times0.50}{50}}} = -2.828$
In a left-tailed test at the 0.05 significance level, the critical value is $-z_\alpha = -z_{.05} = -1.645$.
Since this is a left-tailed test, the P-value is the area to the left of the test statistic. Using Table A-2, we find that the $P-value = 0.0023$.
We reject the null hypothesis.
The sample data support the claim that the proportion of boys born when using this method of gender selection is less than 0.5.

In Exercises 7-18, test the given claim. Identify the null hypothesis, alternative hypothesis, test statistic, P-value or critical value(s), conclusion about the null hypothesis, and final conclusion that addresses the original claim. Use the P-value method unless you instructor specifies otherwise.

7. Cloning Survey
First, we check the requirements. The sample appears to be a simple random sample The conditions for a binomial are satisfied. The sample size is 1012 with the claim that less than 10% of adults say human cloning should be allowed, making p = 0.10, so $np = 1012\times0.10 = 101.2$ and $nq = 1012\times0.90 = 910.8$. The requirements are satisfied.
The claim is that less than 10% of all adults say that human cloning should be allowed, so this is a left-tailed test. The sample proportion is $\hat{p} = 0.09$.
H_0: $p = 0.10$
H_1: $p < 0.10$
The test statistic is $z = \frac{\hat{p}-p}{\sqrt{\frac{pq}{n}}} = \frac{0.09-0.10}{\sqrt{\frac{0.10\times0.90}{1012}}} = -1.060$
In a left-tailed test at the 0.05 significance level, the critical value is $-z_\alpha = -z_{.05} = -1.645$.
Since this is a left tailed test, the P-value is the area to the left of the test statistic. Using Table A-2, we find that the $P-value = 0.1446$.
We fail to reject the null hypothesis.
There is not sufficient sample evidence to support the claim that less than 10% of all adults say that cloning of humans should be allowed. A newspaper could run the headline, but the headline is unsupported by the sample evidence.

9. Cell Phones and Cancer
First, we check the requirements. The sample does not appear to be a simple random sample, but the subjects may comprise a random sample that is representative of the population. The conditions for a binomial are satisfied. The sample size is 420,095 with the claim that the rate at which cell phone users develop cancer of the brain or nervous system is different than 0.0340%, making p = 0.00034, so
$np = 420{,}095\times0.000340 = 142.8323$ and $nq = 420{,}095\times0.999660 = 419{,}952.168$. The requirements are satisfied.

The claim is that the rate at which cell phone users develop cancer of the brain or nervous system is different than 0.0340%, so this is a two-tailed test. In the sample, 135 developed cancer of the brain or nervous system. The sample proportion is $\hat{p} = x/n = 135/420{,}095 = 0.000321$.
H_0: $p = 0.000340$
H_1: $p \neq 0.000340$

The test statistic is $z = \dfrac{\hat{p}-p}{\sqrt{\dfrac{pq}{n}}} = \dfrac{0.000321-0.000340}{\sqrt{\dfrac{0.000340\times 0.999660}{420{,}095}}} = -0.668$

In a two-tailed test at the 0.005 significance level, the critical values are $\pm z_{\alpha/2} = \pm z_{.0025} = \pm 2.81$.
Since this is a two-tailed test and the test statistic is to the left of center, the P-value is twice the area to the left of the test statistic. Using Table A-2, we find that the $P-value = 2\times(0.2514) = 0.5028$.
We fail to reject the null hypothesis.
There is not sufficient sample evidence to support the claim that the rate of cancer of the brain or nervous system is different than the 0.0340% rate for those who do not use cell phones. Users of cell phones need not be more concerned with cancer of the brain or nervous system than those who do not use cell phones.

11. Testing Effectiveness of Nicotine Patches

First, we check the requirements. The sample does not appear to be a simple random sample, but the subjects may comprise a random sample that is representative of the population. The conditions for a binomial are satisfied. The sample size is 39 + 32 = 71 with the claim that the majority of smokers who try to quit with nicotine patch therapy are smoking one year after the treatment, making $p = 0.50$, so $np = 71\times 0.50 = 35.5$ and $nq = 71\times 0.50 = 35.5$. The requirements are satisfied.
The claim is that the majority of smokers who try to quit with nicotine patch therapy are smoking one year after the treatment, so this is a right-tailed test. In the sample, 39 were smoking one year after treatment. The sample proportion is $\hat{p} = x/n = 39/71 = 0.549$.
H_0: $p = 0.50$
H_1: $p > 0.50$

The test statistic is $z = \dfrac{\hat{p}-p}{\sqrt{\dfrac{pq}{n}}} = \dfrac{0.549-0.50}{\sqrt{\dfrac{0.50\times 0.50}{71}}} = 0.826$

In a right-tailed test at the 0.10 significance level, the critical value is $z_\alpha = z_{.10} = 1.28$.
Since this is a right tailed test, the P-value is the area to the right of the test statistic. Using Table A-2, we find that the $P-value = (1-0.7967) = 0.2033$.
We fail to reject the null hypothesis.
There is not sufficient sample evidence to support the claim that the majority of smokers who try to quit with nicotine patch therapy are smoking one year after the treatment. Although a majority return to smoking one year after the treatment, this does not mean the treatment is ineffective. To be certain, the rate at which the therapy patients return to smoking should be compared to others who have quit smoking.

13. Drug Testing for Adverse Reactions

First, we check the requirements. The sample does not appear to be a simple random sample, but the subjects may comprise a random sample that is representative of the population. The conditions for a binomial are satisfied. The sample size is 221 with the claim that the percentage of users who experience dizziness is less than 5%, making $p = 0.05$, so $np = 221\times 0.05 = 11.05$ and $nq = 221\times 0.95 = 209.95$. The requirements are satisfied.
The claim is that the percentage of users who experience dizziness is less than 5%, so this is a left-tailed test. As stated in the exercise, $\hat{p} = .032$.
H_0: $p = 0.05$
H_1: $p < 005$

The test statistic is $z = \dfrac{\hat{p}-p}{\sqrt{\dfrac{pq}{n}}} = \dfrac{0.032-0.05}{\sqrt{\dfrac{0.05\times 0.95}{221}}} = -1.228$

In a left-tailed test at the 0.01 significance level, the critical value is $-z_{\alpha} = -z_{.01} = -2.33$.
Since this is a left tailed test, the P-value is the area to the left of the test statistic. Using Table A-2, we find that the $P-value = 0.1093$.
We fail to reject the null hypothesis.
There is not sufficient sample evidence to support the claim that the percentage of users of Ziac who experience dizziness is less than 5%.

15. Air-Bag Effectiveness

First, we check the requirements. The sample does not appear to be a simple random sample, but the incidents may comprise a random sample that is representative of the population. The conditions for a binomial are satisfied. The sample size is 821 with the claim that the air-bag hospitalization rate is lower than 7.8%, making $p = 0.078$, so $np = 821 \times 0.078 = 64.038$ and $nq = 821 \times 0.922 = 756.962$. The requirements are satisfied.
The claim is that the air-bag hospitalization rate is lower than 7.8%, so this is a left-tailed test. The sample proportion is $\hat{p} = x/n = 46/821 = 0.0560$.
H_0: $p = 0.078$
H_1: $p < 0.078$
The test statistic is $z = \dfrac{\hat{p}-p}{\sqrt{\dfrac{pq}{n}}} = \dfrac{0.0560-0.078}{\sqrt{\dfrac{0.078 \times 0.922}{821}}} = -2.351$
In a left-tailed test at the 0.01 significance level, the critical value is $-z_{\alpha} = -z_{.01} = -2.33$.
Since this is a left tailed test, the P-value is the area to the left of the test statistic. Using Table A-2, we find that the $P-value = 0.0094$.
We reject the null hypothesis.
The sample data support the claim that the air-bag hospitalization rate is lower than the 7.8% rate for crashes of midsized cars equipped with automatic safety belts.

17. Females Underrepresented in Textbooks

First, we check the requirements. The sample does not appear to be a simple random sample, but the illustrations may comprise a random sample that is representative of the population. The conditions for a binomial are satisfied. The sample size is 142 + 554 = 696 with the claim that the majority of nonreproductive illustrations are of males, making $p = 0.50$, so $np = 696 \times 0.50 = 348$ and $nq = 696 \times 0.50 = 348$. The requirements are satisfied.
The claim is that the majority of nonreproductive illustrations are of males, so this is a right-tailed test. In the sample, 554 nonreproductive illustrations depicted males. The sample proportion is $\hat{p} = x/n = 554/696 = 0.796$.
H_0: $p = 0.50$
H_1: $p > 0.50$
The test statistic is $z = \dfrac{\hat{p}-p}{\sqrt{\dfrac{pq}{n}}} = \dfrac{0.796-0.50}{\sqrt{\dfrac{0.50 \times 0.50}{696}}} = 15.618$
In a right-tailed test at the 0.05 significance level, the critical value is $z_{\alpha} = z_{.05} = 1.645$.
Since this is a right tailed test, the P-value is the area to the right of the test statistic. Using Table A-2, we find that the $P-value = (1-0.9999) = 0.0001$.
We reject the null hypothesis.
There is sufficient sample evidence to support the claim that the majority of nonreproductive illustrations are of males.

19. Using Confidence Intervals to Test Hypothesis

a. First, we check the requirements. The sample is a random sample. The conditions for a binomial are satisfied. The sample size is 1000 with the claim that the proportion of zeros is 0.1, making $p = 0.1$, so $np = 1000 \times 0.10 = 100$ and $nq = 1000 \times 0.90 = 900$. The requirements are satisfied.
The claim is that the proportion of zeros is 0.1, so this is a two-tailed test. In the sample, 119 were zero. The sample proportion is $\hat{p} = x/n = 119/1000 = 0.119$.

H_0: $p = 0.1$
H_1: $p \neq 0.1$

The test statistic is $z = \dfrac{\hat{p} - p}{\sqrt{\dfrac{pq}{n}}} = \dfrac{0.119 - 0.1}{\sqrt{\dfrac{0.1 \times 0.9}{1000}}} = 2.003$

In a two-tailed test at the 0.05 significance level, the critical values are $\pm z_{\alpha/2} = \pm z_{.025} = \pm 1.96$.
There is sufficient evidence to warrant the rejection of the claim that the proportion of last digits in weights is 0.1.

b. The preliminaries are the same as in part a.
H_0: $p = 0.1$
H_1: $p \neq 0.1$

The test statistic is $z = \dfrac{\hat{p} - p}{\sqrt{\dfrac{pq}{n}}} = \dfrac{0.119 - 0.1}{\sqrt{\dfrac{0.1 \times 0.9}{1000}}} = 2.003$

Since this is a two-tailed test and the test statistic is to the right of center, the P-value is twice the area to the right of the test statistic. Using Table A-2, we find that the $P - value = 2 \times (1 - 0.9772) = 0.0456$.
There is sufficient evidence to warrant the rejection of the claim that the proportion of last digits in weights is 0.1.

c. We use the information from part a. On page 261, the table of confidence levels and critical values indicates that for a confidence level of 95%, the critical value is 1.96. The margin of error is given by

$$E = z_{\alpha/2}\sqrt{\frac{\hat{p}\hat{q}}{n}} = (1.96)\sqrt{\frac{(0.119)(0.881)}{1000}} = (1.96)(0.010) = 0.020.$$

So the confidence interval is
$(\hat{p} - E, \hat{p} + E) = (0.119 - 0.020, 0.119 + 0.020) = (0.099, 0.139)$
The confidence interval seems to suggest that the proportion of zeros being 0.1 is not out of the question.

d. They do not all lead to the same conclusion. The two hypothesis tests lead to a rejection of the null hypothesis but the confidence interval does not support that result.

21. Power and Probability of Type II Error
The significance level is 5%, this is a right tailed test, so the critical value is $z_{0.05} = 1.645$.
Solving for $\hat{p}$, and substituting the values p=0.5, n = 50, and q = 0.5, we find

$$\hat{p} = p + z_{.05}\sqrt{\frac{pq}{n}} = 0.5 + 1.645\sqrt{\frac{0.5 \cdot 0.5}{50}} = 0.616$$

We now find the area to the right of $\hat{p}$, using the z-score based on a population proportion $p = 0.45$.

$$z = \frac{\hat{p} - p}{\sqrt{\frac{pq}{n}}} = \frac{0.616 - 0.45}{\sqrt{\frac{0.45 \cdot 0.55}{50}}} = 2.359$$

The area to the right of 2.36 is 0.0091, which is the power of the test. In this hypothesis test, it would be very unlikely that we reject H_0 when the true population proportion is p = 0.45. β = 1 – 0.0091 = .9909. There is a high probability we will commit a type II error, in that we would fail to reject an incorrect null hypothesis.

7-4 Testing a Claim About a Mean: σ Known

In Exercises 1-4, determine whether the given conditions justify using the methods of this section when testing a claim about the population mean μ.

Note: It is not known how data was collected, so we assume that the sample is a simple random sample.

1. Since σ is known, and the original population is normally distributed, the conditions are met.

3. Since σ is unknown, the conditions are not met.

In Exercises 5-8, find the test statistic, P-value or critical value(s), and state the final conclusion.

5. The test statistic is $z = \frac{\bar{x}-\mu}{\sigma/\sqrt{n}} = \frac{120-118}{12/\sqrt{50}} = 1.18$

The claim is that the mean IQ is greater than 118, making this a right tailed test. In a right-tailed test, at the 0.05 significance level, the critical value is $z_\alpha = z_{.05} = 1.645$.

Since this is a right tailed test, the P-value is the area to the right of the test statistic. Using Table A-2, we find that the $P-value = 1-0.8810 = 0.1190$.

There is not sufficient evidence to support the claim that the mean IQ of statistics professors is greater than 118.

7. The test statistic is $z = \frac{\bar{x}-\mu}{\sigma/\sqrt{n}} = \frac{5.25-5.00}{2.50/\sqrt{80}} = 0.894$

The claim is that the mean time between uses of a TV remote control by males during commercials is 5.00 seconds, making this a two-tailed test. In a two-tailed test at the 0.01 significance level, the critical values are $\pm z_{\alpha/2} = \pm z_{.005} = \pm 2.575$.

Since this is a two-tailed test and the test statistic is to the right of center, the P-value is twice the area to the right of the test statistic. Using Table A-2, we find that the $P-value = 2\times(1-0.8133) = 0.3734$.

There is not sufficient evidence to warrant the rejection of the claim that the mean time between uses of a TV remote control by males during commercials is 5.00 seconds.

In Exercises 9-12, test the given claim. Identify the null hypothesis, alternate hypothesis, test statistic, P-value or critical value(s), conclusion about the null hypothesis, and final conclusion that addresses the original claim. Use the P-value method unless your instructor specifies otherwise.

9. Everglades Temperatures

First, we check the conditions. The sample size (n = 61) is greater than 30, and we are assuming a value of σ. However, the sample may not be a simple random sample. However, we will continue as if it were. The claim is that the mean temperature is greater than 30.0°C, so this is a right tailed test. The sample mean is 30.4°C and the assumed population standard deviation is 1.7°C. The significance level is 0.05.

H_0: $\mu = 30.0$

H_1: $\mu > 30.0$

The test statistic is $z = \frac{\bar{x}-\mu}{\sigma/\sqrt{n}} = \frac{30.4-30.0}{1.7/\sqrt{61}} = 1.838$

In a right-tailed test at the 0.05 significance level, the critical value is $z_\alpha = z_{.05} = 1.645$.

Since this is a right tailed test, the P-value is the area to the right of the test statistic. Using Table A-2, we find that the $P-value = 1-0.9671 = 0.0329$.

We reject the null hypothesis.

The sample data support the claim that the mean temperature is greater than 30.0°C.

11. Cotinine Levels of Smokers

First, we check the conditions. The sample size (n = 40) is greater than 30, we are assuming a value of σ, and the sample is a random sample. The claim is that the mean cotinine level of all smokers is 200, so this is a two-tailed test. The sample mean is 172.5 and the assumed population standard deviation is 119.5. The significance level is 0.01.

H_0: $\mu = 200$

H_1: $\mu \neq 200$

The test statistic is $z = \frac{\bar{x}-\mu}{\sigma/\sqrt{n}} = \frac{172.5-200}{119.5/\sqrt{40}} = -1.455$

In a two-tailed test at the 0.01 significance level, the critical values are $\pm z_{\alpha/2} = \pm z_{.005} = \pm 2.575$.

Since this is a two-tailed test and the test statistic is to the left of center, the P-value is twice the area to the left of the test statistic. Using Table A-2, we find that the $P-value = 2\times(0.0735) = 0.1470$.

We fail to reject the null hypothesis.

There is not sufficient evidence to warrant the rejection of the claim that the mean cotinine levels of smokers is 200.0.

13. Testing the Assumed σ

a. That the population standard deviation is actually known.

b. The significance level of the test was 0.05, it was a two tailed test, so any test statistic at least as small as $-z_{.025} = -1.96$ would result in the reject of the null hypothesis. Solving the test statistic formula for σ yields

$$\sigma = \frac{\bar{x} - \mu}{z / \sqrt{n}}.$$

Substituting n = 106, $\bar{x} = 98.2$, μ = 98.6, and z = -1.96, we find the largest value for σ that results in a rejection of the null hypothesis.

$$\sigma = \frac{\bar{x} - \mu}{z / \sqrt{n}} = \frac{98.2 - 98.6}{-1.96 / \sqrt{106}} = 2.1011$$

c. Yes, there is a reasonable chance that the population standard deviation is greater than 0.62. Since it was based on a sample, we know that it is highly unlikely that the true value of σ is 0.62. The assumption that σ is 0.62 is not realistic, but that value may be very close.

15. Power of Test

The significance is 0.05 and this is a left-tailed test, so the critical value is $-z_{0.05} = -1.645$. Solving for $\bar{x}$, and substituting the values μ=98.6, n = 106, and σ= 0.62, we find the value for $\bar{x}$

$$\bar{x} = \mu - z_{.05}\frac{\sigma}{\sqrt{n}} = 98.6 - 1.645\frac{0.62}{\sqrt{106}} = 98.501.$$

Since the power is to be 0.8, this means that,under the assumption that the mean is really the alternate value, the z-score for 98.501 has area to its left equal to 0.8. Using table A-2 we find this value, 0.84. Solving the test statistic for μ and substituting in the values gives the following

$$\mu = \bar{x} + z_{.2}\frac{\sigma}{\sqrt{n}} = 98.501 + 0.84\frac{0.62}{\sqrt{106}} = 98.55.$$

7-5 Testing a Claim About a Mean: σ Not Known

In Exercises 1-4, determine whether the hypothesis test involves a sampling distributions of means that is a normal distribution, Student t distribution, or neither.

Note: It is not known how data was collected, so we assume that the sample is a simple random sample.

1. The data appear to come from a normally distributed population, and σ is unknown, so this involves the Student t distribution.

3. The data appear to come from a normally distributed population, and σ is known, so this involves the normal distribution.

In Exercises 5-8, use the given information to find a range of numbers for the P-value..

5. In the row for 11 degrees of freedom, the test statistic falls between 3.106 and 2.718, so the P-value is between 0.005 and 0.01.

7. In the row for 15 degrees of freedom, the test statistic falls to the left of 2.947, so the P-value is less than 0.01.

In Exercises 9-12, assume that a simple random sample has been selected from a normally distributed population. test the given claim. Find the test statistic, P-value or critical value(s), and state the final conclusion.

9. The claim is that the mean IQ score of statistics professors is greater than 118, so this is a right-tailed test. The sample size is n = 20, the sample mean is $\bar{x} = 120$, and the sample standard deviation is s = 12. The significance level is 0.05.

The test statistic is $t = \frac{\bar{x} - \mu}{s/\sqrt{n}} = \frac{120 - 118}{12/\sqrt{20}} = 0.745$

In a right-tailed test at the 0.05 significance level with 19 degrees of freedom, the critical value is $t_\alpha = t_{.05} = 1.729$.

Since the test statistic is smaller than 1.328, the P-value is greater than 0.10.

We fail to reject the null hypothesis.

There is not sufficient sample evidence to support the claim that the mean IQ of statistics professors is greater than 118.

11. The claim is that the mean time between uses of a TV remote control by males during commercials equals 5.00 sec, so this is a two-tailed test. The sample size is n = 81, the sample mean is $\bar{x} = 5.25$, and the sample standard deviation is s = 2.50. The significance level is 0.01.

The test statistic is $t = \frac{\bar{x} - \mu}{s/\sqrt{n}} = \frac{5.25 - 5.00}{2.50/\sqrt{81}} = 0.90$

In a two-tailed test at the 0.01 significance level with 80 degrees of freedom, the critical values are $\pm t_{\alpha/2} = \pm t_{.005} = \pm 2.639$.

Since the test statistic is smaller than 1.292, the P-value is greater than 0.20.

We fail to reject the null hypothesis.

There is not sufficient evidence to warrant the rejection of the claim that the mean time between uses of a TV remote control by males during commercials equals 5.00 sec.

In Exercises 13-24, assume that a simple random sample has been selected from a normally distributed population and test the given claim. Unless specified by your instructor, use either the traditional method or the P-value method for testing hypotheses.

13. Effect of Vitamin Supplement on Birth Weight

The claim is that the mean birth weight for all male babies of mothers given vitamins is 3.39 kg, so this is a two-tailed test. The P-value is provided by the output, and is 0.103, and we would fail to reject the null hypothesis. There is not sufficient evidence to warrant the rejection of the claim that the mean birth weight for all male babies of mothers given vitamins is 3.39 kg. In other words, there isn't enough evidence to say that the vitamin supplements have an effect on birth weight.

15. Heights of Parents

The claim is that women tend to marry men who are taller than themselves, or that the mean difference between husbands and wives is greater than zero, so this is a right-tailed test. The sample size is n = 20 making the degrees of freedom df = 19, the sample mean is $\bar{x} = 4.4$, and the sample standard deviation is s = 4.2. The significance level is 0.01.

H_0: $\mu = 0$

H_1: $\mu > 0$

The test statistic is $t = \frac{\bar{x} - \mu}{s/\sqrt{n}} = \frac{4.4 - 0}{4.2/\sqrt{20}} = 4.685$

In a right-tailed test at the 0.01 significance level with df = 19, the critical value is $t_\alpha = t_{.01} = 2.539$.

In the row for 19 degrees of freedom, the test statistic falls to the left of 2.861 so the P-value is less than 0.005.

We reject the null hypothesis.

The sample data support the claim that women tend to marry men who are taller than themselves.

17. Sugar in Cereal

The claim is that the mean sugar content for all cereals is les than 0.3g, so this is a left-tailed test. The sample size is n = 16 making the degrees of freedom df = 15, the sample mean is $\bar{x} = 0.295$, and the sample standard deviation is s = 0.168. The significance level is 0.05.

H_0: $\mu = 0.3$

H_1: $\mu < 0.3$

The test statistic is $t = \frac{\bar{x} - \mu}{s/\sqrt{n}} = \frac{0.295 - 0.3}{0.168/\sqrt{16}} = -0.119$

In a left-tailed test at the 0.05 significance level with df = 15, the critical value is $-t_\alpha = -t_{.05} = -1.753$.
In the row for 15 degrees of freedom, the absolute value for the test statistic falls to the right of 1.341 so the P-value is greater than 0.10.
We fail to reject the null hypothesis.
There is not sufficient sample evidence to support the claim that the mean sugar content for all cereals is less than 0.3g.

19. Olympic Winners

The claim is that the mean winning time in the Olympic 100 m dash is less than 10.5 sec, so this is a left-tailed test. The sample size is n = 23 making the degrees of freedom df = 22. The significance level is not given so we take the significance level to be 0.05. Calculating $\bar{x}$ from the sample data gives
$\bar{x} = \Sigma x / n = 241.27/23 = 10.49$
Calculating the sample standard deviation from the data gives

$$s = \sqrt{\frac{n\Sigma(x^2) - (\Sigma x)^2}{n(n-1)}} = \sqrt{\frac{23 \cdot 241.27 - (2536.576)^2}{23 \cdot 22}} = 0.507$$

We move on to the test.
H_0: $\mu = 10.5$
H_1: $\mu < 10.5$

The test statistic is $t = \dfrac{\bar{x} - \mu}{s/\sqrt{n}} = \dfrac{10.49 - 10.5}{0.507/\sqrt{23}} = -0.095$

In a left-tailed test at the 0.05 significance level with df = 22, the critical value is $-t_\alpha = -t_{.05} = -1.717$.
In the row for 22 degrees of freedom, the absolute value for the test statistic falls to the right of 1.321 so the P-value is greater than 0.10.
We fail to reject the null hypothesis.
There is not sufficient sample evidence to support the claim that the mean winning time in the Olympic 100 m dash is less than 10.5 sec.
Notice that the precision of the numbers increases in later years.
One important characteristic of the data is that it, generally, decreases in later Olympics. Although the results are that there is no evidence that times will be lower than 10.5, it is an invalid conclusion, as they are clearly going to continue being lower than 10.5 sec.

21. Norwegian Babies Weigh More?

The claim is that Norwegian newborn babies, on average, weigh more than 3420g, which is the mean weight of American newborn babies, so this is a right-tailed test. The sample size is n = 90 making the degrees of freedom df = 89, the sample mean is $\bar{x} = 3570$, and the sample standard deviation is s = 498g. The significance level is 0.05.
H_0: $\mu = 3420$
H_1: $\mu > 3420$

The test statistic is $t = \dfrac{\bar{x} - \mu}{s/\sqrt{n}} = \dfrac{3570 - 3420}{498/\sqrt{90}} = 2.857$

In a right-tailed test at the 0.05 significance level with df = 89, we use the degrees of freedom closest to 89, which is 90, and so the critical value is $t_\alpha = t_{.05} = 1.662$.
In the row for 90 degrees of freedom, the test statistic falls to the left of 2.632 so the P-value is less than 0.005.
We reject the null hypothesis.
The sample data support the claim that Norwegian newborn babies, on average, weigh more than 3420g, which is the mean weight of American newborn babies.

23. Poplar Tree Weights

The claim is that the fertilized trees come from a population with a mean weight less than 0.92kg, so this is a left-tailed test. The sample size is n =5 making the degrees of freedom df = 4. The significance level is not mentioned, and so we use 0.05. Calculating $\bar{x}$ from the sample data gives
$\bar{x} = \Sigma x / n = 1.66/5 = 0.332$
Calculating the sample standard deviation from the data gives

$$s=\sqrt{\frac{n\Sigma(x^2)-(\Sigma x)^2}{n(n-1)}}=\sqrt{\frac{5\cdot 1.8284-(1.66)^2}{5\cdot 4}}=0.565$$

We move on to the test.

H$_0$: $\mu = 0.92$

H$_1$: $\mu < 0.92$

The test statistic is $t=\frac{\bar{x}-\mu}{s/\sqrt{n}}=\frac{0.332-0.92}{0.565/\sqrt{5}}=-2.327$

In a left-tailed test at the 0.05 significance level with df = 4, the critical value is $-t_\alpha=-t_{.05}=-2.132$.

In the row for 4 degrees of freedom, the absolute value for the test statistic falls between 2.776 and 2.132 so the P-value is between 0.025 and 0.05.

We reject the null hypothesis.

The sample data support the claim that the fertilized trees come from a population with a mean weight less than 0.92kg.

The fertilizer appears to have an effect, decreasing the weight of poplars on site 1.

25. **Using the Wrong Distribution**

If one uses the standard normal distribution when the Student t distribution is indicated, there is a greater chance of rejecting the null hypothesis. This is because the critical values found are closer to the center than when dealing with the Student t. This means a test statistic closer to the center could be rejected when, had the Student t be properly employed, the test statistic would not have fallen in the critical region.

7-6 Testing a Claim About a Standard Deviation of Variance

In Exercises 1-4, find the test statistic, then use table A-4 to find the critical value(s) of χ^2 and limits containing the P-value, then determine whether there is sufficient evidence to support the alternate hypothesis.

1. The test statistic is $\chi^2=\frac{(n-1)s^2}{\sigma^2}=\frac{19\cdot 10^2}{15^2}=8.444$.

The alternate hypothesis is H$_1$: $\sigma \neq 15$, so this is a two-tailed test. Since n = 20, df = 19, and the significance level is 0.05. The critical values, from Table A-4 are 8.907 and 32.852. Since the test statistic falls between 7.733 and 8.907, the P-value is between 0.02 and 0.05. There is sufficient evidence to support the alternate hypothesis.

3. The test statistic is $\chi^2=\frac{(n-1)s^2}{\sigma^2}=\frac{29\cdot 30^2}{50^2}=10.44$.

The alternate hypothesis is H$_1$: $\sigma < 50$, so this is a left-tailed test. Since n = 30, df = 29, and the significance level is 0.01. The critical value, from Table A-4 is 14.257. Since the test statistic falls below 13.121, the P-value is below 0.005. There is sufficient evidence to support the alternate hypothesis.

In Exercises 5-12, test the given claim. Assume that a simple random sample is selected from a normally distributed population. Use the traditional method of testing hypotheses unless your instructor indicates otherwise.

5. **Body Temperatures**

The claim is that the population standard deviation of body temperatures is less than 2.11°F, so this is a left-tailed test. The sample size is n = 106 making the degrees of freedom df = 105. The significance level is 0.005. The sample standard deviation is s = 0.62°F. We move on to the test.

H$_0$: $\sigma = 2.11$

H$_1$: $\sigma < 2.11$

The test statistic is $\chi^2=\frac{(n-1)s^2}{\sigma^2}=\frac{105\cdot 0.62^2}{2.11^2}=9.066$

The degrees of freedom table does not contain df = 105, so we use df = 100. In a left-tailed test at the 0.005 significance level with df = 100, the critical value is $\chi^2=67.328$.

In the row for 100 degrees of freedom, the test statistic lies below 67.328, so the P-value is less than 0.005.
We reject the null hypothesis.
The sample data support the claim that the population standard deviation of body temperatures is less than 2.11°F.

7. Supermodel Heights
The claim is that the population standard deviation of supermodel heights vary less than the heights of women in general, who have a standard deviation of 2.5 in, so this is a left-tailed test. The sample size is $n = 18$ making the degrees of freedom df = 17. The significance level is 0.05. Calculating the sample standard deviation from the data gives

$$s = \sqrt{\frac{n\Sigma(x^2) - (\Sigma x)^2}{n(n-1)}} = \sqrt{\frac{18 \cdot 88{,}084 - (1{,}259)^2}{18 \cdot 17}} = 1.187$$

We move on to the test.
H_0: $\sigma = 2.5$
H_1: $\sigma < 2.5$
The test statistic is $\chi^2 = \frac{(n-1)s^2}{\sigma^2} = \frac{17 \cdot 1.187^2}{2.5^2} = 3.832$
In a left-tailed test at the 0.05 significance level with df = 17, the critical value is $\chi^2 = 8.672$.
In the row for 17 degrees of freedom, the test statistic lies below 5.697, so the P-value is less than 0.005.
We reject the null hypothesis.
The sample data support the claim that the population standard deviation of supermodel heights vary less than the heights of women in general.

9. Is the New Machine Better?
The claim is that the population standard deviation of cold medicine poured by the new machine is less than that of the old machine, which was 0.15 oz., so this is a left-tailed test. The sample size is $n = 71$ making the degrees of freedom df = 70. The significance level is 0.05. The sample standard deviation is $s = 0.12$ oz. We move on to the test.
H_0: $\sigma = 0.15$
H_1: $\sigma < 0.15$
The test statistic is $\chi^2 = \frac{(n-1)s^2}{\sigma^2} = \frac{70 \cdot 0.12^2}{0.15^2} = 44.8$
In a left-tailed test at the 0.05 significance level with df = 70, the critical value is $\chi^2 = 51.739$.
In the row for 70 degrees of freedom, the test statistic lies between 43.275 and 45.442, so the P-value is between 0.005 and 0.01.
We reject the null hypothesis.
The sample data support the claim that the population standard deviation of cold medicine poured by the new machine is less than that of the old machine. Its purchase should be considered.

11. Weights of Men
The claim is that the population standard deviation of weights for men is 28.7 lb., so this is a two-tailed test. The sample size is $n = 40$ making the degrees of freedom df = 39. The significance level is 0.05. Calculating the sample standard deviation from the data gives

$$s = \sqrt{\frac{n\Sigma(x^2) - (\Sigma x)^2}{n(n-1)}} = \sqrt{\frac{40 \cdot 1{,}217{,}971.76 - (6{,}902.0)^2}{40 \cdot 39}} = 26.327$$

We move on to the test.
H_0: $\sigma = 28.7$
H_1: $\sigma \neq 28.7$
The test statistic is $\chi^2 = \frac{(n-1)s^2}{\sigma^2} = \frac{39 \cdot 26.327^2}{28.7^2} = 32.817$
The degrees of freedom df = 39 is not displayed in Table A-4,so we use df = 40. In a two-tailed test at the 0.05 significance level with df = 40, the critical values are $\chi^2 = 24.433$ and $\chi^2 = 59.342$.

In the row for 40 degrees of freedom, the test statistic lies between 29.051 and 51.805, so the P-value is larger than 0.20.
We fail to reject the null hypothesis.
There is not sufficient evidence to warrant rejection of the claim that the population standard deviation of weights for men is 28.7 lb. If the standard deviation was larger, the maximum load displayed in an elevator may be inaccurate.

13. Finding Critical Values for χ^2

a. For $\alpha = 0.05$, the critical values found on Table A-2 are z = -1.96 and z = 1.96. Also, k = degrees of freedom = $n - 1 = 100$. Calculating the left critical value for a χ^2, we find

$$\chi^2 = \frac{1}{2}(z+\sqrt{2k-1})^2 = \frac{1}{2}\left(-1.96+\sqrt{2\times 100-1}\right)^2 = 73.772.$$

Calculating the right critical value for χ^2, we find

$$\chi^2 = \frac{1}{2}(z+\sqrt{2k-1})^2 = \frac{1}{2}\left(1.96+\sqrt{2\times 100-1}\right)^2 = 129.070$$

The values from Table A-4 are 74.222 and 129.561. The values found using the formula are fairly accurate (each is off by roughly 0.5).

b. For $\alpha = 0.05$, the critical values found on Table A-2 are z = -1.96 and z = 1.96. Also, k = degrees of freedom = $n - 1 = 149$. Calculating the left critical value for a χ^2, we find

$$\chi^2 = \frac{1}{2}(z+\sqrt{2k-1})^2 = \frac{1}{2}\left(-1.96+\sqrt{2\times 149-1}\right)^2 = 116.643.$$

Calculating the right critical value for χ^2, we find

$$\chi^2 = \frac{1}{2}(z+\sqrt{2k-1})^2 = \frac{1}{2}\left(1.96+\sqrt{2\times 149-1}\right)^2 = 184.199$$

Review Exercises

1.

a. We can conclude that most adult Americans are opposed to estate taxes, but we need to be careful. With such a large sample size, a sample proportion that is only slightly higher than 0.50 would result in that conclusion.

b. This drug should not be used. The actual mean weight loss is insignificant, though it does have statistical significance.

c. The 0.001 significance level would be most convincing. However, I would prefer the 0.01 significance level, as it is a reasonable threshold without being unrealistic.

d. There is not sufficient evidence to warrant the rejection of the claim that the volume of cold medicine is different than 12 oz.

e. "rejecting a true null hypothesis."

3. Effective Treatment

First, we check the requirements. We will assume that the sample is a simple random sample. The conditions for a binomial are satisfied. The sample size is 200 with the claim that the proportion of tree deaths for treated trees is less than 10%, making $p = 0.10$, so $np = 200\times 0.10 = 20$ and $nq = 200\times 0.90 = 180$. The requirements are satisfied.
The claim is that the proportion of tree deaths for treated trees is less than 10%, so this is a left-tailed test. In the sample, 16 of the trees died the following year. The sample proportion is $\hat{p} = x/n = 16/200 = 0.08$.
H_0: $p = 0.10$
H_1: $p < 0.10$

The test statistic is $z = \dfrac{\hat{p}-p}{\sqrt{\dfrac{pq}{n}}} = \dfrac{0.08-0.10}{\sqrt{\dfrac{0.10\times 0.90}{200}}} = -0.943$

In a left-tailed test at the 0.05 significance level, the critical value is $-z_\alpha = -z_{.05} = -1.645$.
Since this is a left tailed test, the P-value is the area to the left of the test statistic. Using Table A-2, we find that the $P-value = 0.1736$.
We fail to reject the null hypothesis.

There is not sufficient sample evidence to support the claim that the proportion of tree deaths for treated trees is less than 10%.

5. Are Consumers Being Cheated?

The claim is that the mean volume of alcohol in the bottles is less than 12 oz, so this is a left-tailed test. The sample size is n = 24, the sample mean is $\bar{x} = 11.4$, and the sample standard deviation is s = 0.62. The significance level is not specified, so we use 0.05.

H_0: $\mu = 12$

H_1: $\mu < 12$

The test statistic is $t = \dfrac{\bar{x}-\mu}{s/\sqrt{n}} = \dfrac{11.4-12.0}{0.62/\sqrt{24}} = -4.741$

In a left-tailed test at the 0.01 significance level with 23 degrees of freedom, the critical value is $t_{\alpha} = t_{.05} = 1.714$.

In the row for 23 degrees of freedom, the absolute value for the test statistic is greater than 2.807 so the P-value is less than 0.005.

We reject the null hypothesis.

The sample data support the claim that the mean volume of alcohol in the bottles is less than 12 oz. However, the claim by Harry Windsor has a little merit. However, the sample size is close to 30, so if the distribution is symmetric, the hypothesis test would still be valid.

7. Controlling Variation

The sample is a random sample. We will assume that the sample is from a normally distributed population. The claim is that the population standard deviation for the drug is greater than 0.5 mg, so this is a two-tailed test. The sample size is n = 15 making the degrees of freedom df = 14. The significance level is 0.05.

Calculating the sample standard deviation from the data gives

$$s = \sqrt{\frac{n\Sigma(x^2)-(\Sigma x)^2}{n(n-1)}} = \sqrt{\frac{15\cdot 9{,}295.07-(373.3)^2}{15\cdot 14}} = 0.590$$

We move on to the test.

H_0: $\sigma = 0.5$

H_1: $\sigma > 0.5$

The test statistic is $\chi^2 = \dfrac{(n-1)s^2}{\sigma^2} = \dfrac{14\cdot 0.590^2}{0.5^2} = 19.494$

In a right-tailed test at the 0.05 significance level with df = 14, the critical value is $\chi^2 = 23.685$.

In the row for 14 degrees of freedom, the test statistic lies between 7.790 and 21.064, so the P-value is greater than 0.10 but less than 0.90.

We fail to reject the null hypothesis.

There is not sufficient evidence to warrant the rejection of the claim that the population standard deviation for the drug is greater than 0.5 mg. If the variance is too great, then the amount of drug in some of the pills may be too large, may exceed safe levels.

9. Effectiveness of Seat Belts

The sample size is larger than 30, so using the Student t distribution is appropriate. The claim is that the mean amount of time spent in an intensive care unit for children, who were wearing safety belts, following an automobile accident is less than 1.39 days, so this is a left-tailed test. The sample size is n = 123 making the degrees of freedom df = 122, the sample mean is $\bar{x} = 0.83$, and the sample standard deviation is s = 0.16. The significance level is 0.01.

H_0: $\mu = 1.39$

H_1: $\mu < 1.39$

The test statistic is $t = \dfrac{\bar{x}-\mu}{s/\sqrt{n}} = \dfrac{0.83-1.39}{0.16/\sqrt{123}} = -38.817$

The critical values for 122 degrees of freedom is not on the table, so we use df = 100. In a left-tailed test at the 0.01 significance level with df = 100, the critical value is $-t_{\alpha} = -t_{.01} = -2.364$.

In the row for 100 degrees of freedom, the absolute value of the test statistic falls far to the left of 2.626 so the P-value is less than 0.005.

We reject the null hypothesis.
The sample data support the claim that the mean amount of time spent in an intensive care unit for children, who were wearing safety belts, following an automobile accident is less than 1.39 days, which is the mean number of days spent following an accident when children are not using a safety belt. There is strong evidence showing that seatbelts seem to help.

Cumulative Review Exercises

1. **Monitoring Dioxin**
Here is the data listed in order, and the necessary sums.

0.018 0.0268 0.0281 0.032 0.044 0.0524 0.161 0.175 0.176

$\Sigma x = 0.7133$ $\Sigma x^2 = 0.09506$

a. The mean is $\bar{x} = \Sigma x / n = 0.7133/9 = 0.07926$

b. The median is 0.044

c. The standard deviation is $s = \sqrt{\dfrac{n\Sigma(x^2) - (\Sigma x)^2}{n(n-1)}} = \sqrt{\dfrac{9 \cdot 0.09506 - (0.7133)^2}{9 \cdot 8}} = 0.0694$

d. The variance is $s^2 = \dfrac{n\Sigma(x^2) - (\Sigma x)^2}{n(n-1)} = \dfrac{9 \cdot 0.09506 - (0.7133)^2}{9 \cdot 8} = 0.00482$

e. The range is (0.176-0.018) = 0.158

f. We assume that the data is from a normally distributed population. For a 95% confidence interval, α=.05, and so $\alpha/2$=.025. Since n=9, the degrees of freedom df=n–1=8. Using Table A-3, we see that $t_{.025} = 2.306$.

The margin of error is $E = t_{.025}\dfrac{s}{\sqrt{n}} = 2.306\dfrac{0.0694}{\sqrt{9}} = 0.0533$

The confidence interval is
$\bar{x} - E < \mu < \bar{x} + E$

$0.07926 - 0.05335 < \mu < 0.07926 + 0.05335$

$0.02591 < \mu < 0.13261$

g. We again assume that the sample data come from a normally distributed population. The claim is that the mean dioxin level is less than 0.16 ng/m^3, so this is a left-tailed test. The sample size is n = 9 making the degrees of freedom df = 8, the sample mean is $\bar{x} = 0.07926$, and the sample standard deviation is s = 0.0694. The significance level is 0.05.

H_0: $\mu = 0.16$
H_1: $\mu < 0.16$

The test statistic is $t = \dfrac{\bar{x} - \mu}{s/\sqrt{n}} = \dfrac{0.07926 - 0.16}{0.0694/\sqrt{9}} = -3.490$

In a left-tailed test at the 0.05 significance level with df = 8, the critical value is $-t_\alpha = -t_{.05} = -1.860$.
In the row for 8 degrees of freedom, the absolute value for the test statistic is greater than 3.355 so the P-value is less than 0.005.
We reject the null hypothesis.
The sample data support the claim that the mean dioxin level is less than 0.16 ng/m^3.

h. One important characteristic of the data that is not addressed is that there was a dramatic drop in values over time, in that the 4th and later measures were all much lower than the initial three measurements.

3. **Blood Pressure Readings**

a. The mean is $\bar{x} = \Sigma x / n = 1987.7/16 = 124.231$

The standard deviation is $s = \sqrt{\dfrac{n\Sigma(x^2) - (\Sigma x)^2}{n(n-1)}} = \sqrt{\dfrac{16 \cdot 254543.43 - (1987.7)^2}{16 \cdot 15}} = 22.523$

b. The sample is a random sample. We will assume that the sample is from a normally distributed population. The claim is that the mean systolic blood pressure for women infected with the new viral strain is 114.8, so this is a two-tailed test. The sample size is n = 16 making the degrees of freedom df = 15. The significance level is 0.05.

H_0: $\mu = 114.8$

H_1: $\mu \neq 114.8$

The test statistic is $t = \dfrac{\bar{x}-\mu}{s/\sqrt{n}} = \dfrac{124.231-114.8}{22.523/\sqrt{16}} = 1.675$

In a two-tailed test at the 0.05 significance level with df = 15, the critical values are $\pm t_{\alpha} = \pm t_{.025} = \pm 2.131$.

In the row for 15 degrees of freedom, the test statistic is between 1.341 and 1.753, so the P-value is between 0.10 and 0.20.

We fail to reject the null hypothesis.

There is not sufficient evidence to warrant the rejection of the claim that the mean systolic blood pressure for women infected with the new viral strain is 114.8.

c. We again assume that the data is from a normally distributed population. For a 95% confidence interval, α=.05, and so $\alpha/2$=.025. Since n=16, the degrees of freedom df=n–1=15. Using Table A-3, we see that $t_{.025} = 2.131$.

The margin of error is $E = t_{.025}\dfrac{s}{\sqrt{n}} = 2.131\dfrac{222.523}{\sqrt{16}} = 11.999$

The confidence interval is

$\bar{x}-E < \mu < \bar{x}+E$

$124.231-11.999 < \mu < 124.231+11.999$

$112.232 < \mu < 136.23$

The confidence interval limits do contain the value 114.8

d. The claim is that the population standard deviation of systolic blood pressure of women infected with the new viral strain is 13.1, so this is a two-tailed test. The sample size is n = 16 making the degrees of freedom df = 15. The significance level is 0.05.

H_0: $\sigma = 13.1$

H_1: $\sigma \neq 13.1$

The test statistic is $\chi^2 = \dfrac{(n-1)s^2}{\sigma^2} = \dfrac{15 \cdot 22.523^2}{13.1^2} = 44.341$

In a two-tailed test at the 0.05 significance level with df = 15, the critical values are $\chi^2 = 6.262$ and $\chi^2 = 27.488$.

In the row for 15 degrees of freedom, the test statistic lies above 32.801, so the P-value is less than 0.01.

We reject the null hypothesis.

There is sufficient evidence to warrant the rejection of the claim that the population standard deviation of systolic blood pressure of women infected with the new viral strain is 13.1.

e. Yes, the variability of the blood pressures has greatly increased.

Chapter 8. Inferences from Two Samples

8-2 Inferences About Two Proportions

In Exercises 1 & 3, find the number of successes x suggested by the given statement.

1. $\hat{p} = 81\% = 0.81 \quad n = 37 \quad x = \hat{p} * n = 0.81 * 37 = 29.97 \approx 30$

3. $\hat{p} = 28.9\% = 0.289 \quad n = 294 \quad x = \hat{p} * n = 0.289 * 294 = 84.96 \approx 85$

In Exercise 5, assume that you plan to use a significance level of $\alpha = 0.05$ to test the claim that $p_1 = p_2$. Use the given sample sizes and numbers of successes to find (a) the pooled estimate $\overline{p}$, (b) the z test statistic, (c) the critical z values, and (d) the P-value.

5. **a.** Pooled estimate $\overline{p}$

$$\hat{p}_1 = \frac{x_1}{n_1} = \frac{192}{436} = 0.440 \quad \hat{p}_2 = \frac{x_2}{n_2} = \frac{40}{121} = 0.330$$

$$\overline{p} = \frac{x_1 + x_2}{n_1 + n_2} = \frac{192 + 40}{436 + 121} = \frac{232}{557} = 0.417 \qquad \overline{q} = 1 - \overline{p} = 1 - 0.417 = 0.583$$

b. z test statistic

$$\hat{p}_1 = \frac{192}{436} = 0.440 \quad \hat{p}_2 = \frac{40}{121} = 0.330$$

$$z = \frac{(\hat{p}_1 - \hat{p}_2) - (p_1 - p_2)}{\sqrt{\frac{\overline{p}\overline{q}}{n_1} + \frac{\overline{p}\overline{q}}{n_2}}} = \frac{(0.440 - 0.330) - (0)}{\sqrt{\frac{0.417 * 0.583}{436} + \frac{0.417 * 0.583}{121}}} = \frac{0.110}{\sqrt{0.0006 + 0.0020}}$$

$$= \frac{0.100}{0.0507} = 2.17$$

c. Critical z values for a non-directional z test with $\alpha = 0.05$ are ± 1.96

d. $z = 2.17$, P-value from Table A2= $2 * P(z > 2.17) = 2 * (1 - 0.9850) = 2 * 0.0150 = 0.03$

7. **Exercise and Heart Disease**

To find the 0.90 or 90% confidence interval, the critical values of z are ± 1.645

$$\hat{p}_1 = \frac{x_1}{n_1} = \frac{101}{10239} = 0.00986 \qquad \hat{q}_1 = 1 - \hat{p}_1 = 1 - 0.00986 = 0.99014$$

$$\hat{p}_2 = \frac{x_2}{n_2} = \frac{56}{9877} = 0.00567 \qquad \hat{q}_1 = 1 - \hat{p}_1 = 1 - 0.00567 = 0.99433$$

$$\hat{p}_1 - \hat{p}_2 = 0.00986 - 0.00567 = 0.00419$$

$$\text{CI}_{90\%} = (\hat{p}_1 - \hat{p}_2) - E < (p_1 - p_2) < (\hat{p}_1 - \hat{p}_2) + E$$

$$E = z_{\alpha/2}\sqrt{\frac{\hat{p}_1\hat{q}_1}{n_1} + \frac{\hat{p}_2\hat{q}_2}{n_2}} = 1.645\sqrt{\frac{0.00986 * 0.99014}{10239} + \frac{0.00567 * 0.99433}{9877}}$$

$$= 1.645\sqrt{0.000000953 + 0.000000571} = 1.645\sqrt{0.000001524} = 1.645 * 0.00123 = 0.00203$$

$$\text{CI}_{90\%} = (0.00986 - 0.00567) - 0.00203 < (p_1 - p_2) < (0.00986 - 0.00567) + 0.00203$$

$$\text{CI}_{90\%} = 0.00419 - 0.00202 < (p_1 - p_2) < 0.00419 + 0.00202$$

$$\text{CI}_{90\%} = 0.00216 < (p_1 - p_2) < 0.00622$$

The confidence interval does not contain zero. Therefore, it appears that physical activity corresponds to a lower rate of coronary artery disease.

9. Effectiveness of Smoking Bans

$H_0: p_1 = p_2 \quad H_1: p_1 \neq p_2$

$\hat{p}_1 = \frac{56}{843} = 0.0664 \quad \hat{p}_2 = \frac{27}{703} = 0.0384$

$\bar{p} = \frac{x_1 + x_2}{n_1 + n_2} = \frac{56+27}{843+703} = 0.0537 \quad \bar{q} = 1 - \bar{p} = 0.9463$

$$z = \frac{(\hat{p}_1 - \hat{p}_2) - (p_1 - p_2)}{\sqrt{\frac{\bar{p}\bar{q}}{n_1} + \frac{\bar{p}\bar{q}}{n_2}}} = \frac{(0.0664 - 0.0384) - (0)}{\sqrt{\frac{0.0537 * 0.9463}{843} + \frac{0.0537 * 0.9463}{703}}} = \frac{0.0280}{\sqrt{0.0000603 + 0.0000723}}$$

$$= \frac{0.0280}{\sqrt{0.0001326}} = \frac{0.0280}{0.0115} = 2.43$$

When α is set at 0.05, there is a significant difference since the observed z of 2.43 is greater than +1.96 and the *P*-value is 2* (1 – 0.9925) = 2* 0.0075= 0.0150, which is lower than $\alpha = 0.05$.
When α is set at 0.01, there is not a significant difference since the observed z of 2.43 is not greater than +2.575. The P-value is the same at 0.0150 and it is higher than α of 0.01, indicating lack of statistical significance at this level of α.

11. Color Blindness in Men and Women

a. Assume group 1 is male and group 2 is female
Will conduct a directional z test to test the hypotheses:
$H_0: p_1 - p_2 = 0 \qquad H_1: p_1 - p_2 > 0$
With $\alpha = 0.01$ and a directional test where a positive z value supports the research (H_1) hypothesis, the critical value (CV) of z is +2.326. A value higher than +2.326 would indicate a significant directional difference

$\hat{p}_1 = \frac{45}{500} = 0.09000 \quad \hat{p}_2 = \frac{6}{2100} = 0.00286$

$\bar{p} = \frac{x_1 + x_2}{n_1 + n_2} = \frac{45 + 6}{500 + 2100} = \frac{51}{2600} = 0.0196 \quad \bar{q} = 1 - \bar{p} = 0.9804$

$$z = \frac{(\hat{p}_1 - \hat{p}_2) - (p_1 - p_2)}{\sqrt{\frac{\bar{p}\bar{q}}{n_1} + \frac{\bar{p}\bar{q}}{n_2}}} = \frac{(0.09000 - 0.00286) - (0)}{\sqrt{\frac{0.0196 * 0.9804}{500} + \frac{0.0196 * 0.9804}{2100}}}$$

$$= \frac{0.08714}{\sqrt{0.0000384 + 0.00000915}} = \frac{0.08714}{\sqrt{0.0000476}} = \frac{0.08714}{0.00690} = 12.64$$

With $\alpha = 0.01$, since the observed z statistic (12.64) is higher than the critical value of +2.326, the male group has a significantly higher rate of red/green color blindness than women. The *P*-value (determined using Excel) is ~ 0, clearly lower than 0.01.

b. Find $CI_{98\%}$

$$CI_{98\%} = (\hat{p}_1 - \hat{p}_2) - E < (p_1 - p_2) < (\hat{p}_1 - \hat{p}_2) + E$$

$$E = z_{\alpha/2}\sqrt{\frac{\hat{p}_1\hat{q}_1}{n_1} + \frac{\hat{p}_2\hat{q}_2}{n_2}} = 2.326\sqrt{\frac{0.0900 * 0.9100}{500} + \frac{0.00286 * 0.99714}{2100}}$$

$$= 2.326\sqrt{0.000163800 + 0.00000136} = 2.326\sqrt{0.00016516} = 2.326 * 0.0129 = 0.0299$$

$$CI_{98\%} = (0.09000 - 0.00286) - 0.0299 < (p_1 - p_2) < (0.09000 - 0.00286) + 0.0299$$

$$CI_{98\%} = 0.0572 < (p_1 - p_2) < 0.1170$$

The confidence interval does not contain zero, thus there seems to be a significant difference between men and women with respect to red/green color blindness.

c. We need a large sample size for women so that the requirements of: $np \geq 5$ and $nq \geq 5$ are both satisfied,so use of the normal approximation is justified

13. Drinking and Crime

Assume group 1 is drinkers convicted of arson and group 2 is drinkers convicted of fraud
Will conduct a directional z test to test the hypotheses:
$H_0 : p_1 - p_2 = 0 \qquad H_1 : p_1 - p_2 > 0$
With $\alpha = 0.01$ and a directional test where a positive z value supports the research (H_1) hypothesis, the critical value (CV) of z is +2.326
A value higher than +2.326 would indicate a significant directional difference

$$\hat{p}_1 = \frac{50}{93} = 0.5376 \quad \hat{p}_2 = \frac{63}{207} = 0.3043$$

$$\overline{p} = \frac{x_1 + x_2}{n_1 + n_2} = \frac{50+63}{93+207} = \frac{113}{300} = 0.3767 \quad \overline{q} = 1 - \overline{p} = 0.6233$$

$$z = \frac{(\hat{p}_1 - \hat{p}_2) - (p_1 - p_2)}{\sqrt{\frac{\overline{p}\overline{q}}{n_1} + \frac{\overline{p}\overline{q}}{n_2}}} = \frac{(0.5376 - 0.3043) - (0)}{\sqrt{\frac{0.3767 * 0.6233}{93} + \frac{0.3767 * 0.6233}{207}}}$$

$$= \frac{0.2333}{\sqrt{0.002525 + 0.001134}} = \frac{0.2333}{\sqrt{0.003659}} = \frac{0.2333}{0.0605} = 3.86$$

Since the observed z statistic is higher than the critical value, we conclude the proportion of drinkers who committed arson was significantly higher than the proportion of drinkers who committed fraud. The difference type of crime committed seems to be related to drinking. Arson would seem to be more of a crime of "passion" or violence than fraud so it would seem that drinking might make a criminal commit more such crimes.

15. Written Survey and Computer Survey

Assume group 1 takes the written survey and group 2 takes a computer survey. We will conduct a non-directional z test to test the hypotheses:
$H_0 : p_1 - p_2 = 0 \qquad H_1 : p_1 - p_2 \neq 0$
Let's set α= 0.01 and a non-directional test, the critical values (CV) of z are ±2.575
A value lower than -2.575 or higher than +2.575 would indicate a significant difference in admitting they carried a gun

a. z test

$$x_1 = n_1 * p_1 = 850 * 0.079 = 67, \quad x_2 = n_2 * p_2 = 850 * 0.124 = 105$$

$$\hat{p}_1 = \frac{67}{850} = 0.0788 \qquad \hat{p}_2 = \frac{105}{850} = 0.1235$$

$$\overline{p} = \frac{x_1 + x_2}{n_1 + n_2} = \frac{67+105}{850+850} = \frac{172}{1700} = 0.1012 \qquad \overline{q} = 1 - \overline{p} = 0.8988$$

$$z = \frac{(\hat{p}_1 - \hat{p}_2) - (p_1 - p_2)}{\sqrt{\frac{\overline{p}\overline{q}}{n_1} + \frac{\overline{p}\overline{q}}{n_2}}} = \frac{(0.0788 - 0.1235) - (0)}{\sqrt{\frac{0.1012 * 0.8988}{850} + \frac{0.1012 * 0.8988}{850}}}$$

$$= \frac{-0.0447}{\sqrt{0.000107 + 0.000107}} = \frac{-0.0447}{\sqrt{0.000214}} = \frac{-0.0447}{0.0146} = -3.06$$

Since the observed z statistic is lower than the CV of -2.575, there is a significant difference in the proportion of respondents who admit to carrying a gun when taking a written rather than a computer survey. The P-value would be 2 * (1 – P(z < 3.06)) = 2 * (1 – 0.9989)= 2 * 0.0011= 0.0022.

b. Find $CI_{99\%}$

$$CI_{99\%} = (\hat{p}_1 - \hat{p}_2) - E < (p_1 - p_2) < (\hat{p}_1 - \hat{p}_2) + E$$

$$E = z_{\alpha/2}\sqrt{\frac{\hat{p}_1\hat{q}_1}{n_1} + \frac{\hat{p}_2\hat{q}_2}{n_2}} = 2.575\sqrt{\frac{0.0788 * 0.9212}{850} + \frac{0.1235 * 0.8765}{850}}$$

$$= 2.575\sqrt{0.0000854 + 0.0001274} = 2.575\sqrt{0.0002128} = 2.575 * 0.0146 = 0.0376$$

$$CI_{99\%} = (0.0788 - 0.1235) - 0.0376 < (p_1 - p_2) < (0.0788 - 0.1235) + 0.0376$$

$$CI_{99\%} = -0.0823 < (p_1 - p_2) < -0.0071$$

Since zero is not in the confidence interval there is a significant difference in the proportion of respondents who admit to carrying a gun when taking a written rather than a computer survey.

17. Testing Adverse Drug Reactions

Assume group 1 is a group taking Viagra and group 2 is a group taking a placebo. The outcome variable is whether one has dyspepsia.

We will conduct a directional z test to test the hypotheses:

$H_0: p_1 - p_2 = 0 \qquad H_1: p_1 - p_2 > 0$

With $\alpha = 0.05$ and a directional test where a positive z value supports the research (H_1) hypothesis, the critical value (CV) of z is +1.645. A value higher than +1.645 would indicate a significant directional difference.

a. z test

$$x_1 = n_1 * p_1 = 734 * 0.07 = 51.38, \quad x_2 = n_2 * p_2 = 725 * 0.02 = 14.5$$

$$\hat{p}_1 = \frac{51}{734} = 0.0695 \quad \hat{p}_2 = \frac{15}{725} = 0.0207$$

$$\overline{p} = \frac{x_1 + x_2}{n_1 + n_2} = \frac{51 + 15}{734 + 725} = \frac{66}{1459} = 0.0452 \qquad \overline{q} = 1 - \overline{p} = 0.9548$$

$$z = \frac{(\hat{p}_1 - \hat{p}_2) - (p_1 - p_2)}{\sqrt{\frac{\overline{p}\overline{q}}{n_1} + \frac{\overline{p}\overline{q}}{n_2}}} = \frac{(0.0695 - 0.0207) - (0)}{\sqrt{\frac{0.0452 * 0.9548}{734} + \frac{0.0452 * 0.9548}{725}}}$$

$$= \frac{0.0488}{\sqrt{0.0000588 + 0.0000595}} = \frac{0.0488}{\sqrt{0.0001183}} = \frac{0.0488}{0.0109} = 4.49$$

Since the observed z statistic is higher than the critical value, we conclude the proportion of Viagra users having the condition of dyspepsia is significantly higher than the proportion of placebo group users who have dyspepsia.

b. Find $CI_{95\%}$

$$CI_{95\%} = (\hat{p}_1 - \hat{p}_2) - E < (p_1 - p_2) < (\hat{p}_1 - \hat{p}_2) + E$$

$$E = z_{\alpha/2}\sqrt{\frac{\hat{p}_1\hat{q}_1}{n_1} + \frac{\hat{p}_2\hat{q}_2}{n_2}} = 1.96\sqrt{\frac{0.0695 * 0.9305}{734} + \frac{0.0207 * 0.9793}{725}}$$

$$= 1.96\sqrt{0.0000881 + 0.0000280} = 1.96\sqrt{0.0001161} = 1.96 * 0.0108 = 0.0211$$

$$CI_{95\%} = (0.0695 - 0.0207) - 0.0211 < (p_1 - p_2) < (0.0695 - 0.0207) + 0.0211$$

$$CI_{95\%} = 0.0277 < (p_1 - p_2) < 0.0699$$

Since zero is not in the confidence interval there is a significant difference in the proportion of Viagra users who experience dyspepsia compared with placebo users who have dyspepsia.

19. Which Treatment Is Better?

Assume group 1 has the surgery treatment and group 2 has the splint treatment. We will conduct a non-directional z test to test the hypotheses:

$H_0 : p_1 - p_2 = 0 \qquad H_1 : p_1 - p_2 \neq 0$

Let's set $\alpha = 0.05$ and a non-directional test, the critical values (CV) of z are ± 1.96

A value lower than -1.96 or higher than +1.96 would indicate a significant difference in treatment results

$$\hat{p}_1 = \frac{67}{73} = 0.9178 \qquad \hat{p}_2 = \frac{60}{83} = 0.7229$$

$$\bar{p} = \frac{x_1 + x_2}{n_1 + n_2} = \frac{67+60}{73+83} = \frac{127}{156} = 0.8141 \qquad \bar{q} = 1 - \bar{p} = 0.1859$$

$$z = \frac{(\hat{p}_1 - \hat{p}_2) - (p_1 - p_2)}{\sqrt{\frac{\bar{p}\bar{q}}{n_1} + \frac{\bar{p}\bar{q}}{n_2}}} = \frac{(0.9178 - 0.7229) - (0)}{\sqrt{\frac{0.8141 * 0.1859}{73} + \frac{0.8141 * 0.1859}{83}}}$$

$$= \frac{0.1949}{\sqrt{0.002073 + 0.001823}} = \frac{0.1949}{\sqrt{0.003896}} = \frac{0.1949}{0.06242} = 3.122$$

With $\alpha = 0.05$, for a non-directional test, since the observed z statistic is higher than + 1.96, there is a significant difference in treatment success rate. The P-value would be

2 * (1 – P(z < 3.12)) = 2 * (1 – 0.9991)= 2 * 0.0009= 0.0018.

It would appear that the success rate from surgery is higher than the use of a splint.

21. Interpreting Overlap of Confidence Intervals

a. Find $CI_{95\%}$ for the difference in the two proportions

$$\hat{p}_1 = \frac{112}{200} = 0.5600 \qquad \hat{p}_2 = \frac{88}{200} = 0.4400$$

$$CI_{95\%} = (\hat{p}_1 - \hat{p}_2) - E < (p_1 - p_2) < (\hat{p}_1 - \hat{p}_2) + E$$

$$E = z_{\alpha/2}\sqrt{\frac{\hat{p}_1\hat{q}_1}{n_1} + \frac{\hat{p}_2\hat{q}_2}{n_2}} = 1.96\sqrt{\frac{0.5600 * 0.4400}{200} + \frac{0.4400 * 0.5600}{200}}$$

$$= 1.96\sqrt{0.001232 + 0.001232} = 1.96\sqrt{0.002464} = 1.96 * 0.0496 = 0.0973$$

$$CI_{95\%} = (0.5600 - 0.4400) - 0.0973 < (p_1 - p_2) < (0.5600 - 0.4400) + 0.0973$$

$$CI_{95\%} = 0.0227 < (p_1 - p_2) < 0.2173$$

The confidence interval does not contain zero. Therefore, there appears to be a significant difference between the two proportions.

b. Individual single group confidence intervals

$CI_{95\%}$ for $\hat{p}_1 = \hat{p} - E < p < \hat{p} + E$

$$E = z_{\alpha/2}\sqrt{\frac{\hat{p}\hat{q}}{n}} = 1.96\sqrt{\frac{0.5600 * 0.4400}{200}} = 1.96\sqrt{0.001232} = 1.96 * 0.0351 = 0.0688$$

$CI_{95\%}$ for $\hat{p}_1 = 0.5600 - 0.0688 < p_1 < 0.5600 + 0.0688 \quad = 0.4912 < p_1 < 0.6288$

$CI_{95\%}$ for $\hat{p}_2 = \hat{p} - E < p < \hat{p} + E$

$$E = z_{\alpha/2}\sqrt{\frac{\hat{p}\hat{q}}{n}} = 1.96\sqrt{\frac{0.4400 * 0.5600}{200}} = 1.96\sqrt{0.001232} = 1.96 * 0.0351 = 0.0688$$

$CI_{95\%}$ for $\hat{p}_2 = 0.4400 - 0.0688 < p_2 < 0.4400 + 0.0688 = 0.3712 < p_2 < 0.5088$

Since there is overlap of the confidence interval (the upper limit of the p_2 confidence interval of 0.5088 is in the p_1 confidence interval). Thus, $H_0 : p_1 - p_2 = 0$ cannot be rejected.

c. Will conduct a non-directional z test to test the hypotheses:

$H_0: p_1 - p_2 = 0 \qquad H_1: p_1 - p_2 \neq 0$

Let's set $\alpha = 0.05$ and a non-directional test, the critical values (CV) of z are ± 1.96. A value lower than -1.96 or higher than +1.96 would indicate a significant difference in the two proportions

$$\hat{p}_1 = \frac{112}{200} = 0.5600 \qquad \hat{p}_2 = \frac{88}{200} = 0.4400$$

$$\bar{p} = \frac{x_1 + x_2}{n_1 + n_2} = \frac{112 + 88}{200 + 200} = \frac{200}{400} = 0.5000 \qquad \bar{q} = 1 - \bar{p} = 0.5000$$

$$z = \frac{(\hat{p}_1 - \hat{p}_2) - (p_1 - p_2)}{\sqrt{\frac{\bar{p}\bar{q}}{n_1} + \frac{\bar{p}\bar{q}}{n_2}}} = \frac{(0.5600 - 0.4400) - (0)}{\sqrt{\frac{0.5000 * 0.5000}{200} + \frac{0.5000 * 0.5000}{200}}}$$

$$= \frac{0.1200}{\sqrt{0.00125 + 0.00125}} = \frac{0.1200}{\sqrt{0.00250}} = \frac{0.1200}{0.0500} = 2.40$$

With $\alpha = 0.05$, for a non-directional test, since the observed z statistic is higher than + 1.96, there is a significant difference in the two proportions. The P-value would be 2 * (1 – P(z < 2.40)) = 2 * (1 – 0.9918)= 2 * 0.0082 = 0.0164. Thus, $H_0: p_1 - p_2 = 0$ is rejected, supporting $H_0: p_1 - p_2 \neq 0$.

d. In order to test the Null hypothesis of $H_0: p_1 - p_2 = 0$, either the z test can be used or the confidence interval of the difference in the proportions can be used, with consistent results. However, comparing the individual group proportion confidence intervals and basing the decision on looking for overlap or lack of overlap of the confidence intervals is the least effective option and it should not be used.

23. Testing for Constant Difference

Will conduct a non-directional z test to test the hypotheses:

$H_0: p_1 - p_2 = 0.1 \qquad H_1: p_1 - p_2 \neq 0.1$

Let's set $\alpha = 0.05$ and a non-directional test, the critical values (CV) of z are ± 1.96

A value lower than -1.96 or higher than +1.96 would indicate a significant difference in the two proportions

$x_1 = n_1 * p_1 = 734 * 0.16 = 117, \quad x_2 = n_2 * p_2 = 725 * 0.04 = 29$

$$\hat{p}_1 = \frac{117}{734} = 0.1594 \qquad \hat{p}_2 = \frac{29}{725} = 0.0400$$

$$\bar{p} = \frac{x_1 + x_2}{n_1 + n_2} = \frac{117 + 29}{734 + 725} = \frac{146}{1459} = 0.1001 \qquad \bar{q} = 1 - \bar{p} = 0.8999$$

$$z = \frac{(\hat{p}_1 - \hat{p}_2) - c}{\sqrt{\frac{\hat{p}_1\hat{q}_1}{n_1} + \frac{\hat{p}_2\hat{q}_2}{n_2}}} = \frac{(0.1594 - 0.0400) - 0.10}{\sqrt{\frac{0.1594 * 0.8406}{734} + \frac{0.0400 * 0.9600}{725}}}$$

$$= \frac{0.0194}{\sqrt{0.000183 + 0.000053}} = \frac{0.0194}{\sqrt{0.000236}} = \frac{0.0194}{0.0154} = 1.26$$

With $\alpha = 0.05$, for a non-directional test, since the observed z statistic is not lower than -1.96 or higher than + 1.96, there is not a significant difference in the two proportions. The P-value would be 2 * (1 – P(z > 1.26)) = 2 * (1 – 0.8962)= 2 * 0.1038= 0.2076. Thus, there is no evidence to indicate the headache rate for Viagra users is 10% different than that of the headache rate for men given a placebo.

8-3 Inferences About Two Means: Independent Samples

In Exercises 1 & 3, determine whether the samples are independent or consist of matched pairs.

1. Independent since groups are comprised of different subjects.

3. Matched pairs since the data are matched on both their estimate and the physician's scale.

In Exercises 5-23, assume that the two samples are independent simple random samples from normally distributed populations. Do not assume that the population standard deviations are equal.

Note: In the 5 – 23 Exercises below, degrees of freedom (*df*) is computed based on the smaller sample size – 1. This approach is used when the assumption that the samples are drawn from normal distributions is not made. Many other examples of this test comparing two means use another approach to finding the degrees of freedom as *df*= $n_1 + n_2 - 2$ whether or not this normality assumption is made. Using smaller sample size – 1 for *df* is considered a conservative approach.

5. **Hypothesis Test for Effect of Marijuana Use on College Students**
Conduct independent *t* test

$$H_0: \mu_1 = \mu_2 \quad H_1: \mu_1 > \mu_2$$

$$n_1 = 64,\ \bar{x}_1 = 53.3,\ s_1 = 3.6$$

$$n_2 = 65,\ \bar{x}_2 = 51.3,\ s_2 = 4.5$$

$$\bar{x}_1 - \bar{x}_2 = 53.3 - 51.3 = 2.000$$

$$\alpha = 0.01,\ df = n_1 - 1 = 63,\ t_{\alpha/2} = +2.390 \text{ (using } df = 60, \text{ from Table A - 3)}$$

$$t = \frac{(\bar{x}_1 - \bar{x}_2) - (\mu_1 - \mu_2)}{\sqrt{\frac{s_1^2}{n_1} + \frac{s_2^2}{n_2}}} = \frac{(53.3 - 51.3) - 0}{\sqrt{\frac{3.6^2}{64} + \frac{4.5^2}{65}}} = \frac{2.00}{\sqrt{0.5140}} = \frac{2.00}{0.7170} = 2.79$$

Since the test statistic (2.79) is greater than the t critical value (+2.390), we reject the null hypothesis and conclude that the population of heavy marijuana users have a lower mean mental ability score than the light users. *P*-value (from Excel)= 0.003, which is less than 0.01.

7. **Confidence Interval for Magnet Treatment of Pain**, Find $CI_{90\%}$

$$n_1 = 20,\ \bar{x}_1 = 0.49,\ s_1 = 0.96$$

$$n_2 = 20,\ \bar{x}_2 = 0.44,\ s_2 = 1.40$$

$$\bar{x}_1 - \bar{x}_2 = 0.49 - 0.44 = 0.050$$

$$\alpha = 0.10, df = n_1 - 1 = 20 - 1 = 19,\ t_{\alpha/2} = \pm 1.729 \text{ (from Table A - 3)}$$

$$E = t_{\alpha/2}\sqrt{\frac{s_1^2}{n_1} + \frac{s_2^2}{n_2}} = 1.729\sqrt{\frac{0.96^2}{20} + \frac{1.40^2}{20}} = 1.729\sqrt{0.1441} = 1.729 * 0.3796 = 0.656$$

$$CI_{90\%} = (\bar{x}_1 - \bar{x}_2) - E < (\mu_1 - \mu_2) < (\bar{x}_1 - \bar{x}_2) + E$$

$$= 0.050 - 0.656 < (\mu_1 - \mu_2) < 0.050 + 0.656$$

$$= -0.606 < (\mu_1 - \mu_2) < 0.706$$

The confidence interval contains zero. Therefore we conclude that there is no difference between the two means and that the magnetic treatment is not effective in reducing pain.

9. Confidence Interval for Identifying Psychiatric Disorders, Find CI_9

$n_1 = 10,\ \bar{x}_1 = 0.45,\ s_1 = 0.08$

$n_2 = 10,\ \bar{x}_2 = 0.34,\ s_2 = 0.08$

$\bar{x}_1 - \bar{x}_2 = 0.45 - 0.34 = 0.11$

$\alpha = 0.01,\ df = n_1 - 1 = 10 - 1 = 9,\ t_{\alpha/2} = \pm 3.250$ (from Table A - 3)

$$E = t_{\alpha/2}\sqrt{\frac{s_1^2}{n_1} + \frac{s_2^2}{n_2}} = 3.250\sqrt{\frac{0.08^2}{10} + \frac{0.08^2}{10}} = 3.250\sqrt{0.00128} = 3.250 * 0.0358 = 0.116$$

$$CI_{99\%} = (\bar{x}_1 - \bar{x}_2) - E < (\mu_1 - \mu_2) < (\bar{x}_1 - \bar{x}_2) + E = 0.110 - 0.116 < (\mu_1 - \mu_2) < 0.110 + 0.116$$
$$= -0.006 < (\mu_1 - \mu_2) < 0.226$$

The confidence interval contains zero, therefore we conclude that there is no difference between the two means. It seems that obsessive compulsive disorders do not have a biological basis.

11. Confidence Interval for Effects of Alcohol, Find $CI_{95\%}$

$n_1 = 22,\ \bar{x}_1 = 4.20,\ s_1 = 2.20$

$n_2 = 22,\ \bar{x}_2 = 1.71,\ s_2 = 0.72$

$\bar{x}_1 - \bar{x}_2 = 4.20 - 1.71 = 2.49$

$\alpha = 0.05, df = n_1 - 1 = 22 - 1 = 21, t_{\alpha/2} = 2.080$ (from Table A - 3)

$$E = t_{\alpha/2}\sqrt{\frac{s_1^2}{n_1} + \frac{s_2^2}{n_2}} = 2.080\sqrt{\frac{2.20^2}{22} + \frac{0.72^2}{22}} = 2.080\sqrt{0.2436} = 2.080 * 0.4935 = 1.027$$

$$CI_{95\%} = (\bar{x}_1 - \bar{x}_2) - E < (\mu_1 - \mu_2) < (\bar{x}_1 - \bar{x}_2) + E = 2.490 - 1.027 < (\mu_1 - \mu_2) < 2.490 + 1.027$$
$$= 1.463 < (\mu_1 - \mu_2) < 3.517$$

The confidence interval does not include zero, therefore, we conclude that there is a significant difference between the two population means, and that significantly more errors are made by those who are treated with alcohol.

13. Poplar Trees, Find $CI_{95\%}$

$n_1 = 5, \bar{x}_1 = 0.184,\ s_1 = 0.127$

$n_2 = 5, \bar{x}_2 = 1.334,\ s_2 = 0.859$

$\bar{x}_1 - \bar{x}_2 = 0.184 - 1.334 = -1.150$

$\alpha = 0.05, df = n_1 - 1 = 5 - 1 = 4, t_{\alpha/2} = \pm 2.776$ (from Table A - 3)

$$E = t_{\alpha/2}\sqrt{\frac{s_1^2}{n_1} + \frac{s_2^2}{n_2}} = 2.776\sqrt{\frac{0.127^2}{5} + \frac{0.859^2}{5}} = 2.776\sqrt{0.1508} = 2.776 * 0.3883 = 1.078$$

$$CI_{95\%} = (\bar{x}_1 - \bar{x}_2) - E < (\mu_1 - \mu_2) < (\bar{x}_1 - \bar{x}_2) + E = -1.150 - 1.078 < (\mu_1 - \mu_2) < -1.150 + 1.078$$
$$= -2.228 < (\mu_1 - \mu_2) < -0.072$$

The confidence interval does not include zero, therefore, we conclude that there is a significant difference between the weights of the two groups of poplar trees given different treatments.

15. Petal Lengths of Irises, Conduct independent t test

$n_1 = 50,\ \bar{x}_1 = 1.46,\ s_1 = 0.17$

$n_2 = 50,\ \bar{x}_2 = 4.26,\ s_2 = 0.47$

$H_0 : \mu_1 = \mu_2 \quad H_1 : \mu_1 \neq \mu_2$

$\alpha = 0.05,\ df = 50 - 1 = 49,\ t_{\alpha/2} = \pm 2.009$ (from Table A – 3)

$$t = \frac{(\bar{x}_1 - \bar{x}_2) - (\mu_1 - \mu_2)}{\sqrt{\frac{s_1^2}{n_1} + \frac{s_2^2}{n_2}}} = \frac{(1.46 - 4.26) - 0}{\sqrt{\frac{0.17^2}{50} + \frac{0.47^2}{50}}} = \frac{-2.8}{\sqrt{0.004996}} = \frac{-2.8}{0.0707} = -39.60$$

The test statistic (-39.6) is lower than -2.009. Thus, we reject the null hypothesis and conclude that there is a significant difference between the mean petal length of irises in the two populations.

17. BMI of Men and Women, Conduct independent t test

$n_1 = 40,\ \bar{x}_1 = 25.998,\ s_1 = 3.431$

$n_2 = 40,\ \bar{x}_2 = 25.740,\ s_2 = 6.166$

$\bar{x}_1 - \bar{x}_2 = 25.998 - 25.740 = 0.258$

$\alpha = 0.05, df = n_1 - 1 = 40 - 1 = 39, t_{\alpha/2} = \pm 2.024$ (from Table A - 3)

$H_0 : \mu_1 = \mu_2 \quad H_1 : \mu_1 \neq \mu_2$

$$t = \frac{(\bar{x}_1 - \bar{x}_2) - (\mu_1 - \mu_2)}{\sqrt{\frac{s_1^2}{n_1} + \frac{s_2^2}{n_2}}} = \frac{(25.998 - 25.740) - 0}{\sqrt{\frac{3.431^2}{40} + \frac{6.166^2}{40}}} = \frac{0.258}{\sqrt{1.2448}} = \frac{0.258}{1.116} = 0.231$$

The test statistic is not greater than +2.024. Therefore, we do not reject the null hypothesis and we conclude that there is not enough evidence to support the claim that the BMI of men and women are not equal.

19. Head Circumferences, Conduct independent t test

$n_1 = 50,\ \bar{x}_1 = 41.098,\ s_1 = 1.498$

$n_2 = 50,\ \bar{x}_2 = 40.048,\ s_2 = 1.640$

$\bar{x}_1 - \bar{x}_2 = 41.098 - 40.048 = 1.050$

$\alpha = 0.05,\ df = n_1 - 1 = 50 - 1 = 49,\ t_{\alpha/2} = \pm 2.009$ (from Table A - 3)

$H_0 : \mu_1 = \mu_2 \quad H_1 : \mu_1 \neq \mu_2$

$$t = \frac{(\bar{x}_1 - \bar{x}_2) - (\mu_1 - \mu_2)}{\sqrt{\frac{s_1^2}{n_1} + \frac{s_2^2}{n_2}}} = \frac{(41.098 - 40.048) - 0}{\sqrt{\frac{1.498^2}{50} + \frac{1.640^2}{50}}} = \frac{1.050}{\sqrt{0.09867}} = \frac{1.050}{0.3141} = 3.343$$

The test statistic (3.343) is higher than +2.009. Thus, we reject the null hypothesis and conclude that there is enough evidence to support the claim that the head circumference of males and females is significantly different.

21. Seat Belts and Hospital Time, Conduct independent t test

$n_1 = 290, \bar{x}_1 = 1.39, s_1 = 3.06$

$n_2 = 123, \bar{x}_2 = 0.83, s_2 = 1.77$

$\bar{x}_1 - \bar{x}_2 = 1.39 - 0.83 = 0.56$

$\alpha = 0.01,\ df = n_2 - 1 = 123 - 1 = 122,\ t_\alpha = +2.364\ (\text{use } df = 100 \text{ from Table A - 3})$

$H_0 : \mu_1 = \mu_2 \quad H_1 : \mu_1 > \mu_2$

$$t = \frac{(\bar{x}_1 - \bar{x}_2) - (\mu_1 - \mu_2)}{\sqrt{\frac{s_1^2}{n_1} + \frac{s_2^2}{n_2}}} = \frac{(1.39 - 0.83) - 0}{\sqrt{\frac{3.06^2}{290} + \frac{1.77^2}{123}}} = \frac{0.56}{\sqrt{0.05776}} = \frac{0.56}{0.2403} = 2.330$$

The test statistic (2.33) is not higher than the critical value of 2.364. Thus, we do not reject the null hypothesis and we conclude that there is not enough evidence to support the claim that seat belt use leads to a lower stay in hospital.

23. Effects of Cocaine on Children, Conduct independent t test

$n_1 = 190, \bar{x}_1 = 7.3, s_1 = 3.0$

$n_2 = 186, \bar{x}_2 = 8.2, s_2 = 3.0$

$\bar{x}_1 - \bar{x}_2 = 7.3 - 8.2 = -0.09$

$\alpha = 0.05, df = n_2 - 1 = 186 - 1 = 185,\ t_\alpha = -1.653\ (\text{use } df = 200 \text{ from Table A - 3})$

$H_0 : \mu_1 = \mu_2 \quad H_1 : \mu_1 < \mu_2$

$$t = \frac{(\bar{x}_1 - \bar{x}_2) - (\mu_1 - \mu_2)}{\sqrt{\frac{s_1^2}{n_1} + \frac{s_2^2}{n_2}}} = \frac{(7.3\text{-}8.2) - 0}{\sqrt{\frac{3.0^2}{190} + \frac{3.0^2}{186}}} = \frac{-0.9}{\sqrt{0.09576}} = \frac{-0.9}{0.3094} = -2.908$$

The test statistic (–2.908) is lower than the critical value of -1.653. Therefore, we reject the null hypothesis and conclude that there is enough evidence to support the claim that prenatal cocaine exposure is associated with lower scores of four-year old children on the test of object assembly.

In Exercises 25 & 27, assume that the two samples are independent simple random samples selected from normally distributed populations. Also assume that the population standard deviations are equal ($\sigma_1 = \sigma_2$). So that the standard error of the differences between means is obtained by pooling the sample variances.

25. Hypothesis Test with Pooling, t test from Exercise 5 conducted with normality assumption

$H_0 : \mu_1 = \mu_2 \quad H_1 : \mu_1 > \mu_2$

$n_1 = 64, \bar{x}_1 = 53.3, s_1 = 3.6$

$n_2 = 65, \bar{x}_2 = 51.3, s_2 = 4.5$

$\bar{x}_1 - \bar{x}_2 = 53.3 - 51.3 = 2.000$

$\alpha = 0.01, df = n_1 + n_2 - 2 = 64 + 65 - 2 = 127,\ t_\alpha = +2.364\ (\text{using } df = 100, \text{ from Table A - 3})$

$$s_p^2 = \frac{(n_1 - 1)s_1^2 + (n_2 - 1)s_2^2}{(n_1 - 1) + (n_2 - 1)} = \frac{(64 - 1) * 3.6^2 + (65 - 1) * 4.5^2}{(64 - 1) + (65 - 1)} = \frac{2112.48}{127} = 16.63$$

$$t = \frac{(\bar{x}_1 - \bar{x}_2) - 0}{\sqrt{\frac{s_p^2}{n_1} + \frac{s_p^2}{n_2}}} = \frac{2.0}{\sqrt{\frac{16.63}{64} + \frac{16.63}{65}}} = \frac{2.0}{\sqrt{0.5157}} = \frac{2.0}{0.718} = 2.785$$

The test statistic (2.785) is higher than the critical value of +2.364. We reject the null hypothesis and conclude that there is enough evidence to support the claim that heavy marijuana users have a lower mean than light users just as we did in Exercise 5. Results are not affected by this additional assumption. However, the critical value in this case is lower than the critical value used in Exercise 5 (2.390).

27. Confidence Interval with Pooling, Confidence interval from Exercise 7 with normality assumption

$n_1 = 20,\ \bar{x}_1 = 0.49,\ s_1 = 0.96$

$n_2 = 20,\ \bar{x}_2 = 0.44,\ s_2 = 1.40$

$\bar{x}_1 - \bar{x}_2 = 0.49 - 0.44 = 0.050$

$\alpha = 0.10, df = n_1 + n_2 - 2 = 20 + 20 - 2 = 38,\ t_{\alpha/2} = \pm 1.686$ (from Table A - 3)

$$s_p^2 = \frac{(n_1 - 1)s_1^2 + (n_2 - 1)s_2^2}{(n_1 - 1) + (n_2 - 1)} = \frac{(20-1)*0.96^2 + (20-1)*1.4^2}{(20-1)+(20-1)} = \frac{54.7504}{38} = 1.4408$$

$$E = t_{\alpha/2}\sqrt{\frac{s_p^2}{n_1} + \frac{s_p^2}{n_2}} = 1.686\sqrt{\frac{0.9216}{20} + \frac{0.9216}{20}} = 1.686\sqrt{0.14408} = 1.686 * 0.3796 = 0.640$$

$$CI_{90\%} = (\bar{x}_1 - \bar{x}_2) - E < (\mu_1 - \mu_2) < (\bar{x}_1 - \bar{x}_2) + E = 0.05 - 0.64 < (\mu_1 - \mu_2) < 0.05 + 0.64$$
$$= -0.59 < (\mu_1 - \mu_2) < 0.69$$

The confidence interval contains zero. Therefore we conclude that there is no difference between the two means and that the magnetic treatment is not effective in reducing pain. This is the same result that was obtained earlier. However, in this case the confidence interval is narrower (in Exercise 7, $CI_{90\%}$ was -0.606 to 0.706).

29. Verifying a Property of Variances

a. Variance of scores: 5, 10, and 15

x	$x - \bar{x}$	$(x - \bar{x})^2$
5	-5	25
10	0	0
15	5	25
$\bar{x} = \frac{30}{3} = 10$		$\Sigma\frac{(x-\bar{x})^2}{n} = \frac{50}{3} = 16.67$

$\sigma_x^2 = 16.67$

b. Variance of scores: 1, 2, and 3

y	$y - \bar{y}$	$(y - \bar{y})^2$
1	-1	1
2	0	0
3	1	1
$\bar{y} = \frac{6}{3} = 2$		$\Sigma\frac{(y-\bar{y})^2}{n} = \frac{2}{3} = 0.67$

$\sigma_y^2 = 0.67$

c. Population of x – y difference values

x - y	d	$d-\bar{d}$	$(d-\bar{d})^2$
5 - 1	4	-4	16
10 - 1	9	1	1
15 - 1	14	6	36
5 - 2	3	-5	25
10 - 2	8	0	0
15 - 2	13	5	25
5 - 3	2	-6	36
10 - 3	7	-1	1
15 - 3	12	4	16
	$\bar{d}=\frac{72}{9}=8$		$\frac{(d-\bar{d})^2}{n}=\frac{156}{9}=17.33$

$\sigma_d^2=17.33$

d. Verify that $\sigma_{x-y}^2=\sigma_x^2+\sigma_y^2=17.33=16.67+0.67$

e. Range of differences x – y is 2 to 14 = 12
Range of x is 5 to 15= 10
Range of y is 1 to 3 = 2
Therefore, range of differences (x – y) is sum of two ranges (range of x + range of y)

8-4 Inferences from Matched Pairs

In Exercise 1, assume that you want to use a 0.05 significance level to test the claim that the paired sample data come from a population for which the mean difference is μ_d= 0. Find (a) $\bar{d}$, (b)s_d , (c) the t test statistic, and (d) the critical values.

1. **a.** $d=-1 \quad \bar{d}=\frac{d}{n}=\frac{-1}{5}=-0.2$

b. $s_d=\sqrt{\frac{(d-\bar{d})^2}{n-1}}=\sqrt{\frac{30.8}{4}}=2.8$

c. $t=\frac{\bar{d}-\mu_d}{s_d/\sqrt{n}}=\frac{-0.2-0}{2.8/\sqrt{5}}=-0.16$

d $.t_{CV}=\pm 2.776 \ (df=n-1=5-1=4)$

3. Find $CI_{95\%}$

$E=t_{\alpha/2}\frac{s_d}{\sqrt{n}}=2.776\frac{2.8}{\sqrt{5}}=3.48$

$CI_{.95\%}=\bar{d}-E<\mu_d<\bar{d}+E=(-0.2-3.48)<\mu_d<(-0.2+3.48)$

$=-3.68<\mu_d<3.28$

5. Testing Corn Seeds

a. Conduct matched pairs t test

$H_0: \mu_d = 0 \qquad H_1: \mu_d \neq 0$

$$d = -12. \quad \bar{d} = \frac{d}{n} = \frac{-12}{11} = -1.09$$

$$s_d = \sqrt{\frac{(d-\bar{d})^2}{n-1}} = \sqrt{\frac{55.79}{10}} = 2.362$$

$$t = \frac{\bar{d}-\mu_d}{s_d/\sqrt{n}} = \frac{-1.09-0}{2.362/\sqrt{11}} = -1.53$$

$t_{CV} = \pm 2.228 \ (df = n-1 = 11-1 = 10)$

The test statistic is not lower than -2.228. Therefore we do not have sufficient evidence to reject the claim that there is no difference between the yields from the two types of seeds.

b. Find $CI_{95\%}$

$$E = t_{\alpha/2}\frac{s_d}{\sqrt{n}} = 2.228 * \frac{2.362}{\sqrt{11}} = 2.228 * \frac{2.362}{3.317} = 1.587$$

$$CI_{95\%} = \bar{d} - E < \mu_d < \bar{d} + E = (-1.09 - 1.587) < \mu_d < (-1.09 + 1.587)$$

$$= -2.677 < \mu_d < 0.497$$

c. No, since zero is in the confidence interval, it does not appear that either type of seed is better.

7. Self-Reported and Measured Male Heights

a. Conduct matched pairs t test

$H_0: \mu_d = 0 \qquad H_1: \mu_d \neq 0$

$$d = -12. \quad \bar{d} = \frac{d}{n} = \frac{-12}{12} = -1.0$$

$$s_d = \sqrt{\frac{(d-\bar{d})^2}{n-1}} = \sqrt{\frac{136.29}{11}} = 3.52$$

$$t = \frac{\bar{d}-\mu_d}{s_d/\sqrt{n}} = \frac{-1.0-0}{3.52/\sqrt{12}} = -0.984$$

$t_{CV} = \pm 2.201 \ (df = n-1 = 12-1 = 11)$

The test statistic is not below -2.201 nor higher than +2.201. Therefore, we do not have sufficient evidence to support the claim that there is a difference between self reported heights and measured heights.

b. Find $CI_{95\%}$

$$E = t_{\alpha/2}\frac{s_d}{\sqrt{n}} = 2.201\frac{3.52}{\sqrt{12}} = 2.24$$

$$CI_{95\%} = \bar{d} - E < \mu_d < \bar{d} + E = (-1.0 - 2.24) < \mu_d < (-1.0 + 2.24)$$

$$= -3.24 < \mu_d < 1.24$$

The confidence interval contains zero. Therefore, we conclude that there is not sufficient evidence to support the claim that there is a difference between self-reported heights and measured heights.

9. Effectiveness of Hypnotism in Reducing Pain

a. Find $CI_{95\%}$

$$d = 25. \quad \bar{d} = \frac{d}{n} = \frac{25}{8} = 3.125$$

$$s_d = \sqrt{\frac{(d-\bar{d})^2}{n-1}} = \sqrt{\frac{59.28}{7}} = 2.91, \quad df = 8-1 = 7$$

$$E = t_{\alpha/2}\frac{s_d}{\sqrt{n}} = 2.365\frac{2.91}{\sqrt{8}} = 2.43$$

$$CI_{95\%} = \bar{d} - E < \mu_d < \bar{d} + E = (3.125 - 2.43) < \mu_d < (3.125 + 2.43)$$

$$= 0.695 < \mu_d < 5.56$$

b. Conduct directional matched pairs t test

$H_0: \mu_d = 0 \quad H_1: \mu_d > 0$

$$t = \frac{\bar{d} - \mu_d}{s_d/\sqrt{n}} = \frac{3.125 - 0}{2.91/\sqrt{8}} = \frac{3.125}{1.029} = 3.037$$

$t_{cv} = +1.895 \ (df = n - 1 = 8 - 1 = 7)$

We reject the null hypothesis and conclude that there is sufficient evidence to support the claim that sensory measurements are lower after hypnotism.

c. Yes, hypnosis appears to be effective in reducing pain.

11. Motion Sickness: Interpreting a Minitab Display

a. Conduct directional matched pairs t test with $\alpha = 0.05$, testing to see an increase in the before and after number of head movements before becoming nauseous.

$H_0: \mu_d = 0 \qquad H_1: \mu_d \neq 0$

$df = 10 - 1 = 9 \qquad t_{CV} = \pm 2.262$

$t = -0.41 \qquad P - \text{value} = 0.691$

Since $P < 0.05$ and the t statistic is not outside of +-2.262, there is no significant difference in the before to after using astemizole mean number of head movements before becoming nauseous.

b. Conduct significance test

$H_0: \mu_d = 0 \qquad H_1: \mu_d < 0$

$df = 10 - 1 = 9 \qquad t_{CV} = -1.833$

$t = -0.41$

$P-\text{value} = 0.691/2 = 0.346$ (for a directional test the P-value is one-half of the non-directional P-value)

Since the observed t is not lower than the critical value of -1.833, we conclude that astemizole has no affect on reducing motion sickness so there is no reason to take it for that purpose.

13. Self-Reporting and Measured Weights of Males: Interpreting an Excel Display

$H_0: \mu_d = 0 \quad H_1: \mu_d \neq 0$

$t = -0.501, \quad P - \text{value} = 0.626, \quad t_{CV} = \pm 2.201 \ (\alpha = 0.05, df = 11)$

We do not reject the null and we conclude that there is not sufficient evidence to support the claim that there is a difference in self reported weights and measured weights.

15. Morning and Night Body Temperatures

a. Find $CI_{95\%}$

$df = 11 - 1 = 10 \quad t_{CV} = \pm 2.228$

$$\Sigma d = -8.6. \quad \bar{d} = \frac{\Sigma d}{n} = \frac{-8.6}{11} = -0.782$$

$$s_d = \sqrt{\frac{\Sigma(d-\bar{d})^2}{n-1}} = \sqrt{\frac{8.336}{10}} = 0.913$$

$$E = t_{\alpha/2}\frac{s_d}{\sqrt{n}} = 2.228\frac{0.913}{\sqrt{11}} = 2.228 * 0.275 = 0.613$$

$$CI_{95\%} = \bar{d} - E < \mu_d < \bar{d} + E = (-0.782 - 0.613) < \mu_d < (-0.782 + 0.613)$$

$$= -1.395 < \mu_d < -0.169$$

b. Conduct matched pairs t test

$H_0 : \mu_d = 0. \quad H_1 : \mu_d \neq 0.$

$$t = \frac{\bar{d} - \mu_d}{s_d/\sqrt{n}} = \frac{-0.78 - 0}{0.913/\sqrt{11}} = -2.83$$

$t_{CV} = \pm 2.228 \ (df = n - 1 = 11 - 1 = 10)$

We reject the null hypothesis and conclude that there is sufficient evidence to reject the claim that the mean difference is zero. Morning and night body temperatures do not appear to be the same.

17. Using the Correct Procedure

a. Conducting matched pairs test

$H_0 : \mu_d = 0 \quad H_1 : \mu_d > 0.$

$$\Sigma d = 5 \quad \bar{d} = \frac{\Sigma d}{n} = \frac{5}{10} = 0.5$$

$$s_d = \sqrt{\frac{\Sigma(d-\bar{d})^2}{n-1}} = \sqrt{\frac{6.50}{9}} = 0.850$$

$$t = \frac{\bar{d} - \mu_d}{s_d/\sqrt{n}} = \frac{0.5 - 0}{0.85/\sqrt{10}} = 1.860$$

$t_{CV} = +1.833 \ (df = n - 1 = 10 - 1 = 9)$

We reject the null hypothesis and conclude that there is sufficient evidence to support $\mu_d > 0$ for the matched pairs test.

b. Conducting as an independent t test

$H_0 : \mu_1 = \mu_2 \quad H_1 : \mu_1 > \mu_2$

$$t = \frac{(\bar{x}_1 - \bar{x}_2) - (\mu_1 - \mu_2)}{\sqrt{\frac{s_1^2}{n_1} + \frac{s_2^2}{n_2}}} = \frac{(2.0 - 1.5) - 0}{\sqrt{\frac{0.82^2}{10} + \frac{0.53^2}{10}}} = \frac{0.5}{\sqrt{0.09533}} = \frac{0.5}{0.3088} = 1.619$$

$df = 10 - 1 = 9, \ t_{cv} = +1.833$ (from Table A – 3)

The test statistic is in the region of retaining the null. We do not reject the null and we conclude that there is not enough evidence to support the claim that $\mu_1 > \mu_2$

c. Yes, the method used affects the results, so it is important to use the correct method. The matched pairs test detected a significant difference and the independent samples test did not detect a significant difference. When an independent test is used with matched or dependent data, statistical power is lost and we may fail to find a difference being significant when it is significant.

8-5 Odds Ratios

In Exercise 1, use the data in the accompanying table (based on data from Pfizer, Inc.). That table describes results from a clinical trial of the well-known drug Viagra.

1. Odds Ratio and $CI_{95\%}$

$$\text{Odds ratio} = \frac{\text{odds for headache in Viagra treatment group}}{\text{odds for headache in placebo group}} = \frac{117/617}{29/696} = 4.55$$

$$CI_{95\%} = \frac{ad}{bc} * e^{-z_{\alpha/2}\sqrt{\frac{1}{a}+\frac{1}{b}+\frac{1}{c}+\frac{1}{d}}} < OR < \frac{ad}{bc} * e^{z_{\alpha/2}\sqrt{\frac{1}{a}+\frac{1}{b}+\frac{1}{c}+\frac{1}{d}}}$$

$$\text{exponent} = z_{\alpha/2}\sqrt{\frac{1}{a}+\frac{1}{b}+\frac{1}{c}+\frac{1}{d}} = 1.96\sqrt{\frac{1}{117}+\frac{1}{617}+\frac{1}{29}+\frac{1}{696}} = 0.4208$$

$$CI_{95\%} = 4.55 * 2.7183^{-0.4208} < OR < 4.55 * 2.7183^{0.4208} = 2.99 < OR < 6.93$$

The confidence interval does not contain 1. Therefore, it appears that the odds of getting a headache when using Viagra are significantly higher than the odds of getting a headache using the placebo when $\alpha = 0.05$.

In Exercise 3, use the data in the accompanying table (based on data from "A Case-Control Study of the Effectiveness of Bicycle Safety Helmets in Preventing Facial Injury," by Thompson, Rivara, and Wolf, American Journal of Public Health, Vol. 80, No. 12).

3. Odds Ratio and $CI_{99\%}$

$$\text{Odds ratio} = \frac{\text{odds for facial injuries in helmet group}}{\text{odds for facial injuries in non-helmet group}} = \frac{ad}{bc} = \frac{30 * 236}{83 * 182} = 0.47$$

$$CI_{99\%} = \frac{ad}{bc} * e^{-z_{\alpha/2}\sqrt{\frac{1}{a}+\frac{1}{b}+\frac{1}{c}+\frac{1}{d}}} < OR < \frac{ad}{bc} * e^{z_{\alpha/2}\sqrt{\frac{1}{a}+\frac{1}{b}+\frac{1}{c}+\frac{1}{d}}}$$

$$\text{exponent} = z_{\alpha/2}\sqrt{\frac{1}{a}+\frac{1}{b}+\frac{1}{c}+\frac{1}{d}} = 2.58\sqrt{\frac{1}{30}+\frac{1}{83}+\frac{1}{182}+\frac{1}{236}} = 0.6057$$

$$CI_{99\%} = 0.469 * 2.7183^{-0.6057} < OR < 0.469 * 2.7183^{0.6057} = 0.26 < OR < 0.86$$

In Exercise 5, use the data in the accompanying table.

5. Odds Ratio and $CI_{95\%}$

$$\text{Odds ratio} = \frac{\text{odds for lung cancer death of smokers}}{\text{odds for lung cancer death of nonsmokers}} = \frac{ad}{bc} = \frac{140 * 1707}{532 * 21} = 21.39$$

$$CI_{95\%} = \frac{ad}{bc} * e^{-z_{\alpha/2}\sqrt{\frac{1}{a}+\frac{1}{b}+\frac{1}{c}+\frac{1}{d}}} < OR < \frac{ad}{bc} * e^{z_{\alpha/2}\sqrt{\frac{1}{a}+\frac{1}{b}+\frac{1}{c}+\frac{1}{d}}}$$

$$\text{exponent} = z_{\alpha/2}\sqrt{\frac{1}{a}+\frac{1}{b}+\frac{1}{c}+\frac{1}{d}} = 1.96\sqrt{\frac{1}{140}+\frac{1}{532}+\frac{1}{21}+\frac{1}{1707}} = 0.4688$$

$$CI_{95\%} = 21.39 * 2.7183^{-0.4688} < OR < 21.39 * 2.7183^{0.4688} = 13.39 < OR < 34.19$$

In Exercises 7 and 8, use the data in the accompanying table (based on data from Parke-Davis). The cholesterol-reducing drug Lipitor consists of atorvastatin calcium.

7. Odds Ratio and $CI_{95\%}$

$$\text{Odds ratio} = \frac{\text{odds for headaches among drug group}}{\text{odds for headaches among placebo group}} = \frac{ad}{bc} = \frac{15*3}{17*65} = 0.04$$

$$CI_{95\%} = \frac{ad}{bc} * e^{-z_{\alpha/2}\sqrt{\frac{1}{a}+\frac{1}{b}+\frac{1}{c}+\frac{1}{d}}} < OR < \frac{ad}{bc} * e^{z_{\alpha/2}\sqrt{\frac{1}{a}+\frac{1}{b}+\frac{1}{c}+\frac{1}{d}}}$$

$$\text{exponent} = z_{\alpha/2}\sqrt{\frac{1}{a}+\frac{1}{b}+\frac{1}{c}+\frac{1}{d}} = 1.96\sqrt{\frac{1}{15}+\frac{1}{17}+\frac{1}{65}+\frac{1}{3}} = 1.3497$$

$$CI_{95\%} = 0.04 * 2.7183^{-1..3497} < OR < 0.04 * 2.7183^{1..3497} = 0.01 < OR < 0.16$$

a. Table

	Mouth or throat soreness	No mouth or throat soreness
Nicorette treatment	43	109
Placebo	35	118

b. $$\text{Odds ratio} = \frac{\text{odds for sore among nicorette}}{\text{odds for sores a mong placebo group}} = \frac{ad}{bc} = \frac{43*118}{109*35} = 1.33$$

c. $$\text{Confidence interval}_{95\%} = \frac{ad}{bc} * e^{-z_{\alpha/2}\sqrt{\frac{1}{a}+\frac{1}{b}+\frac{1}{c}+\frac{1}{d}}} < OR < \frac{ad}{bc} * e^{z_{\alpha/2}\sqrt{\frac{1}{a}+\frac{1}{b}+\frac{1}{c}+\frac{1}{d}}}$$

$$\text{exponent} = z_{\alpha/2}\sqrt{\frac{1}{a}+\frac{1}{b}+\frac{1}{c}+\frac{1}{d}} = 1.96\sqrt{\frac{1}{43}+\frac{1}{109}+\frac{1}{35}+\frac{1}{118}} = 0.5167$$

$$CI_{95\%} = 1.33 * 2.7183^{-0.5167} < OR < 1.33 * 2.7183^{0.5167} = 0.79 < OR < 2.23$$

We are 95% confident that the limits of 0.79 and 2.23 contain the true odds ratio. The confidence interval contains 1. This means that Nicorette does not appear to significantly affect the occurrence of mouth sores.

11. Transposing Table, from Exercise 7

$$\text{Odds ratio} = \frac{ad}{bc} = \frac{15*3}{65*17} = 0.04$$

The results are the same.

13. Entry of Zero

a. Odds Ratio with a zero frequency

$$\text{Odds ratio} = \frac{ad}{bc} = \frac{25*100}{75*0} = \text{ERROR}$$

Division by zero causes a calculation error

b. with a zero frequency

Since $\frac{ad}{bc} = \frac{25*100}{75*0} = \text{ERROR}$, multiplying it by the exponent will result in an error also

c. Making +0.5 adjustment

$$\text{Odds ratio} = \frac{\text{odds for disease among treatment group}}{\text{odds for disease among placebo group}} = \frac{ad}{bc} = \frac{25.5 * 100.5}{75.5 * 0.5} = 67.89$$

$$\text{CI}_{95\%} = \frac{ad}{bc} * e^{-z_{\alpha/2}\sqrt{\frac{1}{a}+\frac{1}{b}+\frac{1}{c}+\frac{1}{d}}} < OR < \frac{ad}{bc} * e^{z_{\alpha/2}\sqrt{\frac{1}{a}+\frac{1}{b}+\frac{1}{c}+\frac{1}{d}}}$$

$$\text{exponent} = z_{\alpha/2}\sqrt{\frac{1}{a}+\frac{1}{b}+\frac{1}{c}+\frac{1}{d}} = 1.96\sqrt{\frac{1}{25.5}+\frac{1}{75.5}+\frac{1}{0.5}+\frac{1}{100.5}} = 2.815$$

$$\text{CI}_{95\%} = 67.89 * 2.7183^{-2.815} < OR < 67.89 * 2.7183^{2.815} = 4.07 < OR < 1133.00$$

The result makes sense, though the confidence interval is very wide. The confidence interval does not

8-6 Comparing Variation in Two Samples

In Exercise 1, test the given claim. Use a significance level of α = 0.05 and assume that all populations are normally distributed. Use the traditional method of testing hypothesis outlined in Figure 7-8.

1. Test of difference between treatment and placebo variances

$H_0 : \sigma_1^2 = \sigma_2^2 \qquad H_1 : \sigma_1^2 \neq \sigma_2^2$

$n_1 = 25, \bar{x}_1 = 98.6, s_1 = 0.78$

$n_2 = 30, \bar{x}_2 = 98.2, s_2 = 0.52$

$$F = \frac{\text{larger variance}}{\text{smaller variance}} = \frac{s_1^2}{s_2^2} = \frac{0.78^2}{0.52^2} = 2.25$$

$df = (n_1 - 1),\ (n_2 - 1) = 24,\ 29$

$F_{CV} = 2.154$

We reject the null hypothesis. There is sufficient evidence to support the claim that the treatment and placebo populations have different variances.

3. Hypothesis Test for Magnet Treatment of Pain

$H_0 : \sigma_1^2 = \sigma_2^2 \qquad H_1 : \sigma_1^2 > \sigma_2^2$

$n_1 = 20, \bar{x}_1 = 0.44, s_1 = 1.40$

$n_2 = 20, \bar{x}_2 = 0.49, s_2 = 0.96$

$$F = \frac{\text{larger variance}}{\text{smaller variance}} = \frac{s_1^2}{s_2^2} = \frac{1.4^2}{0.96^2} = 2.127$$

$df = (n_1 - 1),\ (n_2 - 1) = 19,\ 19$

$F_{CV} = 2.1555$

We do not reject the null hypothesis. There is not sufficient evidence to support the claim that the pain reductions for the sham treatment group vary more that the pain reductions for the magnet treatment group.

5. Cigarette Filters and Nicotine

$H_0 : \sigma_1^2 = \sigma_2^2 \qquad H_1 : \sigma_1^2 > \sigma_2^2$

$n_1 = 21, \bar{x}_1 = 0.94, s_1 = 0.31$

$n_2 = 8, \bar{x}_2 = 1.65, s_2 = 0.16$

$$F = \frac{\text{larger variance}}{\text{smaller variance}} = \frac{s_1^2}{s_2^2} = \frac{0.31^2}{0.16^2} = 3.755$$

$df = (n_1 - 1),\ (n_2 - 1) = 20,\ 7$

$F_{CV} = 3.4445$

We reject the null hypothesis. There is sufficient evidence to support the claim that king-size cigarettes with filters have amounts of nicotine that vary more that the amounts of nicotine in nonfiltered king-size cigarettes.

We reject the null hypothesis. There is sufficient evidence to support the claim that king-size cigarettes with filters have amounts of nicotine that vary more that the amounts of nicotine in nonfiltered king-size cigarettes.

7. **Testing Effects of Zinc**

$H_0 : \sigma_1^2 = \sigma_2^2 \qquad H_1 : \sigma_1^2 < \sigma_2^2$

$n_1 = 294,\ \bar{x}_1 = 3214,\ s_1 = 669$

$n_2 = 286, \bar{x}_2 = 3088, s_2 = 728$

$$F = \frac{\text{larger variance}}{\text{smaller variance}} = \frac{s_2^2}{s_1^2} = \frac{728^2}{669^2} = 1.184$$

$df = (n_1 - 1),\ (n_2 - 1) = 285\ ,\ 293$

$F_{CV} \approx 1.3519$ (from Excel $F_{CV} = 1.214$)

We do not reject the null hypothesis. There is not sufficient evidence to support the claim that the variation of birth weights for the placebo population is greater than the variation for the population treated with zinc supplements.

9. **Blanking Out on Tests,** using SAS printout

$H_0 : \sigma_1^2 = \sigma_2^2 \qquad H_1 : \sigma_1^2 \neq \sigma_2^2$

$F = 2.59$

$df = 24\ ,\ 15$

$P-$ value $= 0.0599$

We do not reject the null hypothesis. There is not sufficient evidence to warrant rejection of the claim that the two samples come from populations with the same variance.

11. Constructing Confidence Intervals

	Sample size	Standard deviation
Placebo	n = 13	$s_1 = 9.4601$
Calcium	n = 15	$s_2 = 8.4689$

For F_R, $df = 12\ ,14$. $F_R = 3.050$

Note: there is no entry in Table A-5 for 14 (numerator) and 12 (denominator). In this case, we usually take the next lowest values for the degrees of freedom. In this case, we would use 12, 12 degrees of freedom which is 3.2773. If we had a table with 14, 12 or a computer program, the actual *F* value would be 3.2062. The value of *F* with 12, 12 *df* will be used in this example since that is the best that can be found using Table A-5.

For F_L, $df = 12, 12$. $F_L = 1/3.2773 = 0.3051$

$$CI_{95\%} = \left(\frac{s_1^2}{s_2^2} * \frac{1}{F_R}\right) < \frac{\sigma_1^2}{\sigma_2^2} < \left(\frac{s_1^2}{s_2^2} * \frac{1}{F_L}\right)$$

$$= \left(\frac{9.4601^2}{8.4689^2} * \frac{1}{F_R}\right) < \frac{\sigma_1^2}{\sigma_2^2} < \left(\frac{9.4601^2}{8.4689^2} * \frac{1}{F_L}\right) =$$

$$(1.2478 * 0.3278) < \frac{\sigma_1^2}{\sigma_2^2} < (1.2478 * 3.2773)$$

$$= 0.41 < \frac{\sigma_1^2}{\sigma_2^2} < 4.09$$

Had we used *df* of 14, 12 and *F*= 3.2062, the CI would have been $0.41 < \frac{\sigma_1^2}{\sigma_2^2} < 4.00$

Review Exercises

1. Warmer Surgical Patients Recover Better?

a. $H_0 : p_1 = p_2 \quad H_1 = p_1 < p_2$

$$p_1 = \frac{6}{104} = 0.06 \quad \hat{p}_2 = \frac{18}{96} = 0.019$$

$$\bar{p} = \frac{x_1 + x_2}{n_1 + n_2} = \frac{6+18}{104+96} = 0.12 \quad \bar{q} = 1 - \bar{p} = 0.88$$

$$z = \frac{(\hat{p}_1 - \hat{p}_2) - (p_1 - p_2)}{\sqrt{\frac{\bar{p}\bar{q}}{n_1} + \frac{\bar{p}\bar{q}}{n_2}}} = \frac{(0.06 - 0.19) - (0)}{\sqrt{\frac{0.12 * 0.88}{104} + \frac{0.12 * 0.88}{96}}}$$

$$= \frac{-0.13}{\sqrt{0.00102 + 0.0011}} = -2.82$$

With $\alpha = 0.05$ for a directional test, the critical value of z is -1.645. Since the observed z statistic is lower than -1.645, there is a significant difference in the direction that supports the claim that warmer surgical patients have fewer wound infections than patients who are not kept warmer. This would certainly indicate that patients should be routinely warmed.

b. A 90% confidence interval should be used since it will cut the distribution into the bottom 5% (the area used in this directional test), the middle 90%, and the top 5%.

c. Find $CI_{90\%}$

$$CI_{90\%} = (\hat{p}_1 - \hat{p}_2) - E < (p_1 - p_2) < (\hat{p}_1 - \hat{p}_2) + E$$

$$E = z_{\alpha/2}\sqrt{\frac{\hat{p}_1\hat{q}_1}{n_1} + \frac{\hat{p}_2\hat{q}_2}{n_2}} = 1.645\sqrt{\frac{0.06 * 0.94}{104} + \frac{0.19 * 0.81}{96}}$$

$$= 1.645\sqrt{0.002145} = 0.0762$$

$$CI_{90\%} = (0.06 - 0.19) - 0.0762 < (p_1 - p_2) < (0.06 - 0.19) + 0.0762$$

$$CI_{90\%} = -0.205 < (p_1 - p_2) < -0.054$$

The confidence interval does not include zero

d. No, the conclusions can be different, but most of the time they will lead to the same conclusion about the Null hypothesis.

e. Odds ratio and $CI_{95\%}$

	Recovered	Did not recover (wound infection)
Warmed	98	6
Cool	78	18

$$\text{Odds ratio } = \frac{ad}{bc} = \frac{98 * 18}{6 * 78} = 3.769$$

$$CI_{95\%} = \frac{ad}{bc} * e^{-z_{\alpha/2}\sqrt{\frac{1}{a}+\frac{1}{b}+\frac{1}{c}+\frac{1}{d}}} < OR < \frac{ad}{bc} * e^{z_{\alpha/2}\sqrt{\frac{1}{a}+\frac{1}{b}+\frac{1}{c}+\frac{1}{d}}}$$

$$\text{exponent } = z_{\alpha/2}\sqrt{\frac{1}{a}+\frac{1}{b}+\frac{1}{c}+\frac{1}{d}} = 1.96\sqrt{\frac{1}{98}+\frac{1}{6}+\frac{1}{78}+\frac{1}{18}} = 0.971$$

$$CI_{95\%} = 3.77 * 2.7183^{-0.971} < OR < \ 3.77 * 2.7183^{0.971} = 1.43 < OR < 9.96$$

The confidence interval does not contain 1. Therefore it seems that warmer surgical patients have higher odds of recovery or lack of infectious wounds compared with patients who are not warmed.

3. Variation of Brain Volumes

$H_0: \sigma_1^2 = \sigma_2^2 \qquad H_0: \sigma_1^2 \neq \sigma_2^2$

$$F = \frac{\text{larger variance}}{\text{smaller variance}} = \frac{s_1^2}{s_2^2} = \frac{156.84^2}{137.97^2} = 1.2922$$

$df = (n_1 - 1),\ (n_2 - 1) = 9,\ 9$

$F_{CV} = 4.0260$

There is not sufficient evidence to support the claim that the two populations have different amounts of variation in brain volume.

5. Zinc for Mothers

$n_1 = 294,\ \bar{x}_1 = 3214,\ s_1 = 669$

$n_2 = 286,\ \bar{x}_2 = 3088,\ s_2 = 728$

$H_0: \mu_1 = \mu_2 \qquad H_1: \mu_1 > \mu_2$

$$t = \frac{(\bar{x}_1 - \bar{x}_2) - (\mu_1 - \mu_2)}{\sqrt{\frac{s^2_1}{n_1} + \frac{s^2_2}{n_2}}} = \frac{(3214 - 3088) - 0}{\sqrt{\frac{669^2}{294} + \frac{728^2}{286}}} = \frac{126}{58.1} = 2.169$$

$df = 285,\quad t \approx 1.65$ (From Table A – 3)

The test statistic (2.169) is higher than the critical value and thus in the rejection region. We reject the null hypothesis and conclude that there is sufficient evidence to support the claim that zinc supplementation is associated with increased birth weight.

7. Drug Solubility

a. Conduct directional matched pairs test

$H_0: \mu_d = 0 \qquad H_1: \mu_d < 0$

$$d = -348. \quad \bar{d} = \frac{d}{n} = \frac{-348}{12} = -29$$

$$s_d = \sqrt{\frac{(d - \bar{d})^2}{n-1}} = \sqrt{\frac{7502}{11}} = 26.12$$

$$t = \frac{\bar{d} - \mu_d}{s_d / \sqrt{n}} = \frac{-29}{26.12 / \sqrt{12}} = -3.847$$

$t_{CV} = -1.796$

The test statistic (-3.8475) is lower than -1.796, in the rejection region. We reject the null hypothesis and conclude that there is sufficient evidence to support the claim that the Dozenol tablets are more soluble after the storage period.

b. Find $CI_{95\%}$

$$E = t_{\alpha/2} \frac{s_d}{\sqrt{n}} = 2.201 \frac{26.12}{\sqrt{12}} = 16.6$$

$$CI_{95\%} = \bar{d} - E < \mu_d < \bar{d} + E = (29 - 16.6) < \mu_d < (29 + 16.6)$$

$$= 12.4 < \mu_d < 45.6$$

Cumulative Review Exercises

1. Gender Difference in Speeding

a. $P(\text{Yes speeding tickets}) = \dfrac{Yes}{Total} = \dfrac{53}{750} = 0.0707$

b. $P(\text{Man or speeding}) = P(\text{Man}) + \text{P(Speeding)} - \text{P(Man and speeding)}$

$= 0.3333 + 0.0707 - 0.0347 = 0.3693$

c. $P(\text{Ticket}|\text{Man}) = \dfrac{26}{250} = 0.1040$

d. $P(\text{Ticket}|\text{Woman}) = \dfrac{27}{500} = 0.0540$

e. $\text{H}_0 : p_1 = p_2 \quad \text{H}_1 = p_1 < p_2$

$\hat{p}_1 = \dfrac{27}{500} = 0.054 \quad \hat{p}_2 = \dfrac{26}{250} = 0.104$

$\overline{p} = \dfrac{x_1 + x_2}{n_1 + n_2} = \dfrac{53}{750} = 0.071 \quad \overline{q} = 1 - \overline{p} = 0.929$

$$z = \frac{(\hat{p}_1 - \hat{p}_2) - (p_1 - p_2)}{\sqrt{\dfrac{\overline{pq}}{n_1} + \dfrac{\overline{pq}}{n_2}}} = \frac{-0.05 - (0)}{\sqrt{\dfrac{0.066}{500} + \dfrac{0.066}{250}}}$$

$$= \frac{-0.05}{\sqrt{0.000396}} = -2.51$$

With $\alpha = 0.05$, since the observed z is lower than the critical value, there is sufficient evidence to support the claim that the percentage of women ticketed for speeding is lower than the proportion of men ticketed for speeding.

3. Clinical Tests of Viagra

a. $x = 734 * 0.04 = 29.34 \approx 29 \qquad \hat{p} = \frac{29}{734} = 0.0395$

$CI_{95\%}$ for $\hat{p}_1 = \hat{p} - E < p < \hat{p} + E$

$$E = z_{\alpha/2}\sqrt{\frac{\hat{p}\hat{q}}{n}} = 1.96\sqrt{\frac{0.03794}{734}} = 0.0141$$

$CI_{95\%}$ for $\hat{p}_1 = 0.0395 - 0.0141 < p < 0.0395 + 0.0141$

$= 0.0254 < p < 0.0536$

b. $x = 725 * 0.021 = 15.225 \approx 15 \qquad \hat{p} = \frac{15}{725} = 0.0207$

$CI_{95\%}$ for $\hat{p}_2 = \hat{p} - E < p < \hat{p} + E$

$$E = z_{\alpha/2}\sqrt{\frac{\hat{p}\hat{q}}{n}} = 1.96\sqrt{\frac{0.02026}{725}} = 0.0104$$

$CI_{95\%}$ for $\hat{p} = 0.0207 - 0.0104 < p < 0.0207 + 0.0104$

$= 0.0103 < p < 0.0311$

c. $\hat{p}_1 = \frac{29}{734} = 0.0395 \qquad \hat{p}_2 = \frac{15}{725} = 0.0207$

$CI_{95\%} = (\hat{p}_1 - \hat{p}_2) - E < (p_1 - p_2) < (\hat{p}_1 - \hat{p}_2) + E$

$$E = z_{\alpha/2}\sqrt{\frac{\hat{p}_1\hat{q}_1}{n_1} + \frac{\hat{p}_2\hat{q}_2}{n_2}} = 1.96\sqrt{\frac{0.03794}{734} + \frac{0.02026}{725}}$$

$= 0.0175$

$CI_{95\%} = 0.0188 - 0.0175 < (p_1 - p_2) < 0.0188 + 0.0175$

$CI_{95\%} = 0.0013 < (p_1 - p_2) < 0.0363$

d. The best method is method iii, to conduct a hypothesis test of the null hypothesis: $p_1 = p_2$, using 0.05 significance level.

Chapter 9. Correlation and Regression

9-2 Correlation

In exercises 1 & 3, use a significance level of $\alpha = 0.05$

1. Chest Sizes and Weights of Bears

a. Checking Table A-6, we see that the critical value for $n = 8$ is 0.707. Since r = 0.993, there is a significant correlation between chest size and weight in bears.

b. The percentage of variation in weight that can be explained by the linear association between weight and chest size is r^2. We find that $r^2 = 0.986$, so 98.6% of the variation in weight can be explained by the linear association between weight and chest size.

3. BMI and Weight of Females

a. Checking Table A-6, we see that the critical value for $n = 40$ is 0.312. Since r = 0.936, there is a significant correlation between BMI and weight in females.

b. The percentage of variation in the weight that can be explained by the variation in BMI is r^2. We find that $r^2 = 0.876$, so 87.6% of variation in the weight can be explained by the variation in BMI.

In exercise 5, use a scatterplot and the linear correlation coefficient r to determine whether there is a correlation between the two variables.

The formula for the linear correlation coefficient (see pg. 432 of text) is $r = \dfrac{n\Sigma xy - (\Sigma x)(\Sigma y)}{\sqrt{n(\Sigma x^2) - (\Sigma x)^2}\sqrt{n(\Sigma y^2) - (\Sigma y)^2}}$

5. The scatterplot for the data is shown. Clearly, there is a correlation between the variables, in the shape of a parabola. We now calculate the linear correlation coefficient, along with the needed sums.

$\Sigma x = 10, \Sigma y = 10, \Sigma x^2 = 30, \Sigma xy = 20, \Sigma y^2 = 34$

$$r = \frac{5 \cdot 20 - 10 \cdot 10}{\sqrt{5 \cdot 30 - (10)^2}\sqrt{5 \cdot 34 - (10)^2}} = 0.00$$

The linear correlation coefficient is 0.00. This does not mean there is no correlation; it indicates that there is no linear correlation.

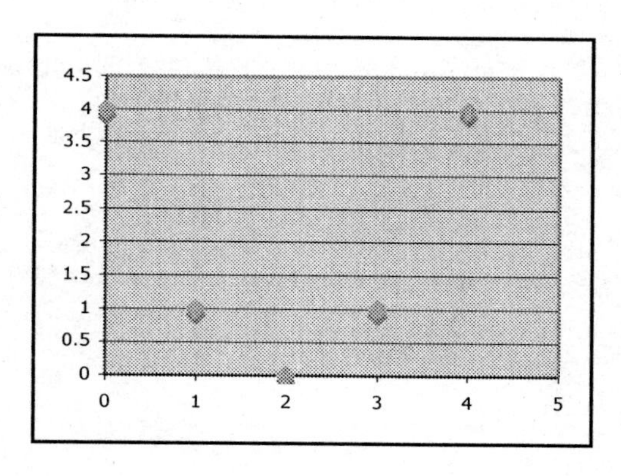

7. Effects of an Outlier

a. There does not appear to be a correlation between the variables x and y.

b. The ten data pairs are given below.

x	1	1	1	2	2	2	3	3	3	10
y	1	2	3	1	2	3	1	2	3	10

We now calculate the linear correlation coefficient, along with the needed sums.

$\Sigma x = 28, \Sigma y = 28, \Sigma x^2 = 142, \Sigma xy = 136, \Sigma y^2 = 142$

$$r = \frac{10 \cdot 136 - 28 \cdot 28}{\sqrt{10 \cdot 142 - (28)^2}\sqrt{10 \cdot 142 - (28)^2}} = 0.906$$

The critical value from Table A-6 with $\alpha = 0.01$ and $n = 10$ is 0.765. There is a significant linear correlation between the variables.

c. The nine data pairs are given below.

x	1	1	1	2	2	2	3	3	3
y	1	2	3	1	2	3	1	2	3

We now calculate the linear correlation coefficient, along with the needed sums.

$\Sigma x = 18, \Sigma y = 18, \Sigma x^2 = 42, \Sigma xy = 36, \Sigma y^2 = 42$

$$r = \frac{9 \cdot 36 - 18 \cdot 18}{\sqrt{9 \cdot 42 - (18)^2}\sqrt{9 \cdot 42 - (18)^2}} = 0.000$$

There is no linear correlation between the variables, as $r = 0.000$.

d. One pair of values can have a large impact of the value of the linear correlation coefficient, creating the impression that there is a strong correlation among the data when one doesn't seem to exist.

In Exercises 9-15, construct a scatterplot, find the value of the linear correlation coefficient and use a significance level of α = 0.05 to determine whether there is a significant linear correlation between the two variables. Save your work because the same data sets will be used in the next section.

Note: It is not explicitly stated that the data was collected via a random sample. This will be assumed. We will determine whether there is a significant linear correlation using both the formal hypothesis test for linear correlation and Table A-6. Additionally, the formulas used are as follows:

$$r = \frac{n\Sigma xy - (\Sigma x)(\Sigma y)}{\sqrt{n(\Sigma x^2) - (\Sigma x)^2}\sqrt{n(\Sigma y^2) - (\Sigma y)^2}}$$ for the linear correlation coefficient, and

$$t = \frac{r}{\sqrt{\frac{1 - r^2}{n - 2}}}$$ as the test statistic for the formal hypothesis test for correlation.

9. Supermodel Heights and Weights

The scatterplot for the data is as shown. The scatterplot approximates a straight-line pattern, and so a hypothesis test is indicated. We now calculate the needed sums and the linear correlation coefficient.

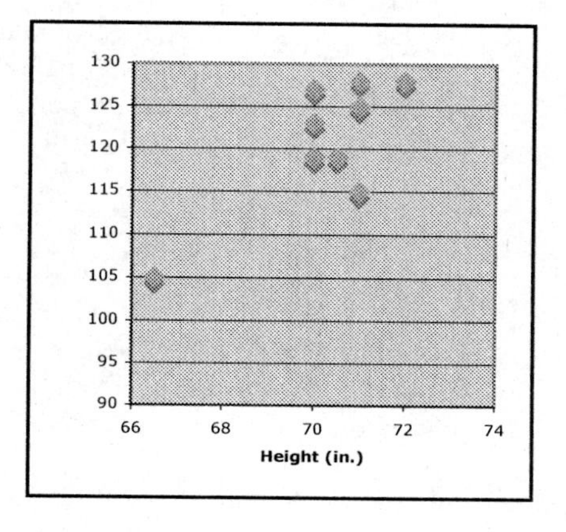

$\Sigma x = 632, \Sigma y = 1089, \Sigma x^2 = 44399.5, \Sigma xy = 76546, \Sigma y^2 = 132223$

$$r = \frac{9 \cdot 76546 - 632 \cdot 1089}{\sqrt{9 \cdot 44399.5 - (632)^2}\sqrt{9 \cdot 132223 - (1089)^2}} = 0.796$$

The critical value from Table A-6 for $n = 9$ and with $\alpha = 0.05$ is 0.666. Since $r = 0.796$, there is a significant linear correlation between heights and weights of supermodels.

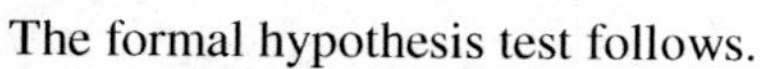

The formal hypothesis test follows.

H_0: $\rho = 0$

H_1: $\rho \neq 0$

The test statistic is

$$t = \frac{r}{\sqrt{\frac{1-r^2}{n-2}}} = \frac{0.796}{\sqrt{\frac{1-(0.796)^2}{9-2}}} = 3.475$$

The critical values are found in Table A-3 using df = n – 2, and are $\pm t_{\alpha/2} = \pm t_{.025} = \pm 2.365$.

Since the test statistic is between greater than 2.365, we reject the null hypothesis. There is sufficient sample evidence to conclude that there is significant linear correlation between the heights and weights of supermodels. Since the data was only from supermodels, the inference cannot be generalize to the general population of all women.

11. Blood Pressure Measurements

The scatterplot for the data is as shown. The scatterplot approximates a straight-line pattern, and so a hypothesis test is indicated. We now calculate the needed sums and the linear correlation coefficient.

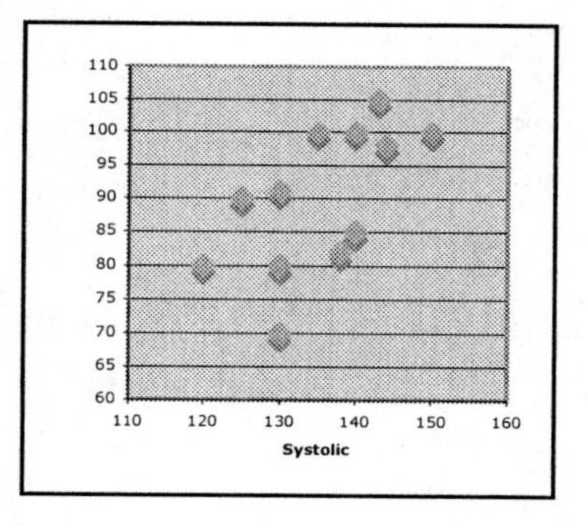

$\Sigma x = 1875, \Sigma y = 1241, \Sigma x^2 = 252179, \Sigma xy = 167023, \Sigma y^2 = 111459$

$$r = \frac{14 \cdot 167023 - 1875 \cdot 1241}{\sqrt{14 \cdot 252179 - (1875)^2}\sqrt{14 \cdot 111459 - (1241)^2}} = 0.658$$

The critical value from Table A-6 for $n = 14$ and with $\alpha = 0.05$ is 0.532. Since $r = 0.658$, there is a significant linear correlation between systolic and diastolic values.

The formal hypothesis test follows.

H_0: $\rho = 0$

H_1: $\rho \neq 0$

The test statistic is

$$t = \frac{r}{\sqrt{\frac{1-r^2}{n-2}}} = \frac{0.658}{\sqrt{\frac{1-(0.658)^2}{14-2}}} = 3.026$$

The critical values are found in Table A-3 using df = n – 2, and are $\pm t_{\alpha/2} = \pm t_{.025} = \pm 2.179$.

Since the absolute value of the test statistic is greater than 2.179, we reject the null hypothesis. There is sufficient sample evidence to conclude that there is significant linear correlation between the systolic and diastolic values. Though there is a linear correlation, there appears to be wide variation in the readings from the same individual. Is this due to fluctuations in the patient, variability in the interpretive abilities of the medical students, or to the tools used by the different students?

13. Smoking and Nicotine

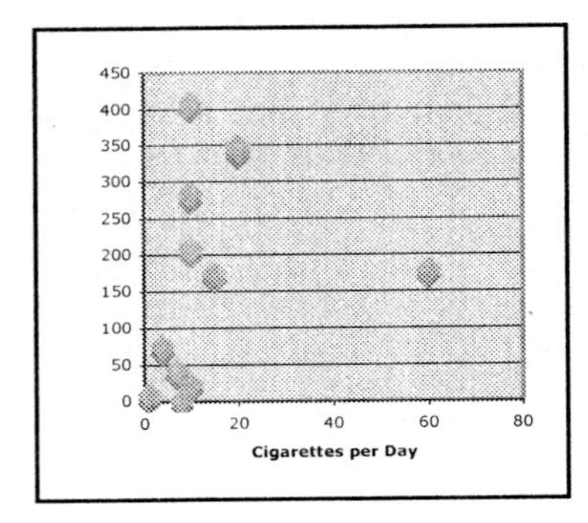

The scatterplot for the data is as shown. The scatterplot does not appear to approximate a straight-line pattern, and so a hypothesis test is not indicated. We will, however, perform the test in order to demonstrate the procedure. We now calculate the needed sums and the linear correlation coefficient.

$\Sigma x = 175, \Sigma y = 2102.46, \Sigma x^2 = 5155, \Sigma xy = 37111.51, \Sigma y^2 = 601709.793$

$$r = \frac{12 \cdot 37111.51 - 175 \cdot 2102.46}{\sqrt{12 \cdot 5155 - (175)^2}\sqrt{12 \cdot 601709.793 - (2102.46)^2}} = 0.262$$

The critical value from Table A-6 for n = 12 and with $\alpha = 0.05$ is 0.576. Since r = 0.262, there is not a significant linear correlation between cigarettes per day and cotinine.

The formal hypothesis test follows.

H_0: $\rho = 0$

H_1: $\rho \neq 0$

The test statistic is

$$t = \frac{r}{\sqrt{\frac{1-r^2}{n-2}}} = \frac{0.262}{\sqrt{\frac{1-(0.262)^2}{14-2}}} = 0.858$$

The critical values are found in Table A-3 using df = n – 2, and are $\pm t_{\alpha/2} = \pm t_{.025} = \pm 2.228$.

Since the absolute value of the test statistic is less than 2.228, we fail to reject the null hypothesis. There is not sufficient sample evidence to conclude that there is significant linear correlation between cigarettes per day and cotinine. The results are affected by the presence of an outlier, (60,179). With that value eliminated, the scatterplot approximates a straight line, and the correlation coefficient is 0.690. It may be that the number of cigarettes was misrecorded, and was 6 instead of 60. This is another example of an outlier affecting the results.

15. Testing Duragesic for Pain Relief

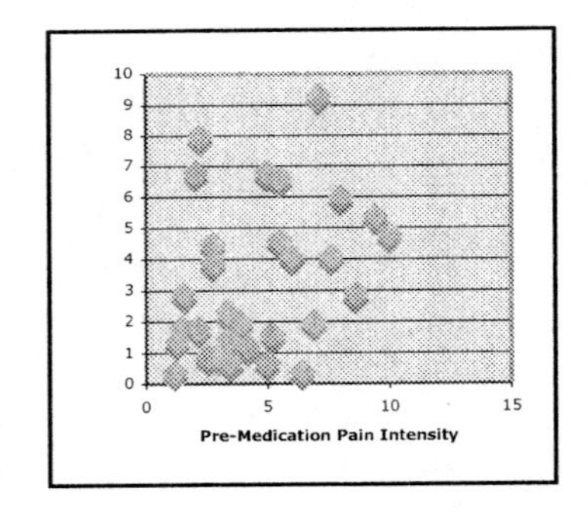

The scatterplot for the data is as shown. The scatterplot does not appear to approximate a straight-line pattern, and so a hypothesis test is not indicated. We will, however, perform the test in order to demonstrate the procedure. We now calculate the needed sums and the linear correlation coefficient.

$\Sigma x = 144.9, \Sigma y = 102.2, \Sigma x^2 = 866.35, \Sigma xy = 533.78, \Sigma y^2 = 516.4$

$$r = \frac{31 \cdot 533.78 - 144.9 \cdot 102.2}{\sqrt{31 \cdot 866.35 - (144.9)^2}\sqrt{31 \cdot 516.4 - (102.2)^2}} = 0.304$$

The critical value from Table A-6 for n = 31 is not shown. We use n = 30 instead. The critical value from Table A-6 for n = 30 with α = 0.05 is 0.361. Since r = 0.304, there is a not a significant linear correlation between the pain measurements.

The formal hypothesis test follows.

H_0: $\rho = 0$

H_1: $\rho \neq 0$

The test statistic is

$$t = \frac{r}{\sqrt{\frac{1-r^2}{n-2}}} = \frac{0.304}{\sqrt{\frac{1-(0.304)^2}{31-2}}} = 1.721$$

The critical values are found in Table A-3 using df = n – 2, and are $\pm t_{\alpha/2} = \pm t_{.025} = \pm 2.045$.

Since the absolute value of the test statistic is less than 2.045, we fail to reject the null hypothesis. There is not sufficient sample evidence to conclude that there is significant linear correlation between the pain measurements. A significant correlation might indicate that the drug is effective, but a placebo effect may occur and the significance could be due primarily to the number of data pairs.

In Exercises 17-21, use the data from Appendix B to construct a scatterplot, find the value of the linear correlation coefficient r, and use a significance level of $\alpha = 0.05$ to determine whether there is a significant linear correlation between the two variables. Save your work because the same data sets will be used in the next section.

17. Cholesterol and Weight

The scatterplot for the data is as shown. The scatterplot does appear to approximate a straight-line pattern (absent the outlier), and so a hypothesis test is indicated. We now calculate the needed sums and the linear correlation coefficient.

$\Sigma x = 9635, \Sigma y = 5848.8, \Sigma x^2 = 3669819, \Sigma xy = 1543475.3, \Sigma y^2 = 910409.9$

$$r = \frac{40 \cdot 1543475.3 - 9635 \cdot 5848.8}{\sqrt{40 \cdot 3669819 - (9635)^2}\sqrt{40 \cdot 910409.9 - (5848.8)^2}} = 0.493$$

The critical value from Table A-6 for $n = 40$ with $\alpha = 0.05$ is 0.312. Since $r = 0.493$, there is a significant linear correlation between the cholesterol and weight.

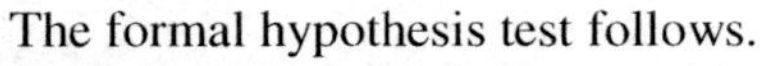

The formal hypothesis test follows.

H_0: $\rho = 0$

H_1: $\rho \neq 0$

The test statistic is

$$t = \frac{r}{\sqrt{\frac{1-r^2}{n-2}}} = \frac{0.493}{\sqrt{\frac{1-(0.493)^2}{40-2}}} = 3.497$$

The critical values are found in Table A-3 using df = n – 2, and are $\pm t_{\alpha/2} = \pm t_{.025} = \pm 2.024$.

Since the absolute value of the test statistic is greater than 2.024, we reject the null hypothesis. There is sufficient sample evidence to conclude that there is significant linear correlation between cholesterol and weight in females.

19. Heights of Mothers and Daughters

The scatterplot for the data is as shown. The scatterplot does appear to approximate a straight-line pattern, and so a hypothesis test is indicated. We now calculate the needed sums and the linear correlation coefficient.

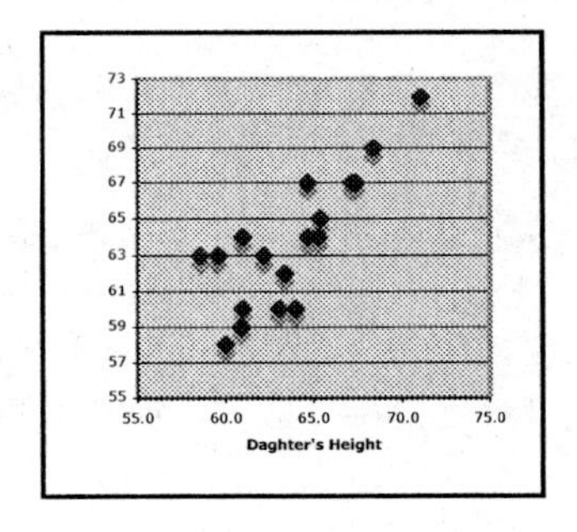

$\Sigma x = 1275.6, \Sigma y = 1275.0, \Sigma x^2 = 81554.7, \Sigma xy = 81491.6, \Sigma y^2 = 81515.0$

$$r = \frac{20 \cdot 81491.6 - 1275.6 \cdot 1275.0}{\sqrt{20 \cdot 81554.7 - (1275.6)^2}\sqrt{20 \cdot 81515.0 - (1275.0)^2}} = 0.802$$

For $n = 20$ with $\alpha = 0.05$ the critical value is 0.444. Since $r = 0.802$, there is a significant linear correlation between the heights of mothers and their daughters.

The formal hypothesis test follows.

H_0: $\rho = 0$

H_1: $\rho \neq 0$

The test statistic is

$$t = \frac{r}{\sqrt{\frac{1-r^2}{n-2}}} = \frac{0.802}{\sqrt{\frac{1-(0.802)^2}{20-2}}} = 5.697$$

The degrees of freedom is df = n – 2 = 18. The critical values are $\pm t_{\alpha/2} = \pm t_{.025} = \pm 2.101$.

Since the absolute value of the test statistic is greater than 2.101, we reject the null hypothesis. There is sufficient sample evidence to conclude that there is significant linear correlation between the heights of mothers and their daughters.

21. Iris Petal Lengths and Widths

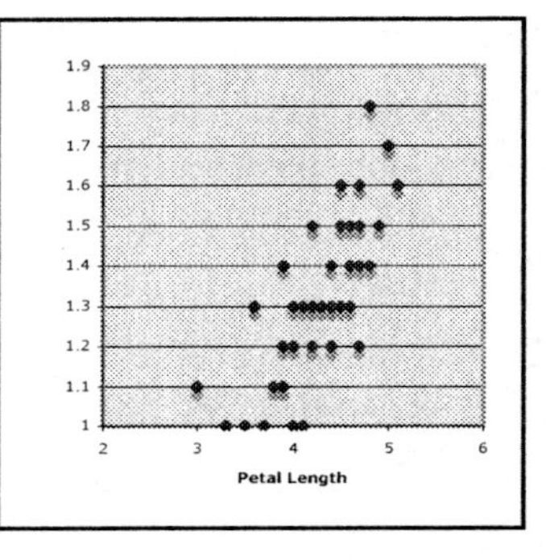

The scatterplot for the data is as shown. The scatterplot appears to approximate a straight-line pattern, and so a hypothesis test is indicated. We now calculate the needed sums and the linear correlation coefficient.

$\Sigma x = 213, \Sigma y = 66.3, \Sigma x^2 = 918.2, \Sigma xy = 286.02, \Sigma y^2 = 89.83$

$$r = \frac{50 \cdot 286.02 - 213 \cdot 66.3}{\sqrt{50 \cdot 918.2 - (213)^2}\sqrt{50 \cdot 89.83 - (66.3)^2}} = 0.787$$

The critical value from Table A-6 for $n = 50$ with $\alpha = 0.05$ is 0.279. Since $r = 0.787$, there is a significant linear correlation between the petal lengths and widths of Versicolor irises.

The formal hypothesis test follows.

H_0: $\rho = 0$

H_1: $\rho \neq 0$

The test statistic is

$$t = \frac{r}{\sqrt{\frac{1-r^2}{n-2}}} = \frac{0.787}{\sqrt{\frac{1-(0.787)^2}{50-2}}} = 8.828$$

Table A-3 does not have an entry for df = 48, so we use df = 50. The critical values found in Table A-3 are $\pm t_{\alpha/2} = \pm t_{.025} = \pm 2.009$.

Since the absolute value of the test statistic is greater than 2.009, we reject the null hypothesis. There is sufficient sample evidence to conclude that there is significant linear correlation between the petal lengths and widths of Versicolor irises. The result is the same as above, the correlation for the petal lengths and widths of the Versicolor class is much stronger.

In Exercises 23 & 25, describe the error in the stated conclusion. (See the list of common sources of errors included in this section.)

23. Correlation does not necessarily imply causality.

25. Averages suppress the variability in individual values, so the linear correlation may not exist on an individual basis.

27. Correlations with Transformed Data

The calculations and scatterplots are given for each case.

a. $\Sigma x = 18.6, \Sigma y = 2.83, \Sigma x^2 = 53.62, \Sigma xy = 8.258, \Sigma y^2 = 1.2817$

$$r = \frac{7 \cdot 8.258 - 18.6 \cdot 2.83}{\sqrt{7 \cdot 53.62 - (18.6)^2}\sqrt{7 \cdot 1.2817 - (2.83)^2}} = 0.972$$

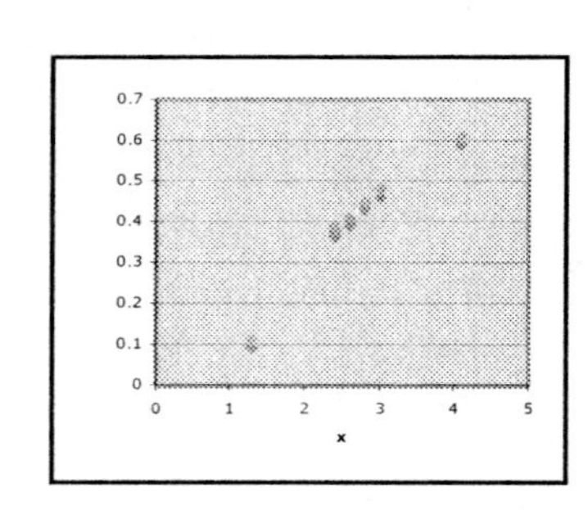

b. $\Sigma x = 53.62, \Sigma y = 2.83, \Sigma x^2 = 539.951, \Sigma xy = 25.495, \Sigma y^2 = 1.282$

$$r = \frac{7 \cdot 25.495 - 53.62 \cdot 2.83}{\sqrt{7 \cdot 539.951 - (53.62)^2}\sqrt{7 \cdot 1.282 - (2.83)^2}} = 0.905$$

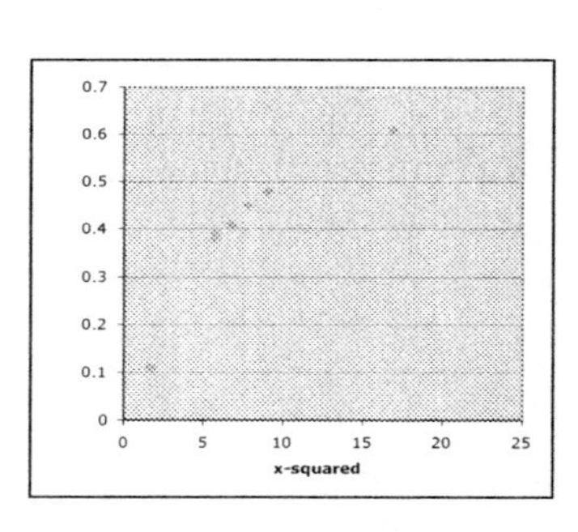

c. $\Sigma x = 2.826, \Sigma y = 2.83, \Sigma x^2 = 1.277, \Sigma xy = 1.279, \Sigma y^2 = 1.282$

$$r = \frac{7 \cdot 1.279 - 2.826 \cdot 2.83}{\sqrt{7 \cdot 1.277 - (2.826)^2}\sqrt{7 \cdot 1.282 - (2.83)^2}} = 0.999$$

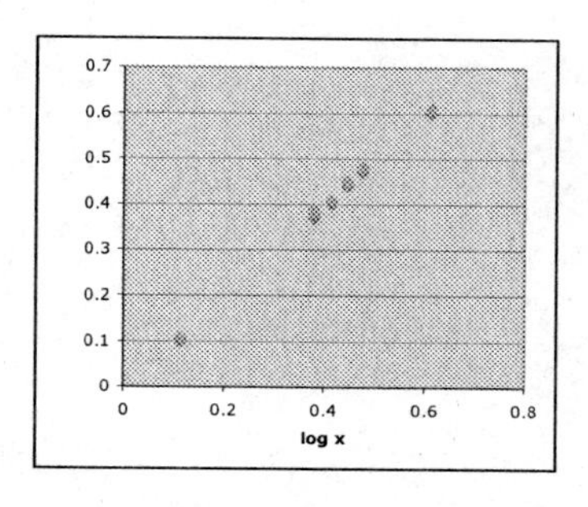

d. $\Sigma x = 11.281, \Sigma y = 2.83, \Sigma x^2 = 18.6, \Sigma xy = 4.799, \Sigma y^2 = 1.282$

$$r = \frac{7 \cdot 4.799 - 11.281 \cdot 2.83}{\sqrt{7 \cdot 18.6 - (11.281)^2}\sqrt{7 \cdot 1.282 - (2.83)^2}} = 0.992$$

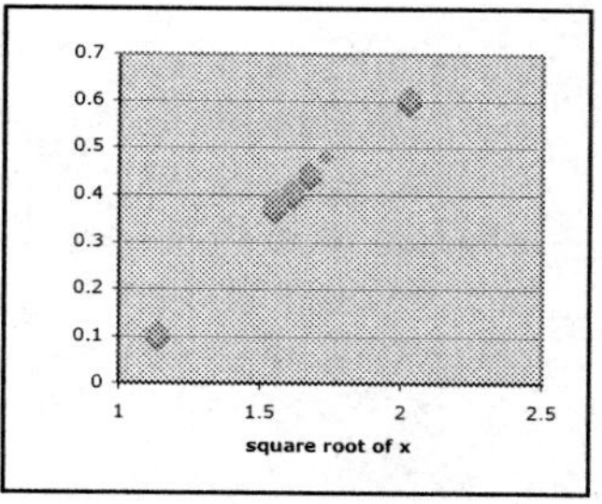

e. $\Sigma x = 2.922, \Sigma y = 2.83, \Sigma x^2 = 1.385, \Sigma xy = 1.033, \Sigma y^2 = 1.282$

$$r = \frac{7 \cdot 1.033 - 2.922 \cdot 2.83}{\sqrt{7 \cdot 1.285 - (2.922)^2}\sqrt{7 \cdot 1.282 - (2.83)^2}} = -0.984$$

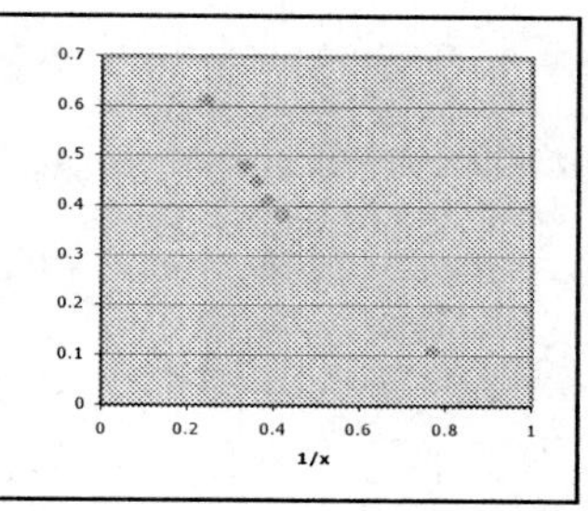

The transformation from **c**, log *x*, yields the largest linear correlation coefficient, 0.999.

29. Including Categorical Data in a Scatterplot

a. It seems that linear relationship exists between weight and remote control use.

b. We calculate *r* and then compare to the value found in Table A-6 for $\alpha = 0.05$.

$\Sigma x = 2408, \Sigma y = 195, \Sigma x^2 = 369474, \Sigma xy = 31884, \Sigma y^2 = 3463$

$$r = \frac{16 \cdot 31884 - 2408 \cdot 195}{\sqrt{16 \cdot 369474 - (2408)^2}\sqrt{16 \cdot 3463 - (195)^2}} = 0.915$$

The critical value from Table A-6 for $n = 16$ with $\alpha = 0.05$ is 0.497. Since $r = 0.915$, there is a significant linear correlation between weight and use of the remote.

c. Restricting the set to the females only, we calculate *r* and then compare to the value found in Table A-6 for $\alpha = 0.05$.

$\Sigma x = 1042, \Sigma y = 34, \Sigma x^2 = 135958, \Sigma xy = 4414, \Sigma y^2 = 164$

$$r = \frac{8 \cdot 4414 - 1042 \cdot 34}{\sqrt{8 \cdot 135958 - (1042)^2}\sqrt{8 \cdot 164 - (34)^2}} = -0.213$$

The critical value from Table A-6 for $n = 8$ with $\alpha = 0.05$ is 0.707. Since $r = -0.213$, the absolute value of which is less than $|-0.707|$, there is not significant linear correlation between weight and use of the remote in females.

d. Restricting the set to the males only, we calculate *r* and then compare to the value found in Table A-6 for $\alpha = 0.05$.

$\Sigma x = 1366, \Sigma y = 161, \Sigma x^2 = 233516, \Sigma xy = 27470, \Sigma y^2 = 3299$

$$r = \frac{8 \cdot 27470 - 1366 \cdot 161}{\sqrt{8 \cdot 233516 - (1366)^2}\sqrt{8 \cdot 3299 - (161)^2}} = -0.164$$

The critical value from Table A-6 for $n = 8$ with $\alpha = 0.05$ is 0.707. Since $r = -0.164$, the absolute value of which is less than |-0.707|, there is not significant linear correlation between weight and use of the remote in males.

e. I would conclude that the correlation seen in part b was due to gender, not weight, since the males were heavier than the females.

9-3 Regression

In exercises 1 & 3, use the given data to find the best predicted value of the dependent variable. Be sure to follow the prediction procedure described in this section.

1. **a.** Checking Table A-6, the critical value for r for a sample size of 20 is 0.444. Since $r = 0.987$, there is a significant linear correlation, so we use the regression equation for the prediction.
$\hat{y} = 6.00 + 4.00x = 6.00 + 4.00 \cdot 3.00 = 18.00$.

b. Checking Table A-6, the critical value for r for a sample size of 20 is 0.444. Since $r = 0.052$, there is not significant linear correlation, so we use the mean, $\bar{y}$ for the prediction. The best prediction is 5.00.

3. **Chest sizes and Weights of Bears**
Checking Table A-6, the critical value for r for a sample size of 8 is 0.707. Since $r = 0.993$, there is a significant linear correlation, so we use the regression equation for the prediction.
$\hat{y} = -187 + 11.3x = -187 + 11.3 \cdot 52 = 400.6$.
The best prediction for the weight of the bear is 400.6 lbs.

In Exercise 5, use the given data to find the equation of the regression line.

5. We need to calculate the required sums and means, and then we find the regression equation for the five data points.

$\Sigma x = 10, \Sigma y = 10, \Sigma x^2 = 30, \Sigma xy = 20$

$\bar{y} = \Sigma y / n = 10/5 = 2$

$\bar{x} = \Sigma x / n = 10/5 = 2$

$$b_1 = \frac{n(\Sigma xy) - (\Sigma x)(\Sigma y)}{n(\Sigma x^2) - (\Sigma x)^2} = \frac{5 \cdot 20 - 10 \cdot 10}{5 \cdot 30 - (10)^2} = 0$$

$b_0 = \bar{y} - b_1\bar{x} = 2 - 0 \cdot 2 = 2$

The regression equation for the data set is $\hat{y} = b_0 + b_1 x = 2 + 0x = 2$

7. **Effects of an Outlier**

a. We need to calculate the required sums and means, and then we find the regression equation for the ten data points, given below as determined from the scatterplot of Exercise 7 in Section 9-2.

x	1	1	1	2	2	2	3	3	3	10
y	1	2	3	1	2	3	1	2	3	10

$\Sigma x = 28, \Sigma y = 28, \Sigma x^2 = 142, \Sigma xy = 136$

$\bar{y} = \Sigma y / n = 28/10 = 2.8$

$\bar{x} = \Sigma x / n = 28/10 = 2.8$

$$b_1 = \frac{n(\Sigma xy) - (\Sigma x)(\Sigma y)}{n(\Sigma x^2) - (\Sigma x)^2} = \frac{10 \cdot 136 - 28 \cdot 28}{10 \cdot 142 - (28)^2} = 0.906$$

$b_0 = \bar{y} - b_1\bar{x} = 2.8 - 0.906 \cdot 2.8 = 0.264$

The regression equation for the data set is $\hat{y} = b_0 + b_1 x = 0.264 + 0.906x$

b. We need to calculate the required sums and means, and then we find the regression equation for the nine data points, given below as determined from the scatterplot of Exercise 7 in Section 9-2.

x	1	1	1	2	2	2	3	3	3
y	1	2	3	1	2	3	1	2	3

$\Sigma x = 18, \Sigma y = 18, \Sigma x^2 = 42, \Sigma xy = 36$

$\bar{y} = \Sigma y / n = 18/9 = 2$

$\bar{x} = \Sigma x / n = 18/9 = 2$

$$b_1 = \frac{n(\Sigma xy) - (\Sigma x)(\Sigma y)}{n(\Sigma x^2) - (\Sigma x)^2} = \frac{9 \cdot 36 - 18 \cdot 18}{9 \cdot 42 - (18)^2} = 0$$

$b_0 = \bar{y} - b_1\bar{x} = 2 - 0 \cdot 2 = 2$

The regression equation for the data set is $\hat{y} = b_0 + b_1 x = 2 + 0x = 2$

c. The removal of the data point (10,10) changed the line of part (a) into a horizontal line.

Exercises 9-21 use the same data sets as the exercises in Section 9-2. In each case, find the regression equation, letting the first variable be the independent (x) variable. Find the indicated predicted values. Caution: When finding predicted values, be sure to follow the prediction procedure described in this section.

9. Supermodel Heights and Weights

We need to calculate the required sums and means, and then we find the regression equation for the data set.

$\Sigma x = 632, \Sigma y = 1089, \Sigma x^2 = 44399.5, \Sigma xy = 76546$

$\bar{y} = \Sigma y / n = 1089/9 = 121$

$\bar{x} = \Sigma x / n = 632/9 = 70.222$

$$b_1 = \frac{n(\Sigma xy) - (\Sigma x)(\Sigma y)}{n(\Sigma x^2) - (\Sigma x)^2} = \frac{9 \cdot 76546 - 632 \cdot 1089}{9 \cdot 44399.5 - (632)^2} = 3.883$$

$b_0 = \bar{y} - b_1\bar{x} = 121 - 3.883 \cdot 70.222 = -151.700$

The regression equation for the data set is $\hat{y} = b_0 + b_1 x = -151.700 + 3.883x$.

In Exercise 9 of Section 9-2, we rejected $\rho = 0$, so the best predicted value for the weight of a supermodel that is 69 inches tall is $\hat{y} = -151.700 + 3.883x = -151.700 + 3.883 \cdot 69 = 116.254$ lbs.

11. Blood Pressure Measurements

We need to calculate the required sums and means, and then we find the regression equation for the data set.

$\Sigma x = 1875, \Sigma y = 1241, \Sigma x^2 = 252179, \Sigma xy = 167023$

$\bar{y} = \Sigma y / n = 1241/14 = 88.643$

$\bar{x} = \Sigma x / n = 1875/14 = 133.929$

$$b_1 = \frac{n(\Sigma xy) - (\Sigma x)(\Sigma y)}{n(\Sigma x^2) - (\Sigma x)^2} = \frac{14 \cdot 167023 - 1875 \cdot 1241}{14 \cdot 252179 - (1875)^2} = 0.769$$

$b_0 = \bar{y} - b_1\bar{x} = 88.643 - 0.769 \cdot 133.929 = -14.380$

The regression equation for the data set is $\hat{y} = b_0 + b_1 x = -14.380 + 0.769x$.

In Exercise 11 of Section 9-2, we rejected $\rho = 0$, so the best predicted value for the diastolic pressure for a systolic pressure of 122 is $\hat{y} = -14.380 + 0.769x = -14.380 + 0.769 \cdot 122 = 79.467$ lbs.

13. Smoking and Nicotine

We need to calculate the required sums and means, and then we find the regression equation for the data set.

$\Sigma x = 175, \Sigma y = 2102.46, \Sigma x^2 = 5155, \Sigma xy = 37111.51$

$\bar{y} = \Sigma y / n = 2102.46/12 = 175.205$

$\bar{x} = \Sigma x / n = 175/12 = 14.583$

$$b_1 = \frac{n(\Sigma xy) - (\Sigma x)(\Sigma y)}{n(\Sigma x^2) - (\Sigma x)^2} = \frac{12 \cdot 37111.51 - 175 \cdot 2102.46}{12 \cdot 5155 - (175)^2} = 2.478$$

$b_0 = \bar{y} - b_1\bar{x} = 175.205 - 2.478 \cdot 14.583 = 139.064$

The regression equation for the data set is $\hat{y} = b_0 + b_1 x = 139.064 + 2.478x$.

In Exercise 13 of Section 9-2, we did not reject that $\rho = 0$, so the best predicted value for the cotinine is its mean, which is 175.205.

15. Testing Duragesic for Pain Relief

We need to calculate the required sums and means, and then we find the regression equation for the data set.

$\Sigma x = 144.9, \Sigma y = 102.2, \Sigma x^2 = 866.35, \Sigma xy = 533.78$

$\bar{y} = \Sigma y / n = 102.2 / 31 = 3.297$

$\bar{x} = \Sigma x / n = 144.9 / 31 = 4.674$

$$b_1 = \frac{n(\Sigma xy) - (\Sigma x)(\Sigma y)}{n(\Sigma x^2) - (\Sigma x)^2} = \frac{31 \cdot 533.78 - 144.9 \cdot 102.2}{31 \cdot 866.35 - (144.9)^2} = 0.297$$

$b_0 = \bar{y} - b_1\bar{x} = 3.297 - 0.297 \cdot 4.674 = 1.910$

The regression equation for the data set is $\hat{y} = b_0 + b_1 x = 1.910 + 0.297x$.

In Exercise 15 of Section 9-2, we did not reject that $\rho = 0$, so the best predicted value for the pain level after the 1 month of Duragesic is its mean, which is 3.297. The equation of the line corresponding to a drug with absolutely no effect would have a slope of 0 and so a y-intercept equal to $\bar{y}$, so its equation would be $\hat{y} = b_0 + b_1 x = \bar{y} + 0x = 3.297$.

17. Cholesterol and Weight

We need to calculate the required sums and means, and then we find the regression equation for the data set.

$\Sigma x = 9635, \Sigma y = 5848.8, \Sigma x^2 = 3669819, \Sigma xy = 1543475.3$

$\bar{y} = \Sigma y / n = 5848.8 / 40 = 146.22$

$\bar{x} = \Sigma x / n = 9635 / 40 = 240.875$

$$b_1 = \frac{n(\Sigma xy) - (\Sigma x)(\Sigma y)}{n(\Sigma x^2) - (\Sigma x)^2} = \frac{40 \cdot 1543475.3 - 9635 \cdot 5848.8}{40 \cdot 3669819 - (9635)^2} = 0.100$$

$b_0 = \bar{y} - b_1\bar{x} = 146.22 - 0.100 \cdot 240.875 = 122.178$

The regression equation for the data set is $\hat{y} = b_0 + b_1 x = 122.178 + 0.100x$.

In Exercise 17 of Section 9-2, we rejected $\rho = 0$, so the best predicted value for the weight of a female with cholesterol of 200 is $\hat{y} = 21.893 + 0.016x = 122.178 + 0.100 \cdot 200 = 142.140$ lbs.

19. Heights of Mothers and Daughters

We need to calculate the required sums and means, and then we find the regression equation for the data set.

$\Sigma x = 1275.0, \Sigma y = 1275.6, \Sigma x^2 = 81515.0, \Sigma xy = 81491.6$

$\bar{y} = \Sigma y / n = 1275.6 / 20 = 63.78$

$\bar{x} = \Sigma x / n = 1275.0 / 20 = 63.75$

$$b_1 = \frac{n(\Sigma xy) - (\Sigma x)(\Sigma y)}{n(\Sigma x^2) - (\Sigma x)^2} = \frac{20 \cdot 81491.6 - 1275.0 \cdot 1275.6}{20 \cdot 81515.0 - (1275.0)^2} = 0.736$$

$b_0 = \bar{y} - b_1\bar{x} = 63.78 - 0.736 \cdot 63.75 = 16.844$

The regression equation for the data set is $\hat{y} = b_0 + b_1 x = 16.844 + 0.736x$.

In Exercise 16 of Section 9-2, we rejected $\rho = 0$, so the best predicted value for the height of the daughter of a mother who is 64 inches tall is $\hat{y} = 16.844 + 0.736x = 16.844 + 0.736 \cdot 64 = 63.964$ in.

21. Iris Petal Lengths and Widths

We need to calculate the required sums and means, and then we find the regression equation for the data set.

$\Sigma x = 213, \Sigma y = 66.3, \Sigma x^2 = 918.2, \Sigma xy = 286.02$

$\bar{y} = \Sigma y / n = 66.3 / 50 = 1.326$

$\bar{x} = \Sigma x / n = 213 / 50 = 4.26$

$$b_1 = \frac{n(\Sigma xy) - (\Sigma x)(\Sigma y)}{n(\Sigma x^2) - (\Sigma x)^2} = \frac{50 \cdot 286.02 - 213 \cdot 66.3}{50 \cdot 918.2 - (213)^2} = 0.331$$

$b_0 = \bar{y} - b_1\bar{x} = 1.326 - 0.331 \cdot 4.26 = -0.084$

The regression equation for the data set is $\hat{y} = b_0 + b_1 x = -0.084 + 0.331x$.

In Exercise 21 of Section 9-2, we rejected $\rho = 0$, so the best predicted value for the petal width of a Versicolor class iris with a petal length of 1.5 mm is $\hat{y} = -0.084 + 0.331x = -0.084 + 0.331 \cdot 1.5 = 0.412$ mm. No, the result is not roughly the same as obtained in Exercise 20. It does seem that when predicting petal length of an iris, it is important to consider the class of the iris.

23. Identifying Outliers and Influential Points

Below are three graphs. The first is simply the scatterplot of the data including the ninth bear, Slim. The second is the scatterplot with the regression line for the data including the ninth bear, Slim. The final graph is the scatterplot with the regression line without the ninth bear. Looking at the first graph, we see that the ninth bear's point is apart from the majority of the data and so could be considered an outlier. In the second and third graphs, we notice that the regression line is quite different, and so the ninth bear is an influential point.

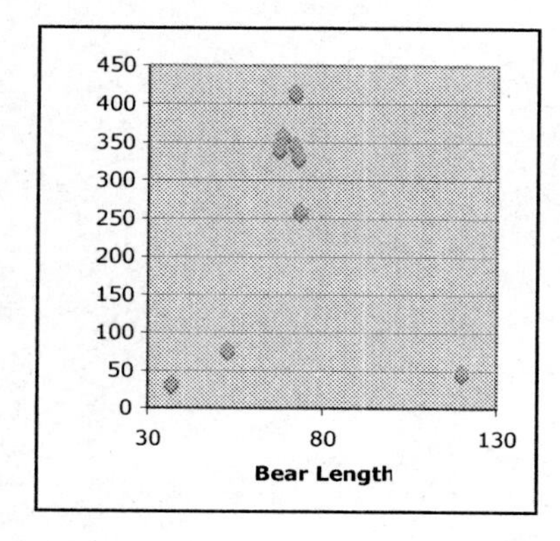

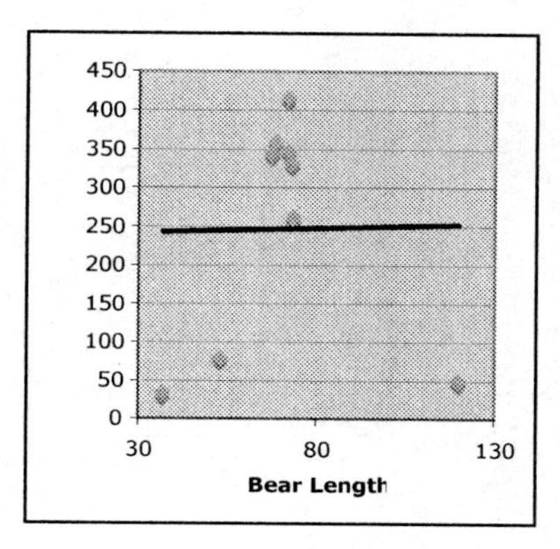

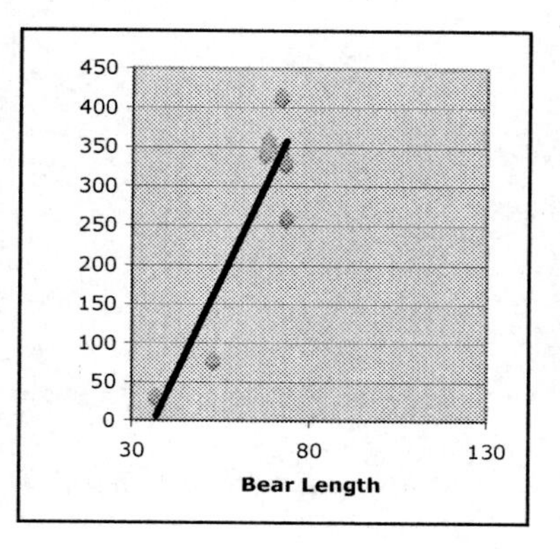

25. Using Logarithms to Transform Data

We need to calculate the required sums and means, and then we find the regression equation for the four data points.

$\Sigma x = 18.7, \Sigma y = 308.7, \Sigma x^2 = 127.89, \Sigma xy = 2543.35, \Sigma y^2 = 53927.69$

$\bar{y} = \Sigma y / n = 308.7/4 = 77.175$

$\bar{x} = \Sigma x / n = 18.7/4 = 4.675$

$$b_1 = \frac{n(\Sigma xy) - (\Sigma x)(\Sigma y)}{n(\Sigma x^2) - (\Sigma x)^2} = \frac{4 \cdot 2543.35 - 18.7 \cdot 308.7}{4 \cdot 127.89 - (18.7)^2} = 27.187$$

$b_0 = \bar{y} - b_1\bar{x} = 77.175 - 27.187 \cdot 4.675 = -49.923$

The regression equation for the data set is $\hat{y} = b_0 + b_1 x = -49.923 + 27.187x$

The correlation coefficient for the data set is

$$r = \frac{n\Sigma xy - (\Sigma x)(\Sigma y)}{\sqrt{n(\Sigma x^2) - (\Sigma x)^2}\sqrt{n(\Sigma y^2) - (\Sigma y)^2}} = \frac{= 4 \cdot 2543.35 - 18.7 \cdot 308.7}{\sqrt{4 \cdot 127.89 - (18.7)^2}\sqrt{4 \cdot 53927.69 - (308.7)^2}} = 0.997$$

We now do the same for the natural logarithm of the x values.

$\Sigma(\ln x) = 5.347, \Sigma y = 308.7, \Sigma(\ln x)^2 = 8.681, \Sigma[(\ln x)y] = 619.594, \Sigma y^2 = 53927.69$

$\bar{y} = \Sigma y / n = 308.7/4 = 77.175$

$\overline{(\ln x)} = \Sigma(\ln x)/n = 5.347/4 = 1.337$

$$b_1 = \frac{n\Sigma[(\ln x)y] - (\Sigma(\ln x))(\Sigma y)}{n(\Sigma(\ln x)^2) - (\Sigma(\ln x))^2} = \frac{4 \cdot 619.594 - 5.347 \cdot 308.7}{4 \cdot 8.681 - (5.347)^2} = 134.938$$

$b_0 = \bar{y} - b_1\overline{\ln x} = 77.175 - 134.938 \cdot 1.337 = -103.208$

The regression equation for the data set is $\hat{y} = b_0 + b_1 \ln x = -103.208 + 134.938 \ln x$

The correlation coefficient for the data set is

$$r = \frac{n\Sigma[(\ln x)y] - (\Sigma(\ln x))(\Sigma y)}{\sqrt{n(\Sigma(\ln x)^2) - (\Sigma(\ln x))^2}\sqrt{n(\Sigma y^2) - (\Sigma y)^2}} = \frac{= 4 \cdot 619.594 - 5.347 \cdot 308.7}{\sqrt{4 \cdot 8.681 - (5.347)^2}\sqrt{4 \cdot 53927.69 - (308.7)^2}} = 0.963$$

It appears the first equation is the better fit, based on the correlation coefficient.

27. Residual Plot
We use the regression equation given on page 452 to generate the table of the residuals for the regression equation.

x	53	67.5	72	72	73.5	68.5	73	37
residual	-79.98	43.95	72.48	4.48	-96.01	50.29	-21.18	28.58

The scatterplot is below.

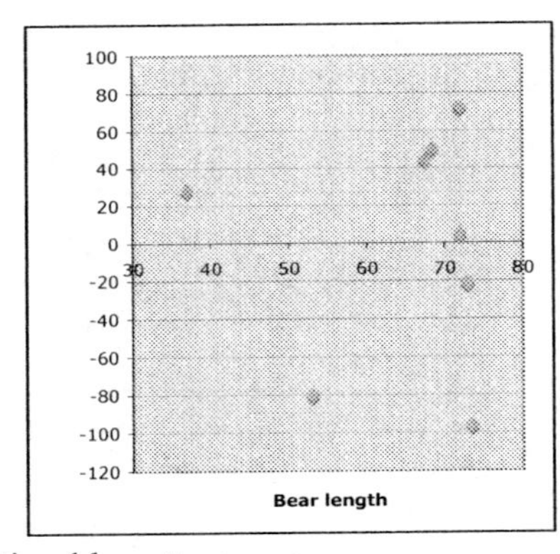

There does not seem to be any noticeable pattern.

9-4 Variation and Prediction Intervals

In exercises 1 & 3, use the value of the linear correlation coefficient r to find the coefficient if determination and the percentage of the total variation that can be explained by the linear association between the two variables.

1. The coefficient of determination is the square of the linear correlation coefficient. Using the value r = 0.8, we find that $r^2 = (0.8)^2 = 0.64$. This means that 64% of the total variation in y can be explained by the linear association between x and y.

3. The coefficient of determination is the square of the linear correlation coefficient. Using the value r = -0.503, we find that $r^2 = (-0.503)^2 = 0.253$. This means that 25.3% of the total variation in y can be explained by the linear association between x and y.

In Exercises 5 & 7, refer to the Minitab display that was obtained by using the paired data consisting of neck size (in inches) and weight (in pounds) for the sample of bears listed in Data Set 6 in Appendix B. Along with the paired sample data, Minitab was also given a neck size of 25.0 in. to be used for predicting the weight.

5. Testing for Correlation
The absolute value of the linear correlation coefficient is the square root of the coefficient of determination, included in the output as R-Sq. Using this value, 87.2%, we find that $|r| = \sqrt{0.872} = 0.934$. Table A-6 does not have an entry for n = 54, so we use n = 50. The critical value is 0.279. Since the absolute value of r is greater than 0.279, we conclude that there is significant linear correlation between bear neck sizes and weight.

7. Predicting Weight
The best prediction for the weight of a bear with a neck size of 25.0 in. is 272.53 lbs., as seen in the output as the Fit for the New Obs.

Exercises 9 & 11 find the (a) explained variation, (b) unexplained variation, (c) total variation, (d) coefficient of determination, and (e) standard error estimate s_e. In each case, there is a significant linear correlation so that it is reasonable to use the regression equation when making predictions.

For Exercises 9-12, we use the following formulas:

Total Variation = $\Sigma(y-\bar{y})^2$

Explained Variation = $\Sigma(\hat{y}-\bar{y})^2$

Unexplained Variation = $\Sigma(y-\hat{y})^2$

Coefficient of Determination = $\dfrac{\text{explained variation}}{\text{total variation}}$

Standard Error of Estimate = $s_e = \sqrt{\dfrac{\Sigma(y-\hat{y})^2}{n-2}}$

Additionally, a table that organizes the data and sums used in the formulas is included.

9. Supermodel Heights and Weights

The linear regression formula used for this problem is $\hat{y} = b_0 + b_1x = -151.700 + 3.883x$, as found in Exercise 9 of Section 9-3 Below is a table of values needed to determine the variations.

x	y	$\hat{y}$	$\bar{y}$	$\hat{y}-\bar{y}$	$(\hat{y}-\bar{y})^2$	$y-\hat{y}$	$(y-\hat{y})^2$	$y-\bar{y}$	$(y-\bar{y})^2$
71.00	125.00	124.02	121.00	3.02	9.12	0.98	0.96	4.00	16.00
70.50	119.00	122.08	121.00	1.08	1.16	-3.08	9.48	-2.00	4.00
71.00	128.00	124.02	121.00	3.02	9.12	3.98	15.84	7.00	49.00
72.00	128.00	127.90	121.00	6.90	47.66	0.10	0.01	7.00	49.00
70.00	119.00	120.14	121.00	-0.86	0.74	-1.14	1.29	-2.00	4.00
70.00	127.00	120.14	121.00	-0.86	0.74	6.86	47.10	6.00	36.00
66.50	105.00	106.55	121.00	-14.45	208.94	-1.55	2.39	-16.00	256.00
70.00	123.00	120.14	121.00	-0.86	0.74	2.86	8.20	2.00	4.00
71.00	115.00	124.02	121.00	3.02	9.12	-9.02	81.37	-6.00	36.00
632.00	1089.00	1089.00	1089.00	0.00	287.37	0.00	166.63	0.00	454.00

a. Explained variation = $\Sigma(\hat{y}-\bar{y})^2 = 287.37$

b. Unexplained variation = $\Sigma(y-\hat{y})^2 = 166.63$

c. Total variation = $\Sigma(y-\bar{y})^2 = 454.00$

d. Coefficient of determination = $r^2 = \dfrac{\text{explained variation}}{\text{total variation}} = \dfrac{287.37}{454.00} = 0.633$

e. Standard error of estimate = $s_e = \sqrt{\dfrac{\Sigma(y-\hat{y})^2}{n-2}} = \sqrt{\dfrac{166.63}{9-2}} = 4.879$

11. Tree Circumference and Height

The linear regression formula used for this problem is $\hat{y} = b_0 + b_1x = 22.463 + 5.341x$, as found in Exercise 14 of Section 9-3 Below is a table of values needed to determine the variations.

x	y	$\hat{y}$	$\bar{y}$	$\hat{y}-\bar{y}$	$(\hat{y}-\bar{y})^2$	$y-\hat{y}$	$(y-\hat{y})^2$	$y-\bar{y}$	$(y-\bar{y})^2$
1.8	21.0	32.08	47.79	-15.72	247.02	-11.08	122.68	-26.79	717.86
1.9	33.5	32.61	47.79	-15.18	230.51	0.89	0.79	-14.29	204.29
1.8	24.6	32.08	47.79	-15.72	247.02	-7.48	55.89	-23.19	537.91
2.4	40.7	35.28	47.79	-12.51	156.56	5.42	29.37	-7.09	50.31
5.1	73.2	49.70	47.79	1.91	3.64	23.50	552.24	25.41	645.52
3.1	24.9	39.02	47.79	-8.77	76.98	-14.12	199.34	-22.89	524.08
5.5	40.4	51.84	47.79	4.04	16.35	-11.44	130.79	-7.39	54.65
5.1	45.3	49.70	47.79	1.91	3.64	-4.40	19.36	-2.49	6.21
8.3	53.5	66.79	47.79	19.00	360.90	-13.29	176.63	5.71	32.57
13.7	93.8	95.63	47.79	47.84	2288.37	-1.83	3.35	46.01	2116.66
5.3	64.0	50.77	47.79	2.98	8.85	13.23	175.08	16.21	262.67
4.9	62.7	48.63	47.79	0.84	0.70	14.07	197.91	14.91	222.22
3.7	47.2	42.22	47.79	-5.57	31.02	4.98	24.77	-0.59	0.35
3.8	44.3	42.76	47.79	-5.04	25.36	1.54	2.38	-3.49	12.20
66.4	669.1	669.10	669.10	0.00	3696.93	0.00	1690.58	0.00	5387.51

a. Explained variation = $\Sigma(\hat{y}-\bar{y})^2 = 3696.93$

b. Unexplained variation = $\Sigma(y-\hat{y})^2 = 1690.58$

c. Total variation = $\Sigma(y-\bar{y})^2 = 5387.51$

d. Coefficient of determination = $r^2 = \dfrac{\text{explained variation}}{\text{total variation}} = \dfrac{3696.93}{5387.51} = 0.686$

e. Standard error of estimate = $s_e = \sqrt{\dfrac{\Sigma(y-\hat{y})^2}{n-2}} = \sqrt{\dfrac{1690.58}{14-2}} = 11.869$

13. Effect of Variation on Prediction Interval

a. The predicted weight for a supermodel who is 69 in. tall is found by substituting x = 69 in the regression equation from Exercise 9, $\hat{y} = -151.700 + 3.883x = -151.700 + 3.883 \cdot 69 = 116.254$.

b. To find the margin of error for the prediction interval, we first calculate the following.

$\Sigma x = 632, \Sigma x^2 = 44399.50, \bar{x} = \Sigma x / n = 632/9 = 70.22$

For a 95% prediction interval estimate, we find $t_{\alpha/2}$ with df = n – 2= 7, which is $t_{\alpha/2} = t_{.025} = 2.365$. From Exercise 9, the standard error of estimate is $s_e = 4.879$. Using these values, we find the margin of error.

$$E = t_{\alpha/2} s_e \sqrt{1 + \frac{1}{n} + \frac{n(x_0-\bar{x})^2}{n(\Sigma x^2)-(\Sigma x)^2}} = 2.365 \cdot 4.879 \sqrt{1 + \frac{1}{9} + \frac{9(69-70.22)^2}{9 \cdot 44399.50 - (632)^2}} = 12.585$$

So the 95% prediction interval estimate for the weight of a supermodel who is 69 inches tall is

$\hat{y} - E < y < \hat{y} + E$

$116.254 - 12.585 < y < 116.254 + 12.585$

$103.669 < y < 128.838$

15. Effect of Variation on Prediction Interval

a. The predicted height of a tree with a circumference of 4.0 ft. is found by substituting x = 4.0 in the regression equation from Exercise 11, $\hat{y} = 22.463 + 5.341x = 22.463 + 5.341 \cdot 4.0 = 43.826$.

b. To find the margin of error for the prediction interval, we first calculate the following.

$\Sigma x = 66.4, \Sigma x^2 = 444.54, \bar{x} = \Sigma x / n = 66.4/14 = 4.743$

For a 99% prediction interval estimate, we find $t_{\alpha/2}$ with df = n – 2= 12, which is $t_{\alpha/2} = t_{.005} = 3.055$. From Exercise 11, the standard error of estimate is $s_e = 11.869$. Using these values, we find the margin of error.

$$E = t_{\alpha/2} s_e \sqrt{1 + \frac{1}{n} + \frac{n(x_0-\bar{x})^2}{n(\Sigma x^2)-(\Sigma x)^2}} = 3.055 \cdot 11.869 \sqrt{1 + \frac{1}{14} + \frac{14(4.0-4.743)^2}{14 \cdot 444.54 - (66.4)^2}} = 37.608$$

So the 95% prediction interval estimate for the height of a tree with circumference 4.0 ft. is

$\hat{y} - E < y < \hat{y} + E$

$43.826 - 37.608 < y < 43.826 + 37.608$

$6.218 < y < 81.434$

In Exercises 17 & 19, refer to the Table 9-1 sample data. Let x represent the length of a bear (in inches) and y represent the weight of a bear (in pounds). Use the given length of a bear and the given confidence level to construct a prediction interval estimate of the weight. (See the example in this section.)

From the example in this chapter, we have the regression equation, $\hat{y} = -352 + 9.66x$, the standard error of estimate, $s_e = 66.5994$, and the following sums, $\Sigma x = 516.5, \Sigma x^2 = 34525.75, \bar{x} = \Sigma x / n = 516.5/8 = 64.5625$. Also, the degrees of freedom for each exercise is df = n-2 = 6. We use this information in Exercises 17-20.

17. The predicted weight of a bear that is 50.0 inches long is found by substituting x = 50.0 in the regression equation, $\hat{y} = -352 + 9.66x = -352 + 9.66 \cdot 50.0 = 131.000$.For a 95% prediction interval estimate, we find $t_{\alpha/2}$, which is $t_{\alpha/2} = t_{.025} = 2.447$. Using these values, we find the margin of error.

$$E = t_{\alpha/2} s_e \sqrt{1 + \frac{1}{n} + \frac{n(x_0-\bar{x})^2}{n(\Sigma x^2)-(\Sigma x)^2}} = 2.447 \cdot 66.5994 \sqrt{1 + \frac{1}{8} + \frac{8(50.0-64.5625)^2}{8 \cdot 34525.75 - (516.5)^2}} = 186.158$$

So the 95% prediction interval estimate for the weight of a bear that is 50.0 inches long is

$\hat{y} - E < y < \hat{y} + E$

$131.000 - 186.158 < y < 131.000 + 186.158$

$-55.158 < y < 317.158$

19. The predicted weight of a bear that is 49.7 inches long is found by substituting $x = 49.7$ in the regression equation, $\hat{y} = -352 + 9.66x = -352 + 9.66 \cdot 49.7 = 128.102$. For a 90% prediction interval estimate, we find $t_{\alpha/2}$, which is $t_{\alpha/2} = t_{.05} = 1.943$. Using these values, we find the margin of error.

$$E = t_{\alpha/2} s_e \sqrt{1 + \frac{1}{n} + \frac{n(x_0 - \bar{x})^2}{n(\Sigma x^2) - (\Sigma x)^2}} = 1.943 \cdot 66.5994 \sqrt{1 + \frac{1}{8} + \frac{8(49.7 - 64.5625)^2}{8 \cdot 34525.75 - (516.5)^2}} = 148.239$$

So the 90% prediction interval estimate for the weight of a bear that is 49.7 inches long is

$\hat{y} - E < y < \hat{y} + E$

$128.102 - 148.239 < y < 128.102 + 148.239$

$-20.137 < y < 276.341$

21. Confidence Intervals for β_0 and β_1.

In order to evaluate the confidence intervals for β_0 and β_1, we need the following, as found in the examples in the text:

$\Sigma x = 516.5$, $\Sigma x^2 = 34525.75$, $\bar{x} = 64.5625$, $s_e = 66.5994$, $b_0 = -352$, and $b_1 = 9.66$

For 95% confidence intervals with $df = 6$, we have $t_{\alpha/2} = t_{.025} = 2.447$.

We start with the confidence interval for the y- intercept, finding the margin of error first.

$$E = t_{\alpha/2} s_e \sqrt{\frac{1}{n} + \frac{\bar{x}^2}{\Sigma x^2 - \frac{(\Sigma x)^2}{n}}} = 2.447 \cdot 66.5994 \sqrt{\frac{1}{8} + \frac{64.5625^2}{34525.75 - \frac{(516.5)^2}{8}}} = 472.305$$

We now find the confidence interval for the y-intercept, β_0

$b_0 - E < \beta_0 < b_0 + E$

$-352 - 472.305 < \beta_0 < -352 + 472.305$

$-824.305 < \beta_0 < 120.305$

We now find the confidence interval for the slope, finding the margin of error first.

$$E = t_{\alpha/2} \frac{s_e}{\sqrt{\Sigma x^2 - \frac{(\Sigma x)^2}{n}}} = 2.447 \frac{66.5994}{\sqrt{34525.75 - \frac{(516.5)^2}{8}}} = 7.189$$

We now find the confidence interval for the slope, β_1

$b_1 - E < \beta_1 < b_1 + E$

$9.66 - 7.189 < \beta_1 < 9.66 + 7.189$

$2.471 < \beta_1 < 16.849$

9-5 Multiple Regression

In exercises 1 & 3, refer to the SPSS display that follows and answer the given questions or identify the indicated terms. The SPSS display is based on the sample of 54 bears listed in Data Set 6 iin Appendix B.

1. Bear Measurements

The multiple regression equation is

$\hat{y} = -271.711 - 0.870x_1 + 0.554x_2 + 12.153x_3$, where $\hat{y}$ is the predicted weight of the bear,

x_1 is the length of the head, x_2 is the length of the body, and x_3 is the distance around the chest.

The coefficients for each variable are found in the Coefficients table of the output, in the column labeled B, under Unstandardized Coefficients.

3. Bear Measurements
Yes, it is usable, because the P-value is 0.000. Also, the Adjusted R^2 is quite large, indicating that the regression is very useful.

In Exercises 5 & 7, refer to the accompanying table, which was obtained by using the data for males in Data Set 1 in Appendix B. The dependent variable is weight (in pounds), and the independent variables are HT (height in inches), WAIST (waist circumference in cm), and CHOL (cholesterol in mg).

5. The best single predictor is WAIST, since the coefficient of determination is 0.790, much larger than the other variables' ones.

7. The best regression equation is $\hat{y} = -206 + 2.66\text{HT} + 2.15\text{WAIST}$, since it has the same adjusted R^2 as using all three variables, and it is significantly higher than the adjusted R^2 using the single variable WAIST.

In Exercises 9 & 11, the results were obtained using JMP.

9. Heights of Parents and Children
- **a.** The regression equation for a child's height is $\hat{y} = 21.560 + 0.690x$, where x is the mother's height.
- **b.** The regression equation for a child's height is $\hat{y} = 45.721 + 0.293x$, where x is the father's height.
- **c.** The regression equation for a child's height is $\hat{y} = 9.804 + 0.658x_1 + 0.200x_2$, where x_1 is the mother's height and x_2 is the father's height.
- **d.** The best equation for predicting height is $\hat{y} = 21.560 + 0.690x$, where x is the mother's height. This is because the adjusted R^2 for using just the mother's height is 0.347 and for using both parent's heights it is 0.366. This is not enough of an increase to justify using both variables.
- **e.** It does not seem to be a good equation for predicting a child's height, because the adjusted R^2 is not very large.

11. Using a Dummy Variable
The regression equation found was $\hat{y} = 3.062 + 2.905x_1 + 82.379x_2$, where x_1 is age of the bear and x_2 is the sex of the bear.
- **a.** For a female bear that is 20 years of age, the predicted weight is
$\hat{y} = 3.062 + 2.905x_1 + 82.379x_2 = 3.062 + 2.905 \cdot 20 + 82.379 \cdot 0 = 61.162$.
- **b.** For a male bear that is 20 years of age, the predicted weight is
$\hat{y} = 3.062 + 2.905x_1 + 82.379x_2 = 3.062 + 2.905 \cdot 20 + 82.379 \cdot 1 = 143.541$.

Review Exercises

1. Crickets and Temperature
- **a.** The scatterplot for the data is as shown. The scatterplot appears to approximate a straight-line pattern, and so a hypothesis test is indicated. We now calculate the needed sums and the linear correlation coefficient.

95
90
85
80
75
70
65
700 900 1100 1300
Chirps in one minute

$\Sigma x = 8130, \Sigma y = 646, \Sigma x^2 = 8391204, \Sigma xy = 663245.4, \Sigma y^2 = 52626.6$

$$r = \frac{n\Sigma xy - (\Sigma x)(\Sigma y)}{\sqrt{n(\Sigma x^2) - (\Sigma x)^2}\sqrt{n(\Sigma y^2) - (\Sigma y)^2}}$$

$$r = \frac{8 \cdot 663245.4 - 8130 \cdot 646}{\sqrt{8 \cdot 8391204 - (8130)^2}\sqrt{8 \cdot 52626.6 - (646)^2}}$$

$r = 0.874$

The critical value from Table A-6 for n = 8 and with α = 0.05 is 0.707. Since r = 0.874, there is a significant linear correlation between chirps per minute and temperature.

The formal hypothesis test follows.

H_0: $\rho = 0$
H_1: $\rho \neq 0$
The test statistic is

$$t = \frac{r}{\sqrt{\frac{1-r^2}{n-2}}} = \frac{0.874}{\sqrt{\frac{1-(0.874)^2}{8-2}}} = 4.399$$

The critical values are found in Table A-3 using df = n – 2, and are $\pm t_{\alpha/2} = \pm t_{.025} = \pm 2.447$.
Since the test statistic is greater than 2.477, we reject the null hypothesis. There is sufficient sample evidence to conclude that there is significant linear correlation between chirps per minute and temperature.

b. We start by finding the means and then find the regression equation.

$\bar{y} = \Sigma y / n = 646/8 = 80.75$

$\bar{x} = \Sigma x / n = 8130/8 = 1016.25$

$$b_1 = \frac{n(\Sigma xy) - (\Sigma x)(\Sigma y)}{n(\Sigma x^2) - (\Sigma x)^2} = \frac{8 \cdot 663245.4 - 8130 \cdot 646}{8 \cdot 8391204 - (8130)^2} = 0.0523$$

$b_0 = \bar{y} - b_1\bar{x} = 80.75 - 0.0523 \cdot 1016.25 = 27.628$

The regression equation for the temperature is $\hat{y} = b_0 + b_1 x = 27.628 + 0.0523x$, where x is the number of chirps for a cricket in one minute.

c. The best predicted value for the temperature when a cricket chirps 1234 times in one minute is $\hat{y} = b_0 + b_1 x = 27.628 + 0.0523x = 27.628 + 0.0523 \cdot 1234 = 92.132$.

d. The coefficient of determination is the square of the linear correlation coefficient. Using the value r = 0.874, we find that $r^2 = (0.874)^2 = 0.763$. This means that 76.3% of the total variation in temperature can be explained by the linear association between the number of chirps of a cricket in one minute and the temperature.

e. It does not show causation. The regression shows association, and may indicate that there is some other variable that influences both.

In Exercises 3 & 5, use the sample data in the accompanying table. The data were collected from Iowa during a 10-year period. The amounts of precipitation are the annual totals (in inches). The average temperatures are annual averages (in degrees Fahrenheit). The corn values are the amounts of corn produced (in millions of bushels). The values of acres harvested are in thousands of acres.

3. **a.** We calculate the needed sums and the linear correlation coefficient.

$\Sigma x = 339.1, \Sigma y = 14059, \Sigma x^2 = 11896.05, \Sigma xy = 483531, \Sigma y^2 = 20770345$

$$r = \frac{n\Sigma xy - (\Sigma x)(\Sigma y)}{\sqrt{n(\Sigma x^2) - (\Sigma x)^2}\sqrt{n(\Sigma y^2) - (\Sigma y)^2}}$$

$$r = \frac{10 \cdot 483531 - 339.1 \cdot 14059}{\sqrt{10 \cdot 11896.05 - (339.1)^2}\sqrt{10 \cdot 20770345 - (14059)^2}}$$

$r = 0.340$

The critical value from Table A-6 for n = 10 and with $\alpha = 0.05$ is 0.632. Since r = 0.340, there is not a significant linear correlation between annual precipitation amounts and corn production amounts.
The formal hypothesis test follows.
H_0: $\rho = 0$
H_1: $\rho \neq 0$
The test statistic is

$$t = \frac{r}{\sqrt{\frac{1-r^2}{n-2}}} = \frac{0.340}{\sqrt{\frac{1-(0.340)^2}{10-2}}} = 1.022$$

The critical values are found in Table A-3 using df = n – 2, and are $\pm t_{\alpha/2} = \pm t_{.025} = \pm 2.306$.
Since the absolute value of the test statistic is less than 2.306, we fail to reject the null hypothesis. There is not sufficient sample evidence to conclude that there is significant linear correlation between annual precipitation amounts and corn production amounts.

b. We start by finding the means and then find the regression equation.

$\bar{y} = \Sigma y / n = 14059/10 = 1405.9$

$\bar{x} = \Sigma x / n = 339.1/10 = 33.91$

$$b_1 = \frac{n(\Sigma xy) - (\Sigma x)(\Sigma y)}{n(\Sigma x^2) - (\Sigma x)^2} = \frac{10 \cdot 483531 - 339.1 \cdot 14059}{10 \cdot 11896.05 - (339.1)^2} = 17.097$$

$b_0 = \bar{y} - b_1\bar{x} = 1405.9 - 17.097 \cdot 33.91 = 826.148$

The regression equation for the corn production amount is $\hat{y} = b_0 + b_1 x = 826.148 + 17.097x$, where x is the precipitation for the year.

c. The best predicted value for the amount of corn is the value $\bar{y} = 1405.9$, because there is not significant linear correlation between annual precipitation amounts and corn production amounts.

5. **a.** We calculate the needed sums and the linear correlation coefficient.

$\Sigma x = 119550, \Sigma y = 14059, \Sigma x^2 = 1454397500, \Sigma xy = 172561550, \Sigma y^2 = 20770345$

$$r = \frac{n\Sigma xy - (\Sigma x)(\Sigma y)}{\sqrt{n(\Sigma x^2) - (\Sigma x)^2}\sqrt{n(\Sigma y^2) - (\Sigma y)^2}}$$

$$r = \frac{10 \cdot 172561550 - 119550 \cdot 14059}{\sqrt{10 \cdot 1454397500 - (119550)^2}\sqrt{10 \cdot 20770345 - (14059)^2}}$$

$r = 0.892$

The critical value from Table A-6 for n = 10 and with α = 0.05 is 0.632. Since r = 0.892, there is a significant linear correlation between acres harvested and corn production amounts.

The formal hypothesis test follows.

H_0: $\rho = 0$

H_1: $\rho \neq 0$

The test statistic is

$$t = \frac{r}{\sqrt{\frac{1-r^2}{n-2}}} = \frac{0.892}{\sqrt{\frac{1-(0.892)^2}{10-2}}} = 5.580$$

The critical values are found in Table A-3 using df = n – 2, and are $\pm t_{\alpha/2} = \pm t_{.025} = \pm 2.306$.

Since the absolute value of the test statistic is greater than 2.306, we reject the null hypothesis. There is sufficient sample evidence to conclude that there is significant linear correlation between acres harvested and corn production amounts.

b. We start by finding the means and then find the regression equation.

$\bar{y} = \Sigma y / n = 14059/10 = 1405.9$

$\bar{x} = \Sigma x / n = 119550/10 = 11955$

$$b_1 = \frac{n(\Sigma xy) - (\Sigma x)(\Sigma y)}{n(\Sigma x^2) - (\Sigma x)^2} = \frac{10 \cdot 172561550 - 119550 \cdot 14059}{10 \cdot 1454397500 - (119550)^2} = 0.178$$

$b_0 = \bar{y} - b_1\bar{x} = 1405.9 - (0.178) \cdot 11955 = -724.300$

The regression equation for the corn production amount is $\hat{y} = b_0 + b_1 x = -724.300 + 0.178x$, where x is the acres harvested.

c. The best predicted value for the amount of corn

$\hat{y} = -724.300 + 0.178x = -724.300 + 0.178 \cdot 13300 = 1645.559$,

because there is significant linear correlation between acres harvested and corn production amounts.

7. **Exercise and Stress**

a. We calculate the needed sums and the linear correlation coefficient.

$\Sigma x = 632.333, \Sigma y = 703.667, \Sigma x^2 = 67820.778, \Sigma xy = 75311, \Sigma y^2 = 83760.556$

$$r = \frac{n\Sigma xy - (\Sigma x)(\Sigma y)}{\sqrt{n(\Sigma x^2) - (\Sigma x)^2}\sqrt{n(\Sigma y^2) - (\Sigma y)^2}}$$

$$r = \frac{6 \cdot 75311 - 632.333 \cdot 703.667}{\sqrt{6 \cdot 67820.778 - (632.333)^2}\sqrt{6 \cdot 83760.556 - (703.667)^2}}$$

$r = 0.954$

The critical value from Table A-6 for n = 6 and with α = 0.05 is 0. 811. Since r = 0.954, there is a significant linear correlation between systolic pressure pre-exercise with no stress and pre-exercise with math stress.

The formal hypothesis test follows.
$H_0: \rho = 0$
$H_1: \rho \neq 0$
The test statistic is

$$t = \frac{r}{\sqrt{\frac{1-r^2}{n-2}}} = \frac{0.954}{\sqrt{\frac{1-(0.954)^2}{6-2}}} = 6.379$$

The critical values are found in Table A-3 using df = n – 2, and are $\pm t_{\alpha/2} = \pm t_{.025} = \pm 2.776$.
Since the absolute value of the test statistic is greater than 2.776, we reject the null hypothesis. There is sufficient sample evidence to conclude that there is significant linear correlation between systolic pressure pre-exercise with no stress and pre-exercise with math stress.

b. We start by finding the means and then find the regression equation.
$\bar{y} = \Sigma y / n = 703.667/6 = 117.278$
$\bar{x} = \Sigma x / n = 632.333/6 = 105.389$

$$b_1 = \frac{n(\Sigma xy) - (\Sigma x)(\Sigma y)}{n(\Sigma x^2) - (\Sigma x)^2} = \frac{6 \cdot 75311 - 632.333 \cdot 703.667}{6 \cdot 67820.778 - (632.333)^2} = 0.977$$

$b_0 = \bar{y} - b_1\bar{x} = 117.278 - 0.977 \cdot 105.389 = 14.347$
The regression equation for the temperature is $\hat{y} = b_0 + b_1 x = 14.347 + 0.977x$, where x is the systolic pressure pre-exercise with no stress.

c. We cannot say that the stress from the math test has an effect of systolic blood pressure, but we can say that they are associated.

Cumulative Review Exercises

1. Effects of Heredity and Environment on IQ

a. $\Sigma x = 1189$ $\quad \Sigma x^2 = 118599$
The mean is $\bar{x} = \Sigma x / n = 1189/12 = 99.083$
The standard deviation is $s = \sqrt{\frac{n\Sigma(x^2) - (\Sigma x)^2}{n(n-1)}} = \sqrt{\frac{12 \cdot 118599 - (1189)^2}{12 \cdot 11}} = 8.469$

b. $\Sigma y = 1234$ $\quad \Sigma y^2 = 127724$
The mean is $\bar{y} = \Sigma y / n = 1234/12 = 102.833$
The standard deviation is $s = \sqrt{\frac{n\Sigma(y^2) - (\Sigma y)^2}{n(n-1)}} = \sqrt{\frac{12 \cdot 127724 - (1234)^2}{12 \cdot 11}} = 8.674$

c. There does not appear to be a difference in the means of the two populations. Even though the difference is more than three points, the standard deviations are both larger than 8, and so a difference as relatively small as 3 is not meaningful. However, this is not the best approach. A formal hypothesis test may be a better approach. A formal hypothesis test is presented below.
We will assume that the sample is a random sample. We will also assume that the matched pairs are from a normally distributed population. The claim is that the mean IQ score for twins raised apart are different, so this is a two-tailed test. The sample size is n = 12 making the degrees of freedom df = 11. The significance level is 0.05.
We now find the mean and standard deviation of the paired differences.
$\Sigma d = -45$ $\quad \Sigma d^2 = 651$
The mean is $\bar{d} = \Sigma d / n = -45/12 = -3.75$
The standard deviation is $s = \sqrt{\frac{n\Sigma(d^2) - (\Sigma d)^2}{n(n-1)}} = \sqrt{\frac{12 \cdot 651 - (-45)^2}{12 \cdot 11}} = 6.621$
We use this information to run the hypothesis test.

H_0: $\mu_d = 0$
H_1: $\mu_d \neq 0$

The test statistic is $t = \dfrac{\overline{d} - \mu_d}{s_d / \sqrt{n}} = \dfrac{-3.75 - 0}{6.621/\sqrt{12}} = -1.962$

In a two-tailed test at the 0.05 significance level with df = 11, the critical values are $\pm t_\alpha = \pm t_{.025} = \pm 2.201$.
Since the test statistic is between -2.201 and 2.201, we fail to reject the null hypothesis.
There is not sufficient sample evidence to support the claim that the mean IQ for twins reared apart are different.

d. We will assume that the sample is a random sample. We will also assume that the sample is from a normally distributed population. The claim is that the mean IQ score for twins raised apart is different than 100, so this is a two-tailed test. The sample size is n = 24 making the degrees of freedom df = 23. The significance level is 0.05.
We now find the mean and standard deviation of the combined IQ scores.
$\Sigma x = 2423$ $\quad \Sigma x^2 = 246323$
The mean is $\overline{x} = \Sigma x / n = 2423/24 = 100.958$

The standard deviation is $s = \sqrt{\dfrac{n\Sigma(x^2) - (\Sigma x)^2}{n(n-1)}} = \sqrt{\dfrac{24 \cdot 246323 - (2423)^2}{24 \cdot 23}} = 8.600$

We use this information to run the hypothesis test.
H_0: $\mu = 100$
H_1: $\mu \neq 100$

The test statistic is $t = \dfrac{\overline{x} - \mu}{s / \sqrt{n}} = \dfrac{100.958 - 100}{8.600/\sqrt{24}} = 0.546$

In a two-tailed test at the 0.05 significance level with df = 23, the critical values are $\pm t_\alpha = \pm t_{.025} = \pm 2.069$.
Since the test statistic is between -2.069 and 2.069, we fail to reject the null hypothesis.
There is not sufficient sample evidence to support the claim that the mean IQ for twins reared apart is different than the mean IQ of 100.

e. We will use the linear correlation coefficient to determine if there is an association between IQs of twins separated at birth. We first calculate the needed sums.
$\Sigma x = 1189, \Sigma y = 1234, \Sigma x^2 = 118599, \Sigma xy = 122836, \Sigma y^2 = 127724$

$$r = \frac{n\Sigma xy - (\Sigma x)(\Sigma y)}{\sqrt{n(\Sigma x^2) - (\Sigma x)^2}\sqrt{n(\Sigma y^2) - (\Sigma y)^2}} = \frac{12 \cdot 122836 - 1189 \cdot 1234}{\sqrt{12 \cdot 118599 - (1189)^2}\sqrt{12 \cdot 127724 - (1234)^2}} = 0.702$$

The linear correlation coefficient is 0.702. The critical value from Table A-6 with $\alpha = 0.05$ and n = 12 is 0.576. There is significant linear correlation between the IQs of twins raised separately. It would seem that environment does not have a statistically significant effect on IQ.

Chapter 10. Multinomial Experiments and Contingency Tables

10-2 Multinomial Experiments: Goodness of Fit

1. Testing Equally Likely Categories

a. The null hypothesis must contain the condition of equality. Therefore, $H_0: p_1 = p_2 = p_3 = p_4$, where p_1, p_2, p_3, p_4 are the probabilities of occurrence of categories 1, 2, 3 and 4, respectively.

b. It is assumed that the probability for occurrence for each category is the same. Since, there are altogether four categories, the probability for each category would be ¼. There are altogether 32 (5 + 6 + 8 + 13) observed frequencies. Therefore, the expected frequency for each of the four categories would be $1/4 * 32 = 8$

c. Calculating the χ^2Test statistic for the four categories

Category	Observed frequency O	Expected frequency E	$O-E$	$(O-E)^2$	$\frac{(O-E)^2}{E}$
1	5	8	-3	9	1.125
2	6	8	-2	4	0.500
3	8	8	0	0	0.000
4	13	8	5	25	3.125

$$\chi^2 = \frac{(O-E)^2}{E} = 1.125 + 0.5 + 0 + 3.125 = 4.750$$

d. $\alpha = 0.05$, $df = 4 - 1 = 3$, critical value (from Table A-4) is $\chi^2 = 7.815$

e. Since the test statistic does not fall within the critical region, we fail to reject the null hypothesis. Therefore, there is not sufficient evidence to support the claim that the probability of at least one category is different from the rest.

3. Testing Fairness of Roulette Wheel

a. $\alpha = 0.10$, $df = 38 - 1 = 37$, critical value (from Table A-4) is $\chi^2 = 51.805$ (approximately) Since $df = 37$ is not in the Table, the value of $df = 40$ at $\alpha = 0.10$ was taken. (From Excel, the actual critical value is found as 48.363)

b. Since the given statistic value of 38.232 falls between the values of 29.051 ($p = 0.90$) and 51.805 ($p = 0.10$), it can be said that the P-value is greater than 0.10 but less than 0.90. $0.10 < P\text{-value} < 0.90$

c. Since the P-value is greater than 0.10, we fail to reject the null hypothesis. Therefore, the claim that the 38 results are equally likely cannot be ruled out.

5. Last Digits of Weights

$H_0: p_0 = p_1 = ... = p_8 = p_9 = 1/10$

H_1 : at least one of the proportions is different from 1/10

It is assumed that the probability for occurrence for each digit is the same. Since, there are altogether ten digits, the probability for each category would be $1/10$. There are altogether 20 + 18 + 19 + 17 + 21 + 22 + 14 + 18 + 20 + 21= 190 observed frequencies. Therefore, the expected frequency for each of the ten digits would be $1/10 * 190 = 19$

Last digit	Observed frequency O	Expected frequency E	$O-E$	$(O-E)^2$	$\frac{(O-E)^2}{E}$
0	20	19	+1	1	0.0526
1	18	19	-1	1	0.0526
2	19	19	0	0	0.0000
3	17	19	-2	4	0.2105
4	21	19	+2	4	0.2105

5	22	19	+3	9	0.4737
6	14	19	-5	25	1.3158
7	18	19	-1	1	0.0526
8	20	19	+1	1	0.0526
9	21	19	+2	4	0.2105

$\chi^2 = \dfrac{(O-E)^2}{E} = 2.631$

$\alpha = 0.05$, $df = 10 - 1 = 9$, critical value (from Table A-4) is $\chi^2 = 16.919$

Since the test statistic (2.635) does not fall within the critical region (above 16.919), we fail to reject the null hypothesis. Therefore, there is not sufficient evidence to support the claim that the probability of at least one digit is different from the rest. The results appear to be as expected from an effective weighing procedure.

7. Do Car Crashes Occur on Different Days with the Same Frequency?

$H_0: p_{Sun} = p_{Mon} = p_{Tues} = p_{Wed} = p_{Thurs} = p_{Fri} = p_{Sat} = 1/7$

H_1 : at least one of the proportions is different from 1/7

It is assumed that the probability for occurrence of accidents is the same on any given day. Since, there are altogether seven days, the probability for each day would be 1/7. There are altogether 31 + 20 + 20 + 22 + 22 + 29 + 36= 180 observed frequencies. Therefore, the expected frequency for each of the seven days would be $1/7 * 180 = 25.71$

Day	Observed frequency O	Expected frequency E	$O-E$	$(O-E)^2$	$\dfrac{(O-E)^2}{E}$
Sun	31	25.71	+5.29	27.9841	1.088
Mon	20	25.71	-5.71	32.6041	1.268
Tue	20	25.71	-5.71	32.6041	1.268
Wed	22	25.71	-3.71	13.7641	0.535
Thu	22	25.71	-3.71	13.7641	0.535
Fri	29	25.71	+3.29	10.8241	0.421
Sat	36	25.71	+10.29	105.8841	4.118

$\chi^2 = \dfrac{(O-E)^2}{E} = 9.233$

$\alpha = 0.05$, $df = 7 - 1 = 6$, critical value (from Table A-4) is $\chi^2 = 12.592$

The test statistic (9.233) does not fall within the critical region (above 12.592). Therefore, we fail to reject the null hypothesis. Since the null hypothesis states that the chances of occurrence of accidents is the same on any given week day, there is not sufficient evidence to warrant rejection of the claim that accidents occur with equal frequency on the different days.

9. Measuring Pulse Rates

$H_0: p_0 = p_1 = ... = p_8 = p_9 = 1/10$

H_1 : at least one of the proportions is different from 1/10

It is assumed that, the probability for occurrence for each digit is the same. Since, there are altogether ten digits, the probability for each category would be 1/10. There are altogether 16 + 0 + 14 + 0 + 17 + 0 + 16 + 0 + 17 + 0= 80 observed frequencies. Therefore, the expected frequency for each of the ten digits would be $1/10 * 80 = 8$

last digit	Observed Frequency O	Expected frequency E	$O-E$	$(O-E)^2$	$\dfrac{(O-E)^2}{E}$
0	16	8	+ 8	64	8.000
1	0	8	- 8	64	8.000
2	14	8	+ 6	36	4.500
3	0	8	- 8	64	8.000

The probability of a randomly selected subject having an IQ of 96 to 110 or from 95.5 –110.5. Given μ = 100, σ = 15; z when x = 110.5

$z = \frac{x - \mu}{\sigma} = \frac{110.5 - 100}{15} = 0.70$. Referring to Table A-2, we find that z = 0.70 corresponds to an area of 0.7580. The probability of getting an IQ score of 110.5 or less is 0.7580. The area in the 95.5 to 110.5 interval is 0.7580 – 0.3821= 0.3759.

The probability of a randomly selected subject having an IQ of 110.5 – 120.5. Given μ = 100, σ = 15; z when x = 120.5

$z = \frac{x - \mu}{\sigma} = \frac{120.5 - 100}{15} = 1.37$. Referring to Table A-2, we find that z = 1.37 corresponds to an area of 0.9147. The probability of getting an IQ score of 120.5 or less is 0.9147. The area in the 110.5 to 120.5 interval is 0.9147 – 0.7580= 0.1567.

The probability of a randomly selected subject having an IQ of more than 120. Given μ = 100, σ = 15; z when x = 120.5

Since this is the area above a z value of 1.37, this would be 1.0000 – 0.9147= 0.0853

b. Find expected frequencies

The sum of the observed frequencies is 20 + 20 + 80 + 40 + 40 = 200. The expected values of all categories are:

Category	Probability	Expected value
Less than 80	0.0853	$0.0853 * 200 = 17.06$
80 – 95	0.2968	$0.2968 * 200 = 59.36$
96 – 110	0.3759	$0.3759 * 200 = 75.18$
111 – 120	0.1567	$0.1567 * 200 = 31.34$
More than 120	0.0853	$0.0853 * 200 = 17.06$

c. Conduct goodness of fit test

Category	Observed frequency O	Expected frequency E	$O-E$	$(O-E)^2$	$\frac{(O-E)^2}{E}$
< 80	20	17.06	2.94	8.6436	0.507
80 – 95	20	59.36	-39.36	1549.2096	26.098
96 – 110	80	75.18	4.82	23.2324	0.309
111 – 120	40	31.34	8.66	75.0000	2.390
> 120	40	17.06	22.94	526.2436	30.850

$$\chi^2 = \frac{(O-E)^2}{E} = 60.154$$

$\alpha = 0.01$, df= 5 – 1= 4, critical value (from Table A-4) is χ^2= 13.277

The test statistic (60.154) falls within the critical region (above 13.277). Therefore, we reject the null hypothesis that the IQ scores come from a normally distributed population with the given mean and standard deviation. There is sufficient evidence to warrant rejection of the claim that the IQ scores were randomly selected from a normally distributed population with mean 100 and standard deviation 15.

10.3 Contingency Tables: Independence and Homogeneity

1. Is there Racial Profiling?

H_0 :ethnicity and being stopped are independent

H_1 : ethnicity and being stopped are not independent

Since the significance level to be tested is 0.05 and the *P*-Value obtained in the Minitab is 0.521(which is greater than 0.05), we fail to reject the hypothesis that being stopped is independent of ethnicity. There is not sufficient evidence to support a claim of racial profiling in being stopped by police. Racial profiling does not seem to be used.

Note for the Exercises in this Section: We need to find the Expected frequencies (*E*).

The expected frequency is found using this equation:

$$E = \frac{(\text{row total}) * (\text{column total})}{(\text{grand total})}$$

The expected frequency is exactly proportional to the row and column marginals or totals and the value represents what would be exactly expected by chance in the state of the Null hypothesis being true. These need to be 5 or greater for the chi-square distribution to be an accurate probability model to use for these types of analyses.

We also need to compute the degrees of freedom. In these two dimensional tables, *r* represents the number of rows and *c* represents the number of columns. We find the degrees of freedom as: $df = (r-1) * (c-1)$.

Rounding at different places in all of the computations in this section may lead to slightly different answers, but there be no difference in the conclusions.

3. Accuracy of Polygraph Tests

H_0 : truthfulness and polygraph result are independent

H_1 : truthfulness and polygraph result are not independent

$\alpha = 0.05$, $df = (2-1)(2-1) = 1$, critical value (from Table A-4) is $\chi^2 = 3.841$

The observed and expected values for the contingency table cells are as given below:

	Polygraph indicated Truth	Polygraph indicated Lie	Total
Subject actually told the truth	65 (54.4)	15 (25.6)	80
Subject actually told a lie	3 (13.6)	17 (6.4)	20
Total	68	32	100

$$\chi^2 = \frac{(O-E)^2}{E} = 2.065 + 4.389 + 8.262 + 17.556 = 32.272$$

Since the test statistic (32.272) falls within the critical region (above 3.841), we reject the null hypothesis. Therefore, there is sufficient evidence to warrant rejection of the claim that whether a subject lies is independent of the polygraph indication. It appears that the polygraph is effective, but it is not 100% effective.

5. Fear of Flying Gender Gap

H_0 : fear of flying and gender of flyer are independent

H_1 : fear of flying and gender of flyer are not independent

$\alpha = 0.05$, $df = (2-1)(2-1) = 1$, critical value (from Table A-4) is $\chi^2 = 3.841$

Total number of adults = 1014

Men makeup 48% of the total = 487

Women = 1014 – 487= 527

12% of the men fear flying = 0.12 * 487= 58

33% of the women fear flying = 0.33 * 527= 174
The observed and expected values for the contingency table cells are as given below:

	Male	Female	Total
Fear flying	58 (111.42)	174 (120.58)	232
Do not fear flying	429 (375.58)	353 (406.42)	782
Total	487	527	1014

$$\chi^2 = \frac{(O-E)^2}{E} = \frac{(58 - 111.42)^2}{111.42} + \frac{(174 - 120.58)^2}{120.58} + \frac{(429 - 375.58)^2}{375.58} + \frac{(353 - 406.42)^2}{406.42}$$

= 25.615 + 23.671 + 7.599 + 7.023= 63.908

Since the test statistic (63.908) falls within the critical region (above 3.841), we reject the null hypothesis. Therefore, there is sufficient evidence to warrant rejection of the claim that the gender is independent of the fear of flying. There appears to be a difference in the male and female proportions that fear flying.

7. **No Smoking**

H_0 : success stopping smoking and type of treatment are independent

H_1 : success stopping smoking and type of treatment are not independent

$\alpha = 0.05$, $df= (2 - 1)(3 - 1)= 2$, critical value (from Table A-4) is $\chi^2= 5.991$

The observed and expected values for the contingency table cells are as given below:

	Nicotine gum	Nicotine Patch	Nicotine inhaler	Total
Smoking	191 (198.34)	263 (253.87)	95 (96.79)	549
Not smoking	59 (51.66)	57 (66.13)	27 (25.21)	143
Total	250	320	122	692

$$\chi^2 = \frac{(O-E)^2}{E} = \frac{(191 - 198.34)^2}{198.34} + \frac{(263 - 253.87)^2}{253.87} + \frac{(95 - 96.79)^2}{96.79} + \frac{(59 - 51.67)^2}{51.67} + \frac{(57 - 66.13)^2}{66.13} + \frac{(27 - 25.21)^2}{25.21}$$

= 0.271 + 0.328 + 0.033 + 1.042 + 1.260 + 0.127 = 3.062

Since the Test statistic (3.062) does not fall within the critical region (less than 5.991), we fail to reject the null hypothesis. Therefore, there is not sufficient evidence to warrant rejection of the claim that success is independent of the method used. The evidence does not suggest that any method is significantly better than others.

9. **Occupational Hazards**

H_0 : death by homicide and occupation are independent

H_1 : death by homicide and occupation are not independent

$\alpha = 0.05$, $df= (2 - 1)(4 - 1)= 3$, critical value (from Table A-4) is $\chi^2= 7.815$

The observed and expected values for the contingency table cells are as given below:

Cause	Police	Cashiers	Taxi drivers	Guards	Total
Homicide	82 (112.92)	107 (75.28)	70 (64.25)	59 (65.55)	318
Other than homicide	92 (61.08)	9 (40.72)	29 (34.75)	42 (35.45)	172
Total	174	116	99	101	490

$$\chi^2 = \frac{(O-E)^2}{E} = \frac{(82 - 112.92)^2}{112.92} + \frac{(107 - 75.28)^2}{75.28} + \frac{(70 - 64.25)^2}{64.25} + \frac{(59 - 65.55)^2}{65.55} + \frac{(92 - 61.08)^2}{61.08} + \frac{(9 - 40.72)^2}{40.72} + \frac{(29 - 34.75)^2}{34.75} + \frac{(42 - 35.45)^2}{35.45}$$
$$= 8.468 + 13.364 + 0.515 + 0.654 + 15.655 + 24.708 + 0.952 + 1.209 = 65.524$$

Since the test statistic (65.524) falls within the critical region (above 7.815), we reject the null hypothesis. Therefore, there is sufficient evidence to warrant rejection of the claim that occupation is independent of whether the cause of death was homicide. Cashiers appear to be most prone to die from homicide.

11. Firearms Training and Safety

H_0 : firearms training and safety practice of locking gun age are independent

H_1 : firearms training and safety practice of locking gun age are not independent

$\alpha = 0.05$, $df = (2-1)(2-1) = 1$, critical value (from Table A-4) is $\chi^2 = 3.841$

The observed and expected values for the contingency table cells are as given below:

	Guns stored loaded and unlocked?		
	Yes	No	Total
Had formal firearm training	122 (96.52)	329 (354.48)	451
Had no formal firearm training	49 (74.48)	299 (273.52)	348
Total	171	628	799

$$\chi^2 = \frac{(O-E)^2}{E} = \frac{(122 - 96.52)^2}{96.52} + \frac{(329 - 354.48)^2}{354.48} + \frac{(49 - 74.48)^2}{74.48} + \frac{(299 - 273.52)^2}{273.52} = 6.726 + 1.832 + 8.717 + 2.374 = 19.649$$

Since the test statistic (19.649) falls within the critical region (above 3.841), we reject the null hypothesis. Therefore, there is sufficient evidence to warrant rejection of the claim that formal firearm training is independent of how firearms are stored. The formal training appears to have a positive effect.

13. Clinical Test of Lipitor

H_0 : getting a headache and dosage of Lipitor are independent

H_1 : getting a headache and dosage of Lipitor are not independent

$\alpha = 0.05$, $df = (2-1)(5-1) = 4$, critical value (from Table A-4) is $\chi^2 = 9.488$

The observed and expected values for the contingency table cells are as given below:

	Placebo	Dosage of Atorvastatin				Total
		10mg	20mg	40mg	80mg	
Headache	19 (16.10)	47 (51.45)	6 (2.15)	2 (4.71)	6 (5.60)	80
No headache	251 (253.90)	816 (811.55)	30 (33.85)	77 (74.29)	88 (97.80)	1262
Total	270	863	36	79	94	1342

Since two of the cells (those shaded above) have expected cell frequencies less than 5, we will combine them. We do this since there is a requirement that every cell have an expected cell frequency of at least five in order for the observed chi-square be distributed as a chi-square with (r – 1) * (c – 1) degrees of freedom.

The observed and expected values for the contingency table cells with 20mg and 40mg combined are as given below:

	Placebo	Dosage of Atorvastatin			Total
		10mg	20 or 40 mg	80mg	
Headache	19 (16.10)	47 (51.45)	8 (6.86)	6 (5.60)	80
No headache	251 (253.90)	816 (811.55)	107 (108.14)	88 (88.40)	1262
Total	270	863	115	94	1342

$$\chi^2 = \frac{(O-E)^2}{E} = \frac{(19 - 16.10)^2}{16.10} + \frac{(47 - 51.45)^2}{51.45} + \frac{(8 - 6.86)^2}{6.86} + \frac{(6-5.60)^2}{5.60} + \frac{(251 - 253.90)^2}{253.90} + \frac{(816 - 811.55)^2}{811.55} + \frac{(107 - 108.14)^2}{108.14} + \frac{(88 - 88.4)^2}{88.4}$$

$= 0.524 + 0.384 + 0.191 + 0.028 + 0.033 + 0.024+ 0.012 + 0.002= 1.199$

Since the test statistic (1.199) does not fall within the critical region (7.815), we fail to reject the null hypothesis. Therefore, there is not sufficient evidence to warrant rejection of the claim that getting a headache is independent of the amount of Atrovastatin used as a treatment.

15. Exercise and Smoking

H_0 : mouth or throat sores and use of Nicorette gum are independent

H_1 : mouth or throat sores and use of Nicorette gum are not independent

$\alpha = 0.05$, $df = (2 - 1)(2 - 1) = 1$, critical value (from Table A-4) is $\chi^2 = 3.841$

The observed and expected values for the contingency table cells are as given below:

	Drug	placebo	Total
Mouth or throat soreness	43 (38.87)	35 (39.13)	78
No mouth or throat soreness	109 (113.13)	118 (113.87)	227
Total	152	153	305

$$\chi^2 = \frac{(O-E)^2}{E} = \frac{(43 - 38.87)^2}{38.87} + \frac{(35 - 39.13)^2}{39.13} + \frac{(109 - 113.13)^2}{113.13} + \frac{(118 - 113.87)^2}{113.87}$$

$= 0.439 + 0.436 + 0.151 + 0.150 = 1.176$

Since the test statistic (1.176) does not fall within the critical region (above 3.841), we fail to reject the null hypothesis. Therefore, there is not sufficient evidence to warrant rejection of the claim that the treatment (drug or placebo) is independent of the reaction (whether or not mouth or throat soreness was experienced). If considering whether to use Nicorette, there is evidence that one should not be concerned about mouth or throat soreness.

17. Using Yates's Correction for Continuity

Chi-Square without Yates' correction:

	Black and Non – Hispanic	White and Non- Hispanic	Total
Stopped by police	24 (21.375)	147 (149.625)	171
Not stopped by police	176 (178.625)	1253 (1250.375)	1429
Total	200	1400	1600

$$\chi^2 = \frac{(O-E)^2}{E} = \frac{(24 - 21.375)^2}{21.375} + \frac{(147 - 149.625)^2}{149.625} + \frac{(176 - 178.625)^2}{178.625} + \frac{(1253 - 1250.375)^2}{1250.375}$$

$= 0.322 + 0.046 + 0.039 + 0.006 = 0.413$

$$\chi^2 = \frac{(|O - E| - 0.5)^2}{E} = \frac{(|24 - 21.375| - 0.5)^2}{21.375} + \frac{(|147 - 149.625| - 0.5)^2}{149.625} + \frac{(|176 - 178.625| - 0.5)^2}{178.625} + \frac{(|1253 - 1250.375| - 0.5)^2}{1250.375}$$

$$= \frac{2.125^2}{21.375} + \frac{2.125^2}{149.625} + \frac{2.125^2}{178.625} + \frac{2.125^2}{1250.375}$$

$= 0.211 + 0.030 + 0.025 + 0.004 = 0.270$

Yates's correction decreases the value of the observed test statistic. This reduces the likelihood of the test being statistically significant. It provides a test statistic that is more consistent with the probability model being used when $df = 1$ (a 2 x 2 Table).

10.4 McNemar Test for Matched Pairs

In Exercises 1 – 7, refer to the following table. The table summarizes results from an experiment in which subjects were first classified as smokers or nonsmokers, then they were given a treatment, then later they were again classified as smokers or nonsmokers.

		Before treatment	
		Smoke	Don't smoke
After Treatment	Smoke	50	6
	Don't smoke	8	80

1. **Sample Size**, sample size, $n = 50 + 6 + 8 + 80 = 144$

3. **Treatment Effectiveness**
From the table it can be said that 50 persons (upper left cell) continued to smoke irrespective of the treatment. Similarly, 80 persons (lower right cell) did not smoke both before and after treatment. Therefore, a total of 50 + 80 = 130 persons' smoking status did not change with the treatment.

5. **Discordant Pairs**
b and c are discordant pairs as they consisted of different groups.

7. **Critical Value**
$df = 1$. The critical value as found from Table A – 4 is 6.635.

9. **Treating Athlete's Foot**
H_0 : treatment and cure are independent
H_1 : treatment and cure are not independent
$\alpha = 0.05$, $df = 1$, critical value (from Table A-4) is $\chi^2 = 3.841$

		Fungicide treatment	
		Cure	No cure
Placebo	Cure	5	12
	No cure	22	55

$$\chi^2 = \frac{(|b - c| - 1)^2}{b + c} = \frac{(|12 - 22| - 1)^2}{12 + 22} = \frac{(|10| - 1)^2}{12 + 22} = \frac{9 * 9}{34} = 2.382$$

Since the Test statistic (2.382) does not fall within the critical region (less than 3.841), we fail to reject the null hypothesis. Therefore, there is not sufficient evidence to warrant rejection of the claim that the efficacy of the fungicidal solution is similar to that of placebo. The fungicide does not appear to be effective.

11. PET/CR Compared with MRI

H_0 : method of tomography and accuracy are independent

H_1 : method of tomography and accuracy are not independent

		PET/CT	
		Correct	Incorrect
MRI	Correct	36	1
	Incorrect	11	2

$$\chi^2 = \frac{(|b - c| - 1)^2}{b + c} = \frac{(|1 - 11| - 1)^2}{1 + 11} = \frac{(|10| - 1)^2}{12} = \frac{9 * 9}{12} = 6.750$$

$\alpha = 0.05$, $df = 1$, critical value (from Table A-4) is $\chi^2 = 3.841$

Since the test statistic (6.750) falls within the critical region (above 3.841), we reject the null hypothesis. Therefore, there is sufficient evidence to warrant rejection of the claim that the two proportions are the same. (1) the proportion of tumors with incorrect staging from MRI and correct staging from PET/CT; (2) the proportion of tumors with correct staging from MRI and incorrect staging from PET/CT. the PET/CT technology appears to be more accurate.

13. Correction for Continuity

		Treatment with Pedacream	
		Cure	No cure
Treatment with Fungacream	Cure	12	8
	No cure	40	20

The test statistic without the correction for continuity:

$$\chi^2 = \frac{(b - c)^2}{b + c} = \frac{(8 - 40)^2}{8 + 40} = \frac{(-32)^2}{48} = \frac{1024}{48} = 21.333$$

The test statistic with the correction for continuity:

$$\chi^2 = \frac{|(b - c) - 1|^2}{b + c} = \frac{(|8 - 40| - 1)^2}{8 + 40} = \frac{(32 - 1)^2}{48} = \frac{31^2}{48} = \frac{961}{48} = 20.021$$

Test statistic with continuity correction as solved in the example is 20.021. Test statistic without continuity correction: $\chi^2 = 21.333$. The test statistic is lower with the continuity correction. The continuity correction is used to compute a test statistic that is based on a more appropriate chi-square probability distribution.

15. Small Sample Case

		Treatment with Pedacream	
		Cure	No cure
Treatment with Fungacream	Cure	12	2
	No cure	6	20

Using a binomial test with $n= 8$ and $p= 0.5$, with values for the probabilities of getting 6 or higher as values for x, Table A-1 indicates the area that corresponds to:

$x \geq 6 = p(6) + p(7) + p(8) = 0.109 + 0.031 + 0.004 = 0.144$.

Since this is for one tail, we double it to get a two-tailed P-Value so the P-Value= 0.288. The chi-square approximation had a P-Value of 0.289. Thus, the chi-square approximation is a very good approximation to the exact probability test.

Review Exercises

1. Do Gunfire Deaths Occur More Often on Weekends?

$H_0 : p_{Sun} = p_{Mon} = p_{Tues} = p_{Wed} = p_{Thurs} = p_{Fri} = p_{Sat} = 1/7$

H_1 : at least one of the proportions is different from 1/7

$\alpha = 0.05$, $df = 7 - 1 = 6$, critical value (from Table A-4) is $\chi^2 = 12.592$

It is assumed that the probability for occurrence of gunfire deaths is the same on any given day. Since, there are altogether seven days, the probability for each day would be 1/7 There are altogether 464 (74 + 60 + 66 + 71 + 51 + 66 + 76= 464) observed frequencies. Therefore, the expected frequency for each of the seven days would be $1/7 * 464 = 66.286$.

Day	Observed frequency O	Expected frequency E	$O - E$	$(O-E)^2$	$\frac{(O-E)^2}{E}$
Mon	74	66.286	+7.714	59.506	0.898
Tue	60	66.286	-6.286	39.514	0.596
Wed	66	66.286	-0.286	0.0818	0.0012
Thu	71	66.286	+4.714	22.222	0.335
Fri	51	66.286	-15.286	233.662	3.525
Sat	66	66.286	-0.286	0.0818	0.001
Sun	76	66.286	+9.714	94.362	1.424

$$\chi^2 = \frac{(O-E)^2}{E} = 6.780$$

The test statistic (6.780) does not fall within the critical region (it is less than 12.592). Therefore, we fail to reject the null hypothesis. Since the null hypothesis states that the chances of occurrence of gun fire deaths is the same on any given week day, there is not sufficient evidence to warrant rejection of the claim that gun fire deaths occur with equal frequency on the different days. Thus, no day appears to have a higher proportion of gunfire deaths than any other day.

3. Testing for Independence Between Early Discharge and Rehospitalization of Newborn

H_0 : early discharge and rehospitalization of newborns are independent

H_1 : early discharge and rehospitalization of newborns are not independent

$\alpha = 0.05$, $df = (2 - 1)(2 - 1) = 1$, critical value (from Table A-4) is $\chi^2 = 3.841$

The observed and expected values for the contingency table cells are as given below:

	Rehospitalized within week of discharge?		Total
	Yes	No	Total
Early discharge (less than 30 hours)	622 (584.017)	3997 (4034.983)	4619
Late discharge (30-78 hours)	631 (668.983)	4660 (4622.017)	5291
Total	1253	8657	9910

$$\chi^2 = \frac{(O-E)^2}{E} = \frac{(622 - 584.017)^2}{584.017} + \frac{(3997 - 4034.983)^2}{4034.983} + \frac{(631 - 668.983)^2}{668.983} +$$

$$\frac{(4660 - 4622.017)^2}{4622.017} = 2.470 + 0.358 + 2.157 + 0.312 = 5.297$$

Since the test statistic (5.297) falls within the critical region (which is above 3.841), we reject the null hypothesis. Therefore, there is sufficient evidence to warrant rejection of the claim that whether the newborn was discharged early or late is independent of whether the newborn was rehospitalized within a week of discharge.

The critical value, if $\alpha = 0.01$ and the degrees of freedom 1 is 6.635. Since, the test statistic (5.297) does not fall within the critical region at $\alpha = 0.01$, we fail to reject null hypothesis. Therefore, the conclusion would be changed at $\alpha = 0.01$. We would be 95% confident that this is a true difference, but not 99% confident.

Cumulative Review Exercises

1. Finding Statistics

In the ascending order the scores are 66, 75, 77, 80, 82, 84, 89, and 94.

$$\text{Mean} = \bar{x} = \frac{x}{n} = \frac{66+80+82+75+77+89+94+84}{8} = \frac{647}{8} = 80.90$$

Median (if even number of scores, the median is the mean of the middle two values when the total subjects are arranged either in the descending or ascending order).

The mean of the two middle scores (80 and 82) is 81. Therefore, the median is (80 + 82)/2= 81.

Range = highest value - the lowest value (94-66) = 28

Subject	Score (x)	$(x-\bar{x})$	$(x-\bar{x})^2$
1	66	66-80.9= -14.9	222.01
2	75	75-80.9= -5.9	34.81
3	77	77-80.9= -3.9	15.21
4	80	80-80.9= -0.9	0.81
5	82	82-80.9= 1.1	1.21
6	84	84-80.9= 3.1	9.61
7	89	89-80.9= 8.1	65.61
8	94	94-80.9= 13.1	171.61
Total	647		520.88

$$\text{Variance} = s^2 = \frac{(x-\bar{x})^2}{n-1} = \frac{520.88}{8-1} = \frac{520.88}{7} = 74.41$$

$$\text{Standard deviation} = s^2 = \sqrt{\frac{(x-\bar{x})^2}{n-1}} = \sqrt{74.41} = 8.63$$

5-number summary, n= 8

x_1 (Minimum score)= 66

Q_1, the first quartile, also known as P_{25} is the mean of the 2nd and 3rd score, which is (75 + 77)/2= 76

The median or Q_2, the second quartile, also known as P_{50} is the average of the 4th and 5th scores or $\tilde{x}$ = (80 + 82)/2= 81

Q_3, the third quartile, also known as P_{75} is the average of the 7th and 8th scores, which is (84 + 89)/2= 86.5

x_8 (Maximum score)= 94

3. Testing for Equal Proportions

H_0 : answer selected and gender of respondent are independent

H_1 : answer selected and gender of respondent are not independent

$\alpha = 0.05$, df= (2 – 1) (4 – 1)= 3, critical value (from Table A-4) is χ^2= 7.815

The observed and expected values for the contingency table cells are as given below:

	Answer Selected				Total
	A	B	C	D	
Men (x)	66 (66.97)	80 (79.15)	82 (82.42)	75 (74.46)	303
Women (y)	77 (76.03)	89 (89.85)	94 (93.58)	84 (84.54)	344

Total	143	169	176	159	647

$$\chi^2 = \frac{(O-E)^2}{E} = \frac{(66 - 66.97)^2}{66.97} + \frac{(80 - 79.15)^2}{79.15} + \frac{(82 - 82.42)^2}{82.42} + \frac{(75-74.46)^2}{74.46} + \frac{(77 - 76.03)^2}{76.03} + \frac{(89 - 89.85)^2}{89.85} + \frac{(94 - 93.58)^2}{93.58} + \frac{(84 - 84.54)^2}{84.54}$$

$= 0.014 + 0.009 + 0.002 + 0.0039 + 0.012 + 0.008 + 0.0019 + 0.0034 = 0.0542$

Since the Test statistic (0.0542) does not fall within the critical region (it is less than 7.815), we fail to reject the null hypothesis. Therefore, there is not sufficient evidence to warrant rejection of the claim that men and women choose the different answers in the same proportions.

5. **Testing for Effectiveness of Training**

$H_0 : \mu_d = 0$ training is not effective, no difference in pre and post memory

$H_1 : \mu_d > 0$ training is effective, difference in favor of post higher than pre score

$\alpha = 0.05$, $df = n - 1 = 3$, critical value (from Table A-6) is $t = +2.353$

$$t = \frac{\bar{d} - \mu_d}{s / \sqrt{n}}$$

where d = individual difference between the two values in a single matched pair

μ_d = mean value of the differences d for the population of all matched pairs

$\bar{d}$ = mean value of the differences d for the paired sample data

s_d = standard deviation of the differences d for the paired sample data

n = number of pairs of data

Subject	x	y	$d = y - x$	$d - \bar{d}$	$(d-\bar{d})^2$
A	66	77	11	11-10.25	0.5625
B	80	89	9	9-10.25	1.5625
C	82	94	12	12-10.25	3.0625
D	75	84	9	9-10.25	1.5625

$$\bar{d} = \frac{d}{n} = \frac{41}{4} = 10.25 \qquad (d-\bar{d})^2 = 6.75$$

$$s_d = \sqrt{\frac{(d-\bar{d})^2}{n-1}} = \sqrt{\frac{6.75}{4-1}} = 1.5$$

$$t = \frac{\bar{d} - \mu_d}{\frac{s_d}{\sqrt{n}}} = \frac{10.25 - 0}{\frac{1.5}{\sqrt{4}}} = \frac{10.25}{0.75} = 13.667$$

Since the test statistic (13.667) is greater than the critical value (2.353), there is sufficient evidence to support the claim that the training session is effective in raising scores.

Chapter 11. Analysis of Variance

11-2 One –Way ANOVA

1. **Weights of Poplar Trees**
   - **a.** $H_0 : \mu_1 = \mu_2 = \mu_3 = \mu_4$
   - **b.** H_1: The means are not all the same.
   - **c.** The test statistic, F, is identified in the output as F, and is $F = 8.448$.
   - **d.** Using Table A-5 and degrees of freedom 3 for the numerator and 15 for the denominator, the critical value is 3.2874.
   - **e.** The *P*-value, labeled as Sig. in the output, is .002.
   - **f.** There is sufficient evidence to conclude that the mean weights of polar trees given different treatments are not all equal.
   - **g.** The apparent outlier of 1.34 kg does not seem to have much of an effect on the results. The conclusion did not change.

3. **Marathon Times**
   - **a.** $H_0 : \mu_1 = \mu_2 = \mu_3$
   - **b.** H_1: The means are not all the same.
   - **c.** The test statistic is identified as *F* in the output, and $F = 0.188679406$.
   - **d.** Using Table A-5 and degrees of freedom 2 for the numerator and 120 for the denominator (the degrees of freedom 108 is not present on the table, so we use the closest value), the critical value is 3.0718.
   - **e.** The *P*-value, labeled as *P-value* in the output, is 0.82832493.
   - **f.** There is not sufficient evidence to conclude that the mean times to finish a marathon for men in different age categories are not all equal.

5. **Weights of Poplar Trees**
   We begin by finding the required values, and then perform the formal hypothesis test.

	None	Fertilizer	Irrigation	Fertilizer and Irrigation
n	5	5	5	5
Σx	2.290	3.150	1.390	6.540
Σx^2	1.796	2.677	0.814	13.056
$\bar{x}$	0.458	0.630	0.278	1.308
s^2	0.187	0.173	0.107	1.125

k = 4 $\quad \bar{\bar{x}} = 0.669$

n = 5 $\quad s_p^2 = \Sigma s^2 / k = 0.398$

$s_{\bar{x}}^2 = 0.202$

$H_0 : \mu_1 = \mu_2 = \mu_3 = \mu_4$

H_1: The means are not all the same.

The significance level is 0.05. The numerator degrees of freedom is $df = k - 1 = 3$, the denominator degrees of freedom is $df = k(n-1) = 16$, so the critical value from Table A-5 is 3.2389.

The test statistic is

$$F = \frac{\text{variance between samples}}{\text{variance within samples}} = \frac{ns_{\bar{x}}^2}{s_p^2} = \frac{5 \times 0.202}{0.398} = 2.5423$$

The test statistic is less than 3.2389 and so we fail to reject H_0. There is not sufficient evidence to warrant the rejection of the claim that the mean weights of poplar trees in year one are equal for different treatments.

7. Head Injury in a Car Crash

We begin by finding the required values, and then perform the formal hypothesis test.

	Subcompact	Compact	Midsize	Full-size
n	5	5	5	5
Σx	3344	2779	2434	2689
Σx^2	2470638	1577659	1297312	1541765
$\bar{x}$	668.8	555.8	486.8	537.8
s^2	58542.7	8272.7	28110.2	23905.2
k = 4	$\bar{\bar{x}} = 562.3$			
n = 5	$s_p^2 = \Sigma s^2 / k = 29707.7$			
$s_{\bar{x}}^2 = 5895$				

$H_0: \mu_1 = \mu_2 = \mu_3 = \mu_4$

H_1: The means are not all the same.

The significance level is 0.05, the numerator degrees of freedom is df = k −1 = 3, the denominator degrees of freedom is $df = k(n-1) = 16$, so the critical value from Table A-5 is 3.2389.

The test statistic is

$$F = \frac{\text{variance between samples}}{\text{variance within samples}} = \frac{ns_{\bar{x}}^2}{s_p^2} = \frac{5 \times 5895}{29707.7} = 0.992$$

The test statistic is less than 3.2389 and so we fail to reject H_0. There is not sufficient evidence to warrant the rejection of the claim that the mean standard head injury condition for different sized cars are different. It does not appear that larger cars are safer.

9. Archaeology: Skull Breadths from Different Epochs

We begin by finding the required values, and then perform the formal hypothesis test.

	4000 B.C.	1850 B.C.	150 A.D.
n	9	9	9
Σx	1194	1210	1243
Σx^2	158544	162768	171853
$\bar{x}$	132.6666667	134.4444444	138.1111111
s^2	17.5	11.27777778	22.61111111
k = 3	$\bar{\bar{x}} = 135.074$		
n = 9	$s_p^2 = \Sigma s^2 / k = 17.130$		
$s_{\bar{x}}^2 = 7.708$			

$H_0: \mu_1 = \mu_2 = \mu_3$

H_1: The means are not all the same.

The significance level is 0.05, the numerator degrees of freedom is df = k −1 = 2, the denominator degrees of freedom is $df = k(n-1) = 24$, so he critical value from Table A-5 is 3.4028.

The test statistic is

$$F = \frac{\text{variance between samples}}{\text{variance within samples}} = \frac{ns_{\bar{x}}^2}{s_p^2} = \frac{9 \times 7.708}{17.130} = 4.050$$

The test statistic is greater than 3.4028 and so we reject H_0. There is sufficient sample evidence to support the claim that the mean skull breadths from the different epochs are not all the same.

11. Iris Sepal Lengths

We begin by finding the required values, and then perform the formal hypothesis test.

	Class Setosa	Class Versiccolor	Class virginica
n	50	50	50
Σx	250.3	296.8	329.4
Σx^2	1259.09	1774.86	2189.9
$\bar{x}$	5.006	5.936	6.588
s^2	0.12424898	0.266432653	0.404342857

k = 3 $\quad \bar{\bar{x}} = 5.843$

n = 50 $\quad s_p^2 = \Sigma s^2 / k = 0.265$

$s_{\bar{x}}^2 = 0.632$

$H_0: \mu_1 = \mu_2 = \mu_3$

H_1: The means are not all the same.

The significance level is 0.05, the numerator degrees of freedom is df = k – 1 = 2, and the denominator degrees of freedom is $df = k(n-1) = 147$. Denominator degrees of freedom of 147 is not on the table, so we use the closest value, which is 120, so the critical value from Table A-5 is 3.0718.

The test statistic is

$$F = \frac{\text{variance between samples}}{\text{variance within samples}} = \frac{ns_{\bar{x}}^2}{s_p^2} = \frac{50 \times 0.632}{0.265} = 119.265$$

The test statistic is greater than 3.0718 and so we reject H_0. There is sufficient sample evidence to support the claim that the mean sepal lengths of irises in the three classes are different.

13. Exercise and Stress

We begin by finding the required values, and then perform the formal hypothesis test.

	Female and Black	Male and Black	Female and White	Male and White
n	6	6	6	6
Σx	632.333	732.667	671.667	692.667
Σx^2	67820.778	89652.889	80800.444	80800.444
$\bar{x}$	105.389	122.111	111.944	115.444
s^2	235.974	37.230	31.352	167.185

k = 4 $\quad \bar{\bar{x}} = 113.722$

n = 6 $\quad s_p^2 = \Sigma s^2 / k = 117.935$

$s_{\bar{x}}^2 = 48.648$

$H_0: \mu_1 = \mu_2 = \mu_3 = \mu_4$

H_1: The means are not all the same.

The significance level is not given, so we use 0.05. The numerator degrees of freedom is df = k – 1 = 3, and the denominator degrees of freedom is $df = k(n-1) = 20$, so the critical value from Table A-5 is 3.0984.

The test statistic is

$$F = \frac{\text{variance between samples}}{\text{variance within samples}} = \frac{ns_{\bar{x}}^2}{s_p^2} = \frac{6 \times 48.648}{117.935} = 2.475$$

The test statistic is less than 3.0984 and so we fail to reject H_0. There is not sufficient evidence to warrant the rejection of the claim that the mean systolic blood pressure for the four groups are the same. The results suggest that the sample is suitable.

15. Using the Tukey Test

The Bonferroni test output from SPSS indicated that the weights of poplar trees treated with both fertilizer and irrigation have a mean that is significantly different from fertilizer only, irrigation only, or no treatment. The Tukey results indicate the same conclusion. However, the *P*-values are lower for the Tukey test.

11-3 Two –Way ANOVA

Some of Exercises 1-7require the given Minitab display, which results from the amounts of the pesticide DDT measured in falcons in three different age categories (young, middle-aged, old) qt three different locations (United States, Canada, Arctic region). The data set is included with the Minitab software as the file FALCON.MTW.

1. **Meaning of Two-way ANOVA**
   It is referred to as *two-way* because the data is classified using two factors. It is referred to as *analysis of variance* because it, like one-way ANOVA, compares the variances between groups of data.

3. **Interaction Effect**
   If there is an interaction effect, we should not consider the effects of one factor without considering the effects of the other.

5. **Interaction Effects**
   H_0: There is no interaction between the two factors, site and age
   H_1: There is interaction between the two factors, site and age
   The test statistic for this hypothesis test is below. The values for MS(interaction) and MS(error) are taken from the Minitab output, from the MS column and the Interaction and Error rows.
   $$F = \frac{\text{MS(interaction)}}{\text{MS(error)}} = \frac{4.43}{3.44} = 1.28$$
   In the column P. we see the P-value for Interaction, which is 0.313. The significance level is not included in the exercise, so we assume a significance level of 0.05. We fail to reject the null hypothesis. There does not appear to be an interaction effect between age and site.

7. **Effect of Age**
   H_0: There are no effects from the factor age
   H_1: There is an effect from the factor age
   The test statistic for this hypothesis test is below. The values for MS(age) and MS(error) are taken from the Minitab output, from the MS column and the Age and Error rows.
   $$F = \frac{\text{MS(age)}}{\text{MS(error)}} = \frac{860.59}{3.44} = 249.85$$
   In the column P. we see the P-value for the factor Age, which is 0.00. The significance level is not included in the exercise, so we assume a significance level of 0.05. We reject the null hypothesis. There appears to be an effect on DDT levels in falcons according to age.

In Exercise 9, use the SPSS display that results from the New York Marathon running times (in seconds) for randomly selected runners who finished.

9. **Effect of Gender**
   H_0: There are no effects from the factor gender
   H_1: There is an effect from the factor gender
   The test statistic for this hypothesis test is below. The values for MS(gender) and MS(error) are taken from the SPSS output, from the Mean Squares column and the GENDER and Error rows.
   $$F = \frac{\text{MS(gender)}}{\text{MS(error)}} = \frac{15225412.80}{9028477.350} = 1.686$$
   In the column labeled Sig., we see the P-value for the factor GENDER, which is 0.206. The significance level is not included in the exercise, so we assume a significance level of 0.05. We fail to reject the null hypothesis. There does not appear to be an effect on time according to gender.

In Exercise 11, refer to the given Minitab display. The display results from a study in which 24 subjects were given hearing tests using four different lists of words. The 24 subjects had normal hearing and the tests were conducted with no background noise. The main objective was to determine whether the four lists are equally difficult to understand. The original data are from* A Study of the Interlist Equivalency of the CID W22 Word List Presented in Quiet and in Noise, *by Faith Loven, University of Iowa. The original data are available on the Internet through DASL (Data and Story Library).

11. Hearing Tests: Effects of Subject

H_0: There are no effects from the factor subject
H_1: There is an effect from the factor subject
The test statistic for this hypothesis test is below. The values for MS(subject) and MS(error) are taken from the Minitab output, from the MS column and the Subject and Error rows.

$$F = \frac{\text{MS(subject)}}{\text{MS(error)}} = \frac{140.5}{36.3} = 3.87$$

In the column P. we see the P-value for the factor Site, which is 0.00. The significance level is not included in the exercise, so we assume a significance level of 0.05. We reject the null hypothesis. There appears to be an effect on hearing scores according to subject. This makes sense since different people have different perceptions and different hearing acuity.

13. Pulse Rates

First, we test for an interaction effect
H_0: There is no interaction between the two factors, gender and age
H_1: There is interaction between the two factors, gender and age
The test statistic for this hypothesis test is below. The values for MS(interaction) and MS(error) are taken from JMP output.

$$F = \frac{\text{MS(interaction)}}{\text{MS(error)}} = \frac{40.667}{112.444} = 0.3617$$

Once more, from the JMP output, we see the P-value for the interaction between gender and age is 0.7015. The significance level is not included in the exercise, so we assume a significance level of 0.05. We fail to reject the null hypothesis. There does not appear to be an interaction effect between gender and age.
We now check for an effect from the factor gender.
H_0: There are no effects from the factor gender
H_1: There is an effect from the factor gender
The test statistic for this hypothesis test is below. The values for MS(gender) and MS(error) are taken from JMP output.

$$F = \frac{\text{MS(gender)}}{\text{MS(error)}} = \frac{10.667}{112.444} = 0.0949$$

Once more, from the JMP output, we see the P-value for the factor gender is 0.7616. The significance level is not included in the exercise, so we assume a significance level of 0.05. We reject the null hypothesis. There appears to be an effect on pulse rate according to gender.
Finally, we check for an effect from the factor age.
H_0: There are no effects from the factor age
H_1: There is an effect from the factor age
The test statistic for this hypothesis test is below. The values for MS(age) and MS(error) are taken from JMP output.

$$F = \frac{\text{MS(age)}}{\text{MS(error)}} = \frac{40.667}{112.444} = 0.3617$$

Once more, from the JMP output, we see the P-value for the factor age is 0.7015. The significance level is not included in the exercise, so we assume a significance level of 0.05. We reject the null hypothesis. There appears to be an effect on pulse rate according to age.

In Exercises 15 & 17, use the poplar tree weights from Year 2, as listed in Data Set 9 in Appendix B. (The example in this section used the data from Year 1.)

15. Effect of Site

H_0: There are no effects from the factor site

H_1: There is an effect from the factor site

The test statistic for this hypothesis test is below. The values for MS(site) and MS(error) are taken from JMP output.

$$F = \frac{\text{MS(site)}}{\text{MS(error)}} = \frac{0.3258025}{0.227360} = 1.4330$$

Once more, from the JMP output, we see the P-value for the factor site is 0.2401. The significance level is 0.05. We fail to reject the null hypothesis. There does not appear to be an effect on poplar weight according to site.

17. Effect of Site

H_0: There are no effects from the factor site

H_1: There is an effect from the factor site

The test statistic for this hypothesis test is below. The values for MS(site) and MS(error) are taken from JMP output.

$$F = \frac{\text{MS(site)}}{\text{MS(error)}} = \frac{0.0018}{0.296367} = 0.0061$$

Once more, from the JMP output, we see the P-value for the factor site is 0.9428. The significance level is 0.05. We fail to reject the null hypothesis. There does not appear to be an effect on poplar weight according to site.

Review Exercises

1. Drinking and Driving

$H_0: \mu_1 = \mu_2 = \mu_3$

H_1: The means are not all the same.

From the output, the test statistic is $F = 46.900$. The reported P-value (presented as Sig. in the output) is .000, so we reject H_0. There is sufficient evidence to warrant the rejection of the claim that the mean blood alcohol levels of the three groups are the same.

3. Iris Sepal Widths

We begin by finding the required values, and then perform the formal hypothesis test.

	setosa	versicolor	virginica
n	50	50	50
Σx	170.900	138.500	148.700
Σx^2	591.250	388.470	447.330
$\bar{x}$	3.418	2.770	2.974
s^2	0.145	0.098	0.104
k = 3	$\bar{\bar{x}} = 3.054$		
n = 50	$s_p^2 = \Sigma s^2 / k = 0.1159$		
$s_{\bar{x}}^2 = 0.1098$			

$H_0: \mu_1 = \mu_2 = \mu_3$
H_1: The means are not all the same.
The significance level is 0.05, the numerator degrees of freedom is $df = k - 1 = 2$, and the denominator degrees of freedom is $df = k(n-1) = 147$. Denominator degrees of freedom of 147 is not on the table, so we use the closest value, which is 120, so the critical value from Table A-5 is 3.0718.

$$F = \frac{\text{variance between samples}}{\text{variance within samples}} = \frac{ns_{\bar{x}}^2}{s_p^2} = \frac{50 \times 0.1098}{0.1159} = 47.364$$

The test statistic is greater than 3.0718 and so we reject H_0. There is sufficient evidence to warrant the rejection of the claim that the mean sepal widths for the three species of iris are the same.

In Exercise 5, use the Minitab display that results from the values listed in the accompanying table. The sample data are student estimates (in feet) of the length of their classroom. The actual length of the classroom is 24 ft. 7 in.

5. Effect of Gender
H_0: There are no effects from the factor gender
H_1: There is an effect from the factor gender
The test statistic for this hypothesis test is below. The values for MS(gender) and MS(error) are taken from the Minitab output, from the MS column and the Gender and Error rows.

$$F = \frac{\text{MS(gender)}}{\text{MS(error)}} = \frac{29.4}{37.8} = 0.78$$

In the column P we see the P-value for the factor Gender, which is 0.395. The significance level is not included in the exercise, so we assume a significance level of 0.05. We fail to reject the null hypothesis. There does not appear to be an effect on estimate of classroom length according to gender.

Cumulative Review Exercises

1. Boston Rainfall Statistics
Summary statistics necessary for the calculations in this problem are presented below.
$\Sigma x = 5.22, \Sigma x^2 = 4.0416, n = 52$

a. $\bar{x} = \Sigma x / n = 5.22 / 52 = 0.100$

b. $s = \sqrt{\dfrac{n\Sigma(x^2) - (\Sigma x)^2}{n(n-1)}} = \sqrt{\dfrac{52 \cdot 4.0416 - (5.22)^2}{52 \cdot 51}} = 0.263$

c. The five-number summary is
$\min = 0,\ Q_1 = 0.00,\ Q_2 = 0.00,\ Q_3 = 0.01,\ \max = 1.41$

d. There is 1 mild outlier, 0.03, and 11 extreme outliers, which are all rainfall values greater than 0.04. They are 0.11, 0.43, 1.41, 0.49, 0.59, 0.05, 0.47, 0.41, 0.12, 0.12, and 0.92.

e.

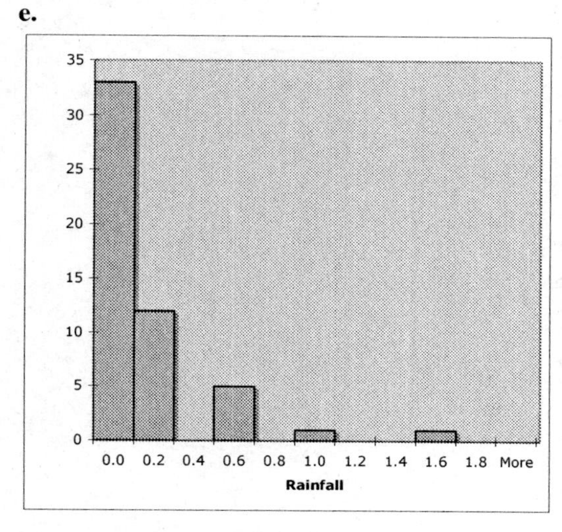

f. No, you should not use one-way ANOVA. Although the data will be categorized using one characteristic, it is fairly clear form the histogram that the data is far from normally distributed.

g. Let x be the number of Mondays on which rain fell in Boston. A good estimate for the probability that rain will fall on a randomly selected Monday in Boston is $\hat{p} = x/n = 19/52 = 0.365$.

3. Weights of Babies: Finding Probabilities

a. This is the area to the right of 8.00 under the normal curve with mean μ = 7.24 and standard deviation σ = 1.09. In order to use table A-2, we must convert into the standard normal distribution,

$$z = \frac{x - \mu}{\sigma} = \frac{8.00 - 7.54}{1.09} = 0.42.$$

The probability is the area to the right of 0.42, and so the probability is P = 1 – .6628 = 0.3372.

b. We use the central limit theorem to convert to the standard normal distribution. The probability will be the area to the right of 8.00 under the normal curve with mean $\mu_{\bar{x}} = \mu = 7.54$ and standard deviation $\sigma_{\bar{x}} = \sigma/\sqrt{n} = 1.09/\sqrt{16} = 0.2725$.

$$z = \frac{x - \mu_{\bar{x}}}{\sigma_{\bar{x}}} = \frac{8.00 - 7.54}{0.2725} = 1.69.$$

The probability is the area to the right of 1.69, and so the probability is P = 1 – .9545= 0.0455.

c. The probability that any baby has a birth weight of over 7.54 lbs. is 0.5000, as the mean is 7.54 lbs. Since the babies are independent of one another, the probability that the next three have weights greater than 7.54 lbs. is $P = (0.5)^3 = 0.125$.

Chapter 12. Nonparametric Statistics

12-2 Sign Test

In Exercises 1 & 3, assume that matched pairs of data result in the given number of signs when the value of the second variable is subtracted from the corresponding value of the first variable. Use the sign test with a 0.05 significance level to test the null hypothesis of no difference.

1. Positive signs= 10, negative signs= 5, ties= 3
H_0: There is no difference.
H_1: There is a difference.
Significance level is α= 0.05, n= 15, x (lower frequency)= 5, CV from Table A-7= 3 Since x= 5 > 3 (CV), there is not sufficient evidence to warrant rejection of the claim that there is no difference between the two variables.

3. Positive signs= 50, negative signs= 40, ties= 5
H_0: There is no difference.
H_1: There is a difference.
Significance level is α= 0.05, n= 90, x (lower frequency)= 40, use z test approximation since n> 25, CV= ±1.96

$$z=\frac{(x+0.5)-\frac{n}{2}}{\frac{\sqrt{n}}{2}}=\frac{(40+0.5)-\frac{90}{2}}{\frac{\sqrt{90}}{2}}=\frac{-4.5}{4.743}=-0.949$$

The test statistic z= −0.949 is not less than −1.96, so there is not sufficient evidence to warrant rejection of the claim that there is no difference between the number of positive and negative signs.

In Exercises 5 – 13, use the sign test.

5. Testing Corn Seeds

Regular	19.25	22.75	23	23	22.5	19.75	24.5	15.5	18	14.25	17
Kiln dried	25	24	24	28	22.5	19.5	22.25	16	17.25	15.75	17.25
Sign of difference	−	−	−	−	0	+	+	−	+	−	−

The null hypothesis is the claim of no difference between the two variables. Therefore,
H_0: There is no difference in the yields
H_1: There is a difference in the yields
n= 10, 3 positive signs, 7 negative sign, 1 tie (not included in n)
Significance level is α= 0.05, n= 10, x (lower frequency)= 3, CV from Table A-7= 1.
Since x=3 > 1 (CV), there is not sufficient evidence to warrant rejection of the claim that there is no difference between the yields of the two types of seeds. No, one type does not appear better than the other.

7. Testing for a Difference Between Reported and Measured Male Heights

11.4	11.8	11.7	11.0	11.9	11.9	11.5	12.0	12.1	11.9	10.9	11.3	11.5	11.5	11.6
–	–	–	–	–	–	–	0	+	–	–	–	–	–	–

The null hypothesis is the claim of no difference between reported and measured heights of male subjects. Therefore,

H_0: There is no difference in the reported and measured heights

H_1: There is a difference in the reported and measured heights

n= 12, 7 positive signs, 5 negative sign, no ties

Significance level is α= 0.05, n= 12, x (lower frequency)= 5, CV from Table A-7= 2.

Since x=5 > 2 (CV), there is not sufficient evidence to warrant rejection of the claim that there is no difference between the reported and measured heights.

9. Testing for Difference Between Prone and Supine Measurements

Prone	2.96	4.65	3.27	2.50	2.59	5.97	1.74	3.51	4.37	4.02
Supine	1.97	3.05	2.29	1.68	1.58	4.43	1.53	2.81	2.70	2.70
Sign of difference	+	+	+	+	+	+	+	+	+	+

The null hypothesis is the claim of no difference between prone and supine measurements. Therefore,

H_0: There is no difference between prone and supine measurements.

H_1: There is a difference between prone and supine measurements.

n= 10, 10 positive signs, 0 negative signs, no ties

Significance level is α= 0.05, n= 10, x (lower frequency)= 0, CV from Table A-7= 1.

Since x=0 < 1 (CV), there is sufficient evidence to warrant rejection of the claim that there is no difference between the reported and measured heights.

11. Testing for Median Underweight

H_0: Median = 12 oz

H_1: Median < 12 oz

We use the negative sign when the value is below 12 oz, and we use the positive sign when it is above 12 oz.

n= 14, 1 positive sign, 13 negative signs, 1 tie (not included in n)

Significance level is α= 0.05 with one-tailed test, n= 14, x (lower frequency)= 1, CV from Table A-7= 3 (directional test)

Since x=1 < 3 (CV), there is sufficient evidence to warrant rejection of the claim that the median is equal to 12 oz. There is sufficient evidence to warrant support of the claim that the company is cheating consumers by giving amounts with a median less than 12 oz.

13. Nominal Data: Smoking and Nicotine Patches

H_0: There is no difference in number who are or are not smoking after one year while using the nicotine patch

H_1: There is a majority of smokers who are smoking after one year while using the nicotine patch

With α= 0.05 and a one tailed test, the critical value is $z = -1.645$ (if using the smaller sign number for the significance test)

Denoting those smokers who are still smoking after using nicotine patch as plus (+) and those who quit smoking as negative (–), we have 41 positives and 30 negatives.

The test statistic x is the smaller of 41 and 30, so, x = 30. Since the value of n = 71 and is above 25, the test statistic x is converted to the test statistic z as follows:

$$z = \frac{(x+0.5) - \frac{n}{2}}{\frac{\sqrt{n}}{2}} = \frac{(30+0.5) - \frac{71}{2}}{\frac{\sqrt{71}}{2}} = \frac{-5}{4.213} = -1.187$$

The test statistic $z = -1.187$ is not less than -1.645, so there is not sufficient evidence to warrant rejection of the claim that there is no difference in number who are or are not smoking after one year while using the nicotine patch.

15. Procedures for Handling Ties

H_0: Median = 100

H_1: Median < 100

Our test is one tailed with $\alpha = 0.05$. Since $n > 25$, the test statistic x is converted to the test statistic z. Critical value is -1.645 (using smaller frequency for testing)

Approach 1, ignore the ties

60 positive signs, 40 negative signs, and 21 ties, $n = 100$

$$z = \frac{(x+0.5) - \frac{n}{2}}{\frac{\sqrt{n}}{2}} = \frac{(40+0.5) - \frac{100}{2}}{\frac{\sqrt{100}}{2}} = \frac{-9.5}{5} = -1.900$$

The test statistic $z = -1.900$ is less than -1.645, so there is sufficient evidence to reject the null hypothesis in favor of the hypothesis than the median is less than 100.

Approach 2, half of ties given + sign and half given – sign

70 positive signs, 50 negative signs, and 1 tie, $n = 120$

$$z = \frac{(x+0.5) - \frac{n}{2}}{\frac{\sqrt{n}}{2}} = \frac{(50+0.5) - \frac{120}{2}}{\frac{\sqrt{120}}{2}} = \frac{-9.5}{5.477} = -1.734$$

The test statistic $z = -1.734$ is less than -1.645, so there is sufficient evidence to reject the null hypothesis in favor of the hypothesis than the median is less than 100.

Approach 3, all ties given sign that supports the null

60 positive signs, 61 negative signs, and 0 ties, $n = 121$

$$z = \frac{(x+0.5) - \frac{n}{2}}{\frac{\sqrt{n}}{2}} = \frac{(60+0.5) - \frac{121}{2}}{\frac{\sqrt{121}}{2}} = \frac{0}{5.50} = 0$$

The test statistic $z = 0$ is not less than -1.645, so there is not sufficient evidence to reject the null hypothesis in favor of the hypothesis than the median is less than 100.

The highest test statistic results from the approach where ties are ignored, the second highest results from splitting have of the ties and distributing them equally as + or – signs, and the lowest, in this case much lower when ties are all given the sign that supports the null hypothesis. In this case, the first two approaches resulted in rejection of the null and the third approach did not.

17. Normal Approximation Error

n= 54, number of men= 36, number of women (x)= 18

H_0: p= 0.5

H_1: p< 0.5

Since n> 25, we use the normal approximation, if $\alpha = 0.01$, the critical value is z= −2.326

$$z = \frac{(x+0.5) - \frac{n}{2}}{\frac{\sqrt{n}}{2}} = \frac{(18+0.5) - \frac{54}{2}}{\frac{\sqrt{54}}{2}} = \frac{-8.5}{3.674} = -2.313$$

The test statistic $z = -2.313$ is not less than −2.326, so there is not sufficient evidence to reject the null hypothesis in favor of the hypothesis than the proportion of women hired is below 0.5. The *P*-Value for this result as a directional or one-tailed test, found using Excel is 0.01036.

If the binomial distribution is used, we need to find and sum all the probabilities from $0 \le x \le 18$ where:

$$P(x) = \frac{n!}{x!(n-x)!} p^x * (1-p)^{n-x} \quad \text{for } x = 0 \text{ through } 18$$

Using the statDisk program that comes with this text, we will find this P-Value to be 0.00992 which is less than 0.01. The *P*-Value of 0.00992 is less than 0.01, so there is sufficient evidence to reject the null hypothesis in favor of the hypothesis than the proportion of women hired is below 0.5.Thus, in this case the normal approximation would be said to be conservative since it did not detect a significant difference that was found using an exact probability test in the form of the binomial test.

12-3 Wilcoxon Signed-Ranks Test for Two Matched Pairs

In Exercise 1, refer to the given paired sample data and use the Wilcoxon signed-rank test to test the claim that the samples come from populations with differences having a median of zero. Use a 0.05 significance level.

1. H_0 :The populations have the same distribution

H_1 :The populations have different distributions

x	12	14	17	19	20	27	29	30
y	12	15	15	14	12	18	19	20
Difference (d)	0	−1	2	5	8	9	10	10
Rank of |differences|	--	1	2	3	4	5	6.5	6.5
Signed ranks	--	−1	2	3	4	5	6.5	6.5

Sum of absolute values of negative ranks= 1, Sum of positive ranks= 27

T will be the smaller of the two sums found, we find that $T = 1$

n is the number of pairs of data for which $d \ne 0$, we have $n = 7$

Since $n \le 30$, we use test statistic of $T = 1$ and we use Table A-8 to find the critical value of T= 2.

The test statistic $T = 1 < 2$, so we have sufficient evidence to warrant rejection of the claim that the populations have the same distributions. Evidence indicates the two distributions are different.

In Exercises 3-7, refer to the sample data for the given Exercises in Section 12-2. Instead of the sign test, use the Wilcoxon signed-rank test to test the claim that both samples come from populations having differences with a median equal to zero.

3. Exercise 5

H_0: There is no difference in the yields
H_1: There is a difference in the yields

Regular	19.25	22.75	23	23	22.5	19.75	24.5	15.5	18	14.25	17
Kiln dried	25	24	24	28	22.5	19.5	22.25	16	17.25	15.75	17.25
Difference (d)	−5.75	−1.25	−1	−5	0	+0.25	+2.25	−0.5	+0.75	−1.5	−0.25
Rank of |differences|	10	6	5	9		1.5	8	3	4	7	1.5
Signed ranks	−10	−6	−5	−9		+1.5	+8	−3	+4	−7	−1.5

Sum of absolute values of negative ranks: 41.5, Sum of positive ranks: 13.5
T is the smaller of the two sums found, $T = 13.5$.
n is the number of pairs of data for which $d \neq 0$, we have $n = 10$
Since $n \leq 30$, we use test statistic of $T = 13.5$ and we use Table A-8 to find the critical value of $T = 8$.
The test statistic $T = 13.5$ is greater than the critical value of 8, so we fail to reject the claim that there is no difference in the distributions between the yields from the two types of seed.

5. Exercise 7

H_0: There is no difference in the reported and measured heights
H_1: There is a difference in the reported and measured heights

Reported height	68	71	63	70	71	60	65	64	54	63	66	72
Measured height	67.9	69.9	64.9	68.3	70.3	60.6	64.5	67.0	55.6	74.2	65	70.8
Difference (d)	+0.1	+1.1	−1.9	+1.7	+0.7	−0.6	+0.5	−3	−1.6	−11.2	+1	+1.2
Rank of |difference|	1	6	10	9	4	3	2	11	8	12	5	7
Signed ranks	+1	+6	−10	+9	+4	−3	+2	−11	−8	−12	+5	+7

Sum of absolute values of negative ranks: 44, Sum of positive ranks: 34
T is the smaller of the two sums found, we find that $T = 34$.
n is the number of pairs of data for which $d \neq 0$, we have $n = 12$.
Test is two tailed with $\alpha = 0.05$.
Since $n \leq 30$, we use test statistic of $T = 34$ and we use Table A-8 to find the critical value of 14.
The test statistic $T = 34$ is greater than the critical value of 14, so we fail to reject the claim that there is a difference between self-reported and measured heights.

7. Exercise 9

H_0: There is no difference between prone and supine measurements.
H_1: There is a difference between prone and supine measurements.

Prone	2.96	4.65	3.27	2.50	2.59	5.97	1.74	3.51	4.37	4.02
Supine	1.97	3.05	2.29	1.68	1.58	4.43	1.53	2.81	2.70	2.70
Difference (d)	0.99	1.6	0.98	0.82	1.01	1.54	0.21	0.7	1.67	1.32
Rank of |difference|	5	9	4	3	6	8	1	2	10	7
Signed ranks	5	9	4	3	6	8	1	2	10	7

Sum of absolute values of negative ranks: 0, Sum of positive ranks: 55
T is the smaller of the two sums found, we find that $T = 0$.
n is the number of pairs of data for which $d \neq 0$, we have $n = 10$.
Test is two tailed with $\alpha = 0.05$.
Since $n \leq 30$, we use test statistic of $T = 0$ and we use Table A-8 to find the critical value of 8.
The test statistic $T = 0$ is less than the critical value of 8, so we reject the claim that there is a difference between prone and supine measurements.

9. Using the Wilcoxon Signed-Ranks Test for Claims About a Median

H_0: Median = 98.6^0F

H_1: Median $\neq 98.6^0$F

1	98.6	98.6	0	-	-
2	98.6	98.6	0	-	-
3	98	98.6	-0.6	48.5	-48.5
4	98	98.6	-0.6	48.5	-48.5
5	99	98.6	0.4	37	37
6	98.4	98.6	-0.2	16.5	-16.5
7	98.4	98.6	-0.2	16.5	-16.5
8	98.4	98.6	-0.2	16.5	-16.5
9	98.4	98.6	-0.2	16.5	-16.5
10	98.6	98.6	0	-	-
11	98.6	98.6	0	-	-
12	98.8	98.6	0.2	26	26
13	98.6	98.6	0	-	-
14	97	98.6	-1.6	88	-88
15	97	98.6	-1.6	88	-88
16	98.8	98.6	0.2	26	26
17	97.6	98.6	-1	71	-71
18	97.7	98.6	-0.9	66.5	-66.5
19	98.8	98.6	0.2	26	26
20	98	98.6	-0.6	48.5	-48.5
21	98	98.6	-0.6	48.5	-48.5
22	98.3	98.6	-0.3	30.5	-30.5
23	98.5	98.6	-0.1	5.5	-5.5
24	97.3	98.6	-1.3	81	-81
25	98.7	98.6	0.1	5.5	5.5
26	97.4	98.6	-1.2	78	-78
27	98.9	98.6	0.3	32.5	32.5
28	98.6	98.6	0	-	-
29	99.5	98.6	0.9	66.5	66.5
30	97.5	98.6	-1.1	75.5	-75.5

31	97.3	98.6	-1.3	81	-81
32	97.6	98.6	-1	71	-71
33	98.2	98.6	-0.4	37	-37
34	99.6	98.6	1	71	71
35	98.7	98.6	0.1	5.5	5.5
36	99.4	98.6	0.8	64.5	64.5
37	98.2	98.6	-0.4	37	-37
38	98	98.6	-0.6	48.5	-48.5
39	98.6	98.6	0	-	-
40	98.6	98.6	0	-	-
41	97.2	98.6	-1.4	83	-83
42	98.4	98.6	-0.2	16.5	-16.5
43	98.6	98.6	0	-	-
44	98.2	98.6	-0.4	37	-37
45	98	98.6	-0.6	48.5	-48.5
46	97.8	98.6	-0.8	61	-61
47	98	98.6	-0.6	48.5	-48.5
48	98.4	98.6	-0.2	16.5	-16.5
49	98.6	98.6	0	-	-
50	98.6	98.6	0	-	-
51	97.8	98.6	-0.8	61	-61
52	99	98.6	0.4	37	37
53	96.5	98.6	-2.1	91	-91
54	97.6	98.6	-1	71	-71
55	98	98.6	-0.6	48.5	-48.5
56	96.9	98.6	-1.7	90	-90
57	97.6	98.6	-1	71	-71
58	97.1	98.6	-1.5	85	-85
59	97.9	98.6	-0.7	57	-57
60	98.4	98.6	-0.2	16.5	-16.5
61	97.3	98.6	-1.3	81	-81
62	98	98.6	-0.6	48.5	-48.5
63	97.5	98.6	-1.1	75.5	-75.5
64	97.6	98.6	-1	71	-71
65	98.2	98.6	-0.4	37	-37
66	98.5	98.6	-0.1	5.5	-5.5
67	98.8	98.6	0.2	26	26
68	98.7	98.6	0.1	5.5	5.5
69	97.8	98.6	-0.8	61	-61
70	98	98.6	-0.6	48.5	-48.5
71	97.1	98.6	-1.5	85	-85
72	97.4	98.6	-1.2	78	-78
73	99.4	98.6	0.8	64.5	64.5
74	98.4	98.6	-0.2	16.5	-16.5
75	98.6	98.6	0	-	-
76	98.4	98.6	-0.2	16.5	-16.5
77	98.5	98.6	-0.1	5.5	-5.5
78	98.6	98.6	0	-	-
79	98.3	98.6	-0.3	30.5	-30.5
80	98.7	98.6	0.1	5.5	5.5
81	98.8	98.6	0.2	26	26
82	99.1	98.6	0.5	41	41
83	98.6	98.6	0	-	-

84	97.9	98.6	-0.7	57	-57
85	98.8	98.6	0.2	26	26
86	98	98.6	-0.6	48.5	-48.5
87	98.7	98.6	0.1	5.5	5.5
88	98.5	98.6	-0.1	5.5	-5.5
89	98.9	98.6	0.3	32.5	32.5
90	98.4	98.6	-0.2	16.5	-16.5
91	98.6	98.6	0	-	-
92	97.1	98.6	-1.5	85	-85
93	97.9	98.6	-0.7	57	-57
94	98.8	98.6	0.2	26	26
95	98.7	98.6	0.1	5.5	5.5
96	97.6	98.6	-1	71	-71
97	98.2	98.6	-0.4	37	-37
98	99.2	98.6	0.6	48.5	48.5
99	97.8	98.6	-0.8	61	-61
100	98	98.6	-0.6	48.5	-48.5
101	98.4	98.6	-0.2	16.5	-16.5
102	97.8	98.6	-0.8	61	-61
103	98.4	98.6	-0.2	16.5	-16.5
104	97.4	98.6	-1.2	78	-78
105	98	98.6	-0.6	48.5	-48.5
106	97	98.6	-1.6	88	-88

Sum of absolute values of negative ranks: 3476
Sum of positive ranks: 710
Letting T be the smaller of the two sums found, we find that $T = 710$
Letting n be the number of pairs of data for $d \neq 0$, $n = 91$
Because $n = 91$, we calculate a z test statistic).

$$z = \frac{T - \frac{n(n+1)}{4}}{\sqrt{\frac{n(n+1)(2n+1)}{24}}} = \frac{710 - \frac{91(91+1)}{4}}{\sqrt{\frac{91(91+1)(182+1)}{24}}} = \frac{710-2093}{\sqrt{\frac{1532076}{24}}} = \frac{-1383}{\sqrt{63836.5}} = \frac{-1383}{252.66} = -5.474$$

With $\alpha = 0.05$ in a two tailed test, the critical value is $z = -1.96$. The test statistic $z = -5.474$ is less than -1.96, so there is sufficient evidence to warrant rejection of the claim that healthy adults have a median body temperature that is equal to 98.6° F.

12-4 Wilcoxon Rank-Sum Test for Two Independent Samples

Identifying Rank Sums. In Exercise 1, use a 0.05 significance level with the methods of this section to identify the rank sums R_1 and R_2, μ_R, σ_R, the test statistic z, the critical values, and then state the conclusion.

1. H_0 : The populations have the same median
H_1 : The populations have different medians
$\alpha = 0.05$, two-tailed test using z statistic, CV= ±1.96

Sample 1		Sample 2	
x	Rank	x	Rank
1	1	2	2
3	3	5	5
4	4	7	7
6	6	9	9
8	8	11	10
12	11	13	12
15	14	14	13
16	15	18	17
17	16	19	18
22	20	20	19
26	22.5	25	21
		26	22.5
$n_1 = 11$ $R_1 = 120.5$		$n_2 = 12$ $R_2 = 155.5$	

$$\mu_R = \frac{n_1(n_1+n_2+1)}{2} = \frac{11(11+12+1)}{2} = \frac{264}{2} = 132$$

$$\sigma_R = \sqrt{\frac{n_1 n_2(n_1+n_2+1)}{12}} = \sqrt{\frac{11*12(11+12+1)}{12}} = \sqrt{\frac{3168}{12}} = \sqrt{264} = 16.248$$

$$z = \frac{R-\mu_R}{\sigma_R} = \frac{120.5-132}{16.248} = \frac{-11.5}{16.248} = -0.708$$

Since it is a two tailed test and $\alpha = 0.05$, critical value of z would be ±1.96. The test statistic of $z = -0.71$ does not fall within the critical region. So, we fail to reject the null hypothesis that the populations have the same distribution.

Using the Wilcoxon Rank-Sum Test. In Exercises 3-9, use the Wilcoxon rank-sum test

3. Are Severe Psychiatric Disorders Related to Biological Factors?

H_0 : The populations have the same median

H_1 : The populations have different medians

$\alpha = 0.01$, two-tailed test using z statistic, CV= ±2.575

Obsessive – Compulsive Patients		Control group	
x	Rank	x	Rank
0.308	6	0.519	23
0.407	13	0.501	22
0.463	19	0.334	7.5
0.210	1	0.476	20
0.455	17	0.402	12
0.334	7.5	0.483	21
0.304	4	0.413	14
0.287	2	0.349	11
0.340	9	0.460	18
0.344	10	0.429	15
0.288	3	0.594	24
0.305	5	0.445	16
$n_1 = 12$, $R_1 = 96.5$		$n_2 = 12$, $R_2 = 203.5$	

$$\mu_R = \frac{n_1(n_1+n_2+1)}{2} = \frac{12(12+12+1)}{2} = \frac{300}{2} = 150$$

$$\sigma_R = \sqrt{\frac{n_1 n_2 (n_1+n_2+1)}{12}} = \sqrt{\frac{12*12(12+12+1)}{12}} = \sqrt{\frac{3600}{12}} = \sqrt{300} = 17.321$$

$$z = \frac{R-\mu_R}{\sigma_R} = \frac{96.5-150}{17.321} = \frac{-53.5}{17.321} - 3.089$$

Since it is a two tailed test and $\alpha = 0.01$, critical value of z would be $\pm$ 2.575. The test statistic of $z = -3.09$ falls within the critical region. So, we reject the claim that the distributions of the obsessive compulsive patients and healthy persons have the same median or that have the same brain volumes. We could conclude that obsessive-compulsive disorders have a biological basis.

5. Cholesterol Levels

H_0 : The populations have the same median

H_1 : The populations have different medians

$\alpha = 0.05$, two-tailed test using z statistic, CV= ±1.96

Cholesterol levels of men		Cholesterol levels of women	
x	Rank	x	Rank
522	21	264	15
127	9	181	12
740	25	267	17
49	2	384	19
230	13	98	6
316	18	62	3
590	24	126	8
466	20	89	5
121	7	531	22
578	23	130	10
78	4	175	11
265	16	44	1
250	14		
$n_1 = 13$ $R_1 = 196$		$n_2 = 12$ $R_2 = 129$	

$$\mu_R = \frac{n_1(n_1+n_2+1)}{2} = \frac{13(13+12+1)}{2} = \frac{338}{2} = 169$$

$$\sigma_R = \sqrt{\frac{n_1 n_2 (n_1+n_2+1)}{12}} = \sqrt{\frac{13*12(13+12+1)}{12}} = \sqrt{\frac{4056}{12}} = \sqrt{338} = 18.385$$

$$z = \frac{R-\mu_R}{\sigma_R} = \frac{196-169}{18.385} = \frac{27}{18.385} = 1.469$$

Since it is a two tailed test and $\alpha = 0.05$, critical value of z would be ± 1.96. The test statistic of $z = 1.47$ does not fall within the critical region. So, there is not sufficient evidence to reject the null of the populations having the same medians. We cannot support the claim that two populations of men and women have different median cholesterol levels.

7. Pulse Rates

H_0 : The populations have the same median

H_1 : The populations have different medians

$\alpha = 0.05$, two-tailed test using z statistic, CV= ± 1.96

Pulse rates for men		Pulse rates for women	
x	Rank	x	Rank
68	31.5	76	53.5
64	22	72	42.5
88	72	88	72
72	42.5	60	11.5
64	22	72	42.5
72	42.5	68	31.5
60	11.5	80	60.5
88	72	64	22
76	53.5	68	31.5
60	11.5	68	31.5
96	77.5	80	60.5
72	42.5	76	53.5
56	3.5	68	31.5
64	22	72	42.5
60	11.5	96	77.5
64	22	72	42.5
84	65.5	68	31.5
76	53.5	72	42.5
84	65.5	64	22
88	72	80	60.5
72	42.5	64	22
56	3.5	80	60.5
68	31.5	76	53.5
64	22	76	53.5
60	11.5	76	53.5
68	31.5	80	60.5
60	11.5	104	79
60	11.5	88	72
56	3.5	60	11.5
84	65.5	76	53.5
72	42.5	72	42.5
84	65.5	72	42.5
88	72	88	72
56	3.5	80	60.5
64	22	60	11.5
56	3.5	72	42.5
56	3.5	88	72
60	11.5	88	72
64	22	124	80
72	42.5	64	22
$n_1 = 40$		$n_2 = 40$	
$R_1 = 1339.5$		$R_2 = 1900.5$	

$$\mu_R = \frac{n_1(n_1+n_2+1)}{2} = \frac{40(40+40+1)}{2} = \frac{3240}{2} = 1620$$

$$\sigma_R = \sqrt{\frac{n_1 n_2(n_1+n_2+1)}{12}} = \sqrt{\frac{40*40(40+40+1)}{12}} = \sqrt{\frac{129600}{12}} = \sqrt{10800} = 103.923$$

$$z = \frac{R-\mu_R}{\sigma_R} = \frac{1339.5-1620}{103.923} = \frac{280.5}{103.923} = -2.699$$

Since it is a two tailed test and $\alpha = 0.05$, critical value of z would be $\pm$ 1.96. The test statistic of $z = -2.699$ falls within the critical region. So, there is sufficient evidence to warrant rejection of the claim that two samples of pulse rates come from populations with the same median. We conclude that the two groups have different medians.

9. Systolic Blood Pressure

H_0 : The populations have the same median

H_1 : The populations have different medians

$\alpha = 0.05$, two-tailed test using z statistic, CV= ± 1.96

Systolic blood pressures of males		Systolic blood pressures of females	
x	Rank	x	Rank
125	64.5	104	15
107	22	99	10
126	68.5	102	13
110	31.5	114	44.5
110	31.5	94	5
107	22	101	12
113	41.5	108	25.5
126	68.5	104	15
137	77	123	59
110	31.5	93	3.5
109	27.5	89	1
153	78	112	37
112	37	107	22
119	54.5	116	49.5
113	41.5	181	80
125	64.5	98	9
131	72.5	100	11
121	57.5	127	70.5
132	74.5	107	22
112	37	116	49.5
121	57.5	97	8
116	49.5	155	79
95	6.5	106	18.5
110	31.5	110	31.5
110	31.5	105	17
125	64.5	118	52.5
124	60.5	133	76
131	72.5	113	41.5
109	27.5	113	41.5
112	37	107	22
127	70.5	95	6.5
132	74.5	108	25.5
116	49.5	114	44.5
125	64.5	104	15

112	37	125	64.5
125	64.5	124	60.5
120	56	92	2
118	52.5	119	54.5
115	46.5	93	3.5
115	46.5	106	18.5
$n_1 = 40$ $R_1 = 2003.5$		$n_2 = 40$ $R_2 = 1236.5$	

$$\mu_R = \frac{n_1(n_1+n_2+1)}{2} = \frac{40(40+40+1)}{2} = \frac{3240}{2} = 1620$$

$$\sigma_R = \sqrt{\frac{n_1 n_2(n_1+n_2+1)}{12}} = \sqrt{\frac{40*40(40+40+1)}{12}} = \sqrt{\frac{129600}{12}} = \sqrt{10800} = 103.923$$

$$z = \frac{R-\mu_R}{\sigma_R} = \frac{2003.5-1620}{103.923} = \frac{383.5}{103.923} = 3.690$$

Since it is a two tailed test and $\alpha = 0.05$, critical value of z would be $\pm$ 1.96. The test statistic of $z = 3.690$ falls within the critical region. So, there is sufficient evidence to warrant rejection of the claim that the two populations have the same median pulse rates. We conclude that the two groups have different medians.

11. Using the Mann-Whitney U Test

H_0 : The populations have the same median

H_1 : The populations have different medians

$\alpha = 0.05$, two-tailed test using z statistic, CV= ± 1.96

From Table 12-5, we have the following:

$n_1 = 13$, $R_1 = 187 = R$

$n_2 = 12$, $R_2 = 138$

z test using data from Table 12-5 in text= 0.98

Using the Mann-Whitney U:

$$z = \frac{U - \frac{n_1 n_2}{2}}{\sqrt{\frac{n_1 n_2(n_1+n_2+1)}{12}}} \quad \text{where } U = n_1 n_2 + \frac{n_1(n_1+1)}{2} - R$$

$$U = 13*12 + \frac{13(13+1)}{2} - 187 = 156 + \frac{182}{2} - 187 = 156 + 91 - 187 = 60$$

$$z = \frac{60 - \frac{13*12}{2}}{\sqrt{\frac{13*12\,(13+12+1)}{12}}} = \frac{60-78}{\sqrt{\frac{4056}{12}}} = \frac{-18}{\sqrt{338}} = \frac{-18}{18.385} = -0.979 \approx -0.98$$

The z test statistic found by calculating Mann-Whitney U test is same as that found using Wilcoxon rank-sum test but sign is opposite. Since the test is a two-tailed test, the sign of the test statistic is not relevant.

12.5 Kruskal-Wallis Test

In Exercise 1, interpret the given Kruskal-Wallis test results and address the given question.

1. Weights of Poplar Trees

a. H_0 : The four populations have the same median weights

b. H_1 : There is a difference in the medians weights of at least two of the four populations

c. In the Kruskal-Wallis test, the test statistic is represented by '*H*'. In this exercise, *H*= 8.37

d. The critical value can be obtained from table A-4 since *H* is distributed as Chi-square with *df*= number of groups – 1= 4 – 1= 3 The critical value of *H*, with $\alpha = 0.05$ is 7.815

e. *P*-value (from printout)= 0.039

f. Since the test statistic, *H* = 8.37 falls within the critical region and the *P*-value is less than 0.05, the null hypothesis can be rejected. It can be concluded that at least one of the four treatments has a median weight different from at least one other treatment group.

3. Weights of Poplar Trees, Year 1, Site 2

H_0 : The four populations have the same median weights

H_1 : There is a difference in the medians weights of at least two of the four populations

No treatment		Fertilizer		Irrigation		Fertilizer and irrigation	
x	Rank	*x*	Rank	*x*	Rank	*x*	Rank
0.60	12	1.16	17	0.65	14	0.22	8
1.11	16	0.93	15	0.08	5	2.13	19
0.07	3.5	0.30	9	0.62	13	2.33	20
0.07	3.5	0.59	11	0.01	1	1.74	18
0.44	10	0.17	7	0.03	2	0.12	6
$n_1 = 5$		$n_2 = 5$		$n_3 = 5$		$n_4 = 5$	
$R_1 = 45$		$R_2 = 59$		$R_3 = 35$		$R_4 = 71$	

$$H = \frac{12}{N(N+1)}\left(\frac{R_1^2}{n_1} + \frac{R_2^2}{n_2} + \cdots + \frac{R_K^2}{n_K}\right) - 3(N+1)$$

Where N = total number of observations in all samples combined

K= number of samples, k is the group number from 1 to K

R_k= sum of ranks for sample k

n_k =number of observations in sample k

$$H = \frac{12}{N(N+1)}\left(\frac{R_1^2}{n_1} + \frac{R_2^2}{n_2} + \frac{R_3^2}{n_3} + \frac{R_4^2}{n_4}\right) - 3(N+1)$$

$$= \frac{12}{20(20+1)}\left(\frac{45^2}{5} + \frac{59^2}{5} + \frac{35^2}{5} + \frac{71^2}{5}\right) - 3(20+1)$$

$$= \frac{12}{420} * \left(\frac{2025}{5} + \frac{3481}{5} + \frac{1225}{5} + \frac{5041}{5}\right) - 63 = 0.0285714 * \frac{11772}{5} - 63$$

$$= 0.0285714 * 2354.40 - 63 = 4.269$$

The number of samples is K = 4, so, *df*= 4 – 1 = 3. Table A – 4 shows that for 3 degrees of freedom and 0.05 significance, the critical value is 7.815. The test statistic *H* = 4.269 is not in the critical region, so we fail to reject the null hypothesis of equal medians. There is not sufficient evidence to support the claim that the medians are not all equal. It does not appear that the weights are different for the various treatments at the drier site for Year 1.

5. Skull Breadths from Different Epochs

H_0 : The three populations have the same median skull breadths

H_1 : There is a difference in the median skull breadths of at least two of the three populations

4000 B.C.		1850 B.C.		150 A.D.	
x	Rank	x	Rank	x	Rank
131	6	129	4	128	2
138	21.5	134	10	138	21.5
125	1	136	14	136	14
129	4	137	17.5	139	24
132	7.5	137	17.5	141	25
135	12	129	4	142	26
132	7.5	136	14	137	17.5
134	10	138	21.5	145	27
138	21.5	134	10	137	17.5
$n_1 = 9$, $R_1 = 91$		$n_2 = 9$, $R_2 = 112.5$		$n_3 = 9$, $R_3 = 174.5$	

$$H = \frac{12}{N(N+1)}\left(\frac{R_1^2}{n_1}+\frac{R_2^2}{n_2}+\frac{R_3^2}{n_3}\right) - 3(N+1)$$

$$= \frac{12}{27(27+1)}\left(\frac{91^2}{9}+\frac{112.5^2}{9}+\frac{174.5^2}{9}\right) - 3(27+1)$$

$$= \frac{12}{756} * \left(\frac{8281}{9}+\frac{12656.25}{9}+\frac{30450.25}{9}\right) - 84$$

$$= 0.015873 * \frac{51387.5}{9} - 84 = 0.015873 * 5709.722 - 84 = 6.630$$

The number of samples is K = 3, so, df= 3 – 1 = 2. Table A – 4 shows that for 2 degrees of freedom and 0.05 significance, the critical value is 5.991. The test statistic H = 6.630 is in the critical region, so we reject the null hypothesis of equal medians. There is sufficient evidence to support the claim that the medians are not all equal. It does not appear that the skull breadths are different for the various epochs.

In Exercise 7, use the listed sample data from the car crash experiments conducted by the National Transportation Safety Administration. New cars were purchased and crashed into a fixed barrier at 35 mi/h, and the listed measurements were recorded for the dummy in the driver's seat. The subcompact cars were the Ford Escort, Honda Civic, Hyundai Accent, Nissan Sentra, and Saturn SL4. The compact cars are Chevrolet Cavalier, Dodge Neon, Mazda 626 DX, Pontiac Sunfire, and Subaru Legacy. The midsize cars are Chevrolet Camaro, Dodge Intrepid, Ford Mustang, Honda Accord, and Volvo S70. The full-size cars are Audi A8, Cadillac Deville, Ford Crown Victoria, Oldsmobile Aurora, and Pontiac Bonneville.

7. Head Injury in a Car Crash

H_0 : The four populations have the same median head injury condition

H_1 : There is a difference in the median head injury condition of at least two of the four populations

Subcompact		Compact		Midsize		Full-size	
x	Rank	x	Rank	x	Rank	x	Rank
681	16	643	13	469	8	384	3
428	5	655	14	727	18	656	15
917	20	442	6	525	10.5	602	12
898	19	514	9	454	7	687	17
420	4	525	10.5	259	1	360	2
$n_1 = 5$		$n_2 = 5$		$n_3 = 5$		$n_4 = 5$	
$R_1 = 64$		$R_2 = 52.5$		$R_3 = 44.5$		$R_4 = 49$	

$$H = \frac{12}{N(N+1)}\left(\frac{R_1^2}{n_1}+\frac{R_2^2}{n_2}+\frac{R_3^2}{n_3}+\frac{R_4^2}{n_4}\right) - 3(N+1)$$

$$= \frac{12}{20(20+1)}\left(\frac{64^2}{5}+\frac{52.5^2}{5}+\frac{44.5^2}{5}+\frac{49^2}{5}\right) - 3(20+1)$$

$$= \frac{12}{420} * \left(\frac{4096}{5}+\frac{2756.25}{5}+\frac{1980.25}{5}+\frac{2401}{5}\right) - 63 = 0.0285714 * \frac{11233.5}{5} - 63$$

$$= 0.0285714 * 2246.7 - 63 = 1.191$$

The number of samples is K = 4, so, df= 4 – 1 = 3. Table A – 4 shows that for 3 degrees of freedom and 0.05 significance, the critical value is 7.815. The test statistic H = 1.191 is not in the critical region, so we fail to reject the null hypothesis of equal medians. There is not sufficient evidence to support the claim that the medians are not all equal. It does not appear that the head injury medians are different for the different class of cars. There is no evidence that larger cars are safer relative to head injury condition.

9. Secondhand Smoke in Different Groups

H_0 : The three populations have the same median cotinine level

H_1 : There is a difference in the median cotinine level of at least two of the three populations

Smokers		Nonsmokers Exposed to Secondhand Smoke		Nonsmokers Not Exposed to Secondhand Smoke	
x	Rank	x	Rank	x	Rank
1	52	384	116	0	22.5
35	74	4	66	0	22.5
130	88	0	22.5	1	52
123	87	0	22.5	0	22.5
0	22.5	0	22.5	0	22.5
112	85	0	22.5	9	67
234	102	3	63.5	0	22.5
167	92	551	120	0	22.5
131	89	69	79	0	22.5
477	117	543	119	0	22.5
164	91	1	52	0	22.5
250	106	2	60.5	0	22.5
173	93.5	19	72	0	22.5
289	112	17	70.5	0	22.5
198	97	45	76	0	22.5
245	105	1	52	0	22.5
265	108	1	52	0	22.5

227	101	1	52	0	22.5
17	70.5	13	68.5	90	83
48	77	1	52	0	22.5
210	99	0	22.5	0	22.5
103	84	0	22.5	0	22.5
253	107	3	63.5	1	52
86	81	1	52	0	22.5
44	75	178	95	0	22.5
222	100	51	78	0	22.5
87	82	1	52	0	22.5
284	111	0	22.5	0	22.5
277	110	2	60.5	0	22.5
149	90	0	22.5	0	22.5
121	86	1	52	309	114
1	52	74	80	0	22.5
32	73	13	68.5	0	22.5
313	115	197	96	244	104
266	109	1	52	0	22.5
208	98	1	52	0	22.5
3	63.5	1	52	0	22.5
491	118	3	63.5	0	22.5
290	113	0	22.5	0	22.5
173	93.5	241	103	0	22.5
$n_1 = 40$		$n_2 = 40$		$n_3 = 40$	
$R_1 = 3629.5$		$R_2 = 2393.5$		$R_3 = 1237.0$	

$$H = \frac{12}{N(N+1)}\left(\frac{R_1^2}{n_1} + \frac{R_2^2}{n_2} + \frac{R_3^2}{n_3}\right) - 3(N+1)$$

$$= \frac{12}{120(120+1)}\left(\frac{3629.5^2}{40} + \frac{2393.5^2}{40} + \frac{1237^2}{40}\right) - 3(120+1)$$

$$= \frac{12}{14520} * \left(\frac{13173270.25}{40} + \frac{5728842.25}{40} + \frac{1530169}{40}\right) - 363$$

$$= 0.0008264463 * \frac{20432281.5}{40} - 363 = 0.0008264463 * 510807.04 - 363 = 59.155$$

The number of samples is K = 3, so, $df = 3 - 1 = 2$. Table A – 4 shows that for 3 degrees of freedom and 0.05 significance, the critical value is 5.991. The test statistic $H = 59.155$ is in the critical region, so we reject the null hypothesis of equal medians. There is sufficient evidence to support the claim that the medians are not all equal. It does appear that the cotinine level medians are different for three secondhand smoking conditions. Cotinine levels are higher for the smoking group.

11. Iris Petal Widths

H_0 : The three populations have the same median iris petal width

H_1 : There is a difference in the median iris petal width of at least two of the three populations

Setosa		Versicolor		Virginica	
x	Rank	x	Rank	x	Rank
0.2	20.5	1.4	82.5	2.5	149
0.2	20.5	1.5	92.5	1.9	119
0.2	20.5	1.5	92.5	2.1	130.5
0.2	20.5	1.3	72	1.8	110.5
0.2	20.5	1.5	92.5	2.2	135
0.4	45	1.3	72	2.1	130.5

0.3	38	1.6	100.5	1.7	103.5
0.2	20.5	1.0	54	1.8	110.5
0.2	20.5	1.3	72	1.8	110.5
0.1	3.5	1.4	82.5	2.5	149
0.2	20.5	1.0	54	2.0	124.5
0.2	20.5	1.5	92.5	1.9	119
0.1	3.5	1.0	54	2.1	130.5
0.1	3.5	1.4	82.5	2.0	124.5
0.2	20.5	1.3	72	2.4	146
0.4	45	1.4	82.5	2.3	140.5
0.4	45	1.5	92.5	1.8	110.5
0.3	38	1.0	54	2.2	135
0.3	38	1.5	92.5	2.3	140.5
0.3	38	1.1	59	1.5	92.5
0.2	20.5	1.8	110.5	2.3	140.5
0.4	45	1.3	72	2.0	124.5
0.2	20.5	1.5	92.5	2.0	124.5
0.5	49	1.2	63	1.8	110.5
0.2	20.5	1.3	72	2.1	130.5
0.2	20.5	1.4	82.5	1.8	110.5
0.4	45	1.4	82.5	1.8	110.5
0.2	20.5	1.7	103.5	1.8	110.5
0.2	20.5	1.5	92.5	2.1	130.5
0.2	20.5	1.0	54	1.6	100.5
0.2	20.5	1.1	59	1.9	119
0.4	45	1.0	54	2.0	124.5
0.1	3.5	1.2	63	2.2	135
0.2	20.5	1.6	100.5	1.5	92.5
0.1	3.5	1.5	92.5	1.4	82.5
0.2	20.5	1.6	100.5	2.3	140.5
0.2	20.5	1.5	92.5	2.4	146
0.1	3.5	1.3	72	1.8	110.5
0.2	20.5	1.3	72	1.8	110.5
0.2	20.5	1.3	72	2.1	130.5
0.3	38	1.2	63	2.4	146
0.3	38	1.4	82.5	2.3	140.5
0.2	20.5	1.2	63	1.9	119
0.6	50	1.0	54	2.3	140.5
0.4	45	1.3	72	2.5	149
0.3	38	1.2	63	2.3	140.5
0.2	20.5	1.3	72	1.9	119
0.2	20.5	1.3	72	2.0	124.5
0.2	20.5	1.1	59	2.3	140.5
0.2	20.5	1.3	72	1.8	110.5
$n_1 = 50$ $R_1 = 1275$		$n_2 = 50$ $R_2 = 3824$		$n_3 = 50$ $R_3 = 6226$	

$$H = \frac{12}{N(N+1)}\left(\frac{R_1^2}{n_1}+\frac{R_2^2}{n_2}+\frac{R_3^2}{n_3}\right) - 3(N+1)$$

$$= \frac{12}{150(150+1)}\left(\frac{1275^2}{50}+\frac{3824^2}{50}+\frac{6226^2}{50}\right) - 3(150+1)$$

$$= \frac{12}{22650}*\left(\frac{1625625}{50}+\frac{14622976}{50}+\frac{38763076}{50}\right) - 453$$

$$= 0.0005298013*\frac{55011677}{50} - 453 = 0.0005298013*1100233.54 - 453 = 129.905$$

The number of samples is K = 3, so, df= 3 – 1 = 2. Table A – 4 shows that for 3 degrees of freedom and 0.05 significance, the critical value is 5.991. The test statistic H = 129.905 is in the critical region, so we reject the null hypothesis of equal medians. There is sufficient evidence to support the claim that the medians are not all equal. It does appear that the iris petal length medians are different for three iris species.

12.6 Rank Correlation

In Exercises 1 & 3, sketch a scatter diagram, use it to estimate the value of r_s, then calculate the value of r_s, and state whether there appears to be a correlation between x and y.

1. Scatter Diagram

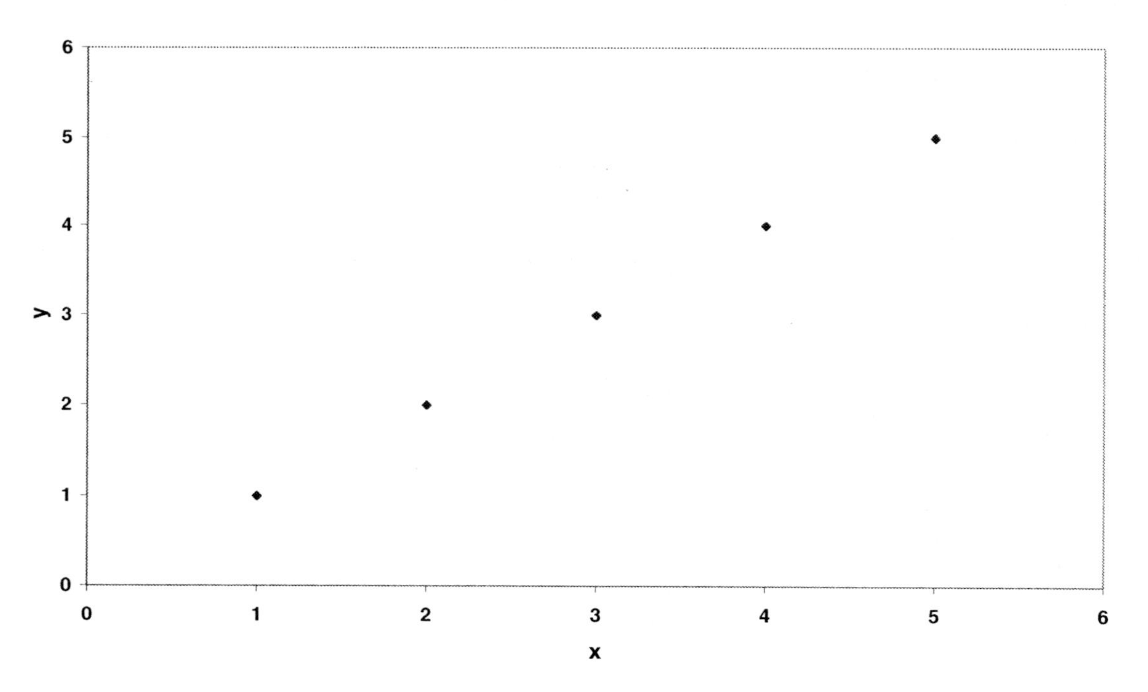

From the scatter diagram, there appears to be a perfect positive correlation between x and y.
In the table below the x and y scores also correspond to the ranks the scores would have received. Thus, $R_x = x$ and $R_y = y$, and d= $R_x - R_y$

R_x	R_y	|Difference| (d)	d^2
1	1	0	0
3	3	0	0
5	5	0	0
4	4	0	0
2	2	0	0
			Total = 0

$$r_s = 1 - \frac{6 \quad d^2}{n(n^2-1)}$$

Where r_s = rank correlation coefficient for sample paired data
n = number of pairs of sample data
d = difference between ranks for the two values within a pair
Substituting 5 for n and 0 for $\quad d^2$

$$r_s = 1 - \frac{6\Sigma d^2}{n(n^2-1)} = 1 - \frac{6 * 0}{5(5^2-1)} = 1 - \frac{0}{120} = 1$$

Since $r_s = 1$, there appears to be a perfect positive correlation between x and y

3. Scatter Diagram

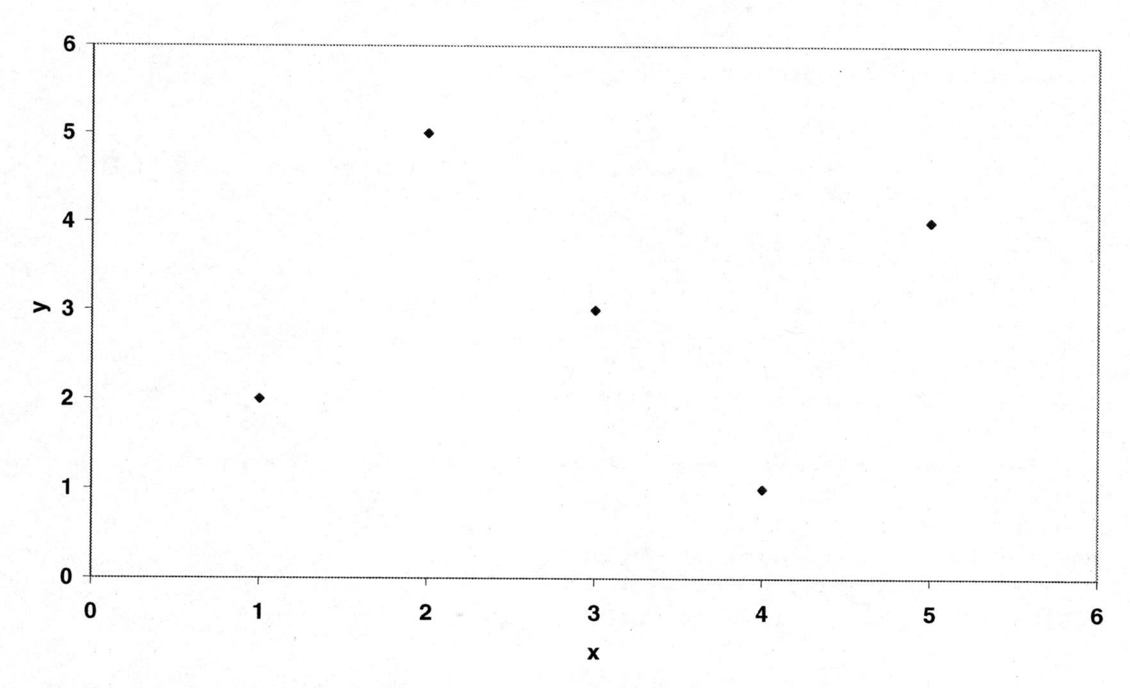

From the scatter diagram, there appears to be no correlation between x and y.
In the table below the x and y scores also correspond to the ranks the scores would have received. Thus, $R_x = x$ and $R_y = y$, and d= $R_x - R_y$

R_x	R_y	|Difference| (d)	d^2
1	2	1	1
2	5	3	9
3	3	0	0
4	1	3	9
5	4	1	1
			Total= 20

$$r_s = 1 - \frac{6\ d^2}{n(n^2-1)}$$

Where r_s = rank correlation coefficient for sample paired data

n = number of pairs of sample data

d = difference between ranks for the two values within a pair

Substituting 5 for n and 20 for d^2

$$r_s = 1 - \frac{6\ d^2}{n(n^2-1)} = 1 - \frac{6*20}{5(5^2-1)} = 1 - \frac{120}{120} = 1 - 1 = 0$$

Since $r_s = 0$, there appears to be no correlation between x and y

In Exercises 5 & 7, find the critical value(s) for r_s by using either Table A – 9 or Formula 12-1, as appropriate. Assume two-tailed cases, where α represents the level of significance and n represents the number of pairs of data.

5. $n = 50,\ \alpha = 0.05$

Since n is more than 30, the following formula should be used.

$$r_s = \frac{\pm z}{\sqrt{n-1}}$$

where the value of z corresponds to the significance level, and the value of z is found in Table A-2.

Substituting 50 for n and 1.96 for z,

$$r_s = \frac{\pm z}{\sqrt{n-1}} = \frac{\pm 1.96}{\sqrt{50-1}} = \frac{\pm 1.96}{7} = \pm 0.280$$

7. $n = 15,\ \alpha = 0.01$

Since n is less than 30, Table A-9 should be used to obtain the critical value for r_s. According to Table A-9, the critical value when n=15 and $\alpha = 0.01$ is ± 0.654

In Exercises 9 – 15, use the rank correlation coefficient to test for a correlation between the two variables. Use a significance level of $\alpha = 0.05$.

9. Correlation Between Salary and Physical Demand

$H_0: \rho_s = 0$ There is no correlation between salary and physical demand

$H_1: \rho_s \neq 0$ There is a correlation between salary and physical demand

Salary Rank	Physical Demand Rank	|Difference| (d)	d^2
2	5	3	9
6	2	4	16
3	3	0	0
5	8	3	9
7	10	3	9
10	9	1	1
9	1	8	64
8	7	1	1
4	6	2	4
1	4	3	9
			Total =122

$$r_s = 1 - \frac{6\ d^2}{n(n^2-1)}$$

Where r_s = rank correlation coefficient for sample paired data
n = number of pairs of sample data
d = difference between ranks for the two values within a pair
Substituting 10 for n and 122 for d^2

$$r_s = 1 - \frac{6\ d^2}{n(n^2-1)} = 1 - \frac{6*122}{10(10^2-1)} = 1 - \frac{732}{990} = 1 - 0.739 = 0.261$$

Since n is less than 30, Table A-9 should be used to obtain the critical value for r_s. According to Table A-9, the critical value when n=10 and $\alpha = 0.05$ is ± 0.648. Because the test statistic r_s =0.261 does not exceed the critical value of 0.648, we fail to reject the null hypothesis. There does not appear to be a correlation between salary and physical demand.

11. **Correlation Between Heights and Weights of Supermodels**

$H_0 : \rho_s = 0$ There is no correlation between heights and weights of supermodels
$H_1 : \rho_s \neq 0$ There is a correlation between heights and weights of supermodels

Height			Weight			
x	$x = R_x$	$x^2 = R_x^2$	y	$y = R_y$	$y^2 = R_y^2$	$xy = R_x * R_y$
71	7	49	125	6	36	42
70.5	5	25	119	3.5	12.25	17.5
71	7	49	128	8.5	72.25	59.5
72	9	81	128	8.5	72.25	76.5
70	3	9	119	3.5	12.25	10.5
70	3	9	127	7	49	21
66.5	1	1	105	1	1	1
70	3	9	123	5	25	15
71	7	49	115	2	4	14
	x= 45	x^2= 281		y= 45	y^2=284	xy= 257

Since there are ties among ranks for both variables, the exact value of test statistic can be calculated using the formula:

$$r_s = \frac{n\sum xy - (\sum x)(\sum y)}{\sqrt{n(\sum x^2) - (\sum x)^2}\sqrt{n(\sum y^2) - (\sum y)^2}}$$

Where r_s = rank correlation coefficient for sample paired data
n = number of pairs of sample data
x = rank of first variable
y = rank of second variable

$$r_s = \frac{9*257-(45)(45)}{\sqrt{9(281)-(45)^2}\sqrt{9(284)-(45)^2}} = \frac{288}{\sqrt{504}\sqrt{531}} = \frac{288}{22.450*23.043} = \frac{288}{517.325} = 0.557$$

Since n is less than 30, Table A-9 should be used to obtain the critical value for r_s. According to Table A-9, the critical value when n=9 and $\alpha = 0.05$ is ± 0.700. Because the test statistic r_s =0.557 does not exceed the critical value of 0.700, we fail to reject the null hypothesis. There does not appear to be a correlation between weight and height of supermodels.

13. Systolic and Diastolic Blood Pressure

$H_0: \rho_s = 0$ There is no correlation between systolic and diastolic blood pressure
$H_1: \rho_s \neq 0$ There is a correlation between systolic and diastolic blood pressure

Systolic Blood Pressure			Diastolic Blood Pressure			
x	$x = R_x$	$x^2 = R_x^2$	y	$y = R_y$	$y^2 = R_y^2$	$xy = R_x * R_y$
125	6	36	78	6	36	36
107	1.5	2.25	54	1	1	1.5
126	7.5	56.25	81	7	49	52.5
110	3.5	12.25	68	3	9	10.5
110	3.5	12.25	66	2	4	7
107	1.5	2.25	83	8	64	12
113	5	25	71	4	16	20
126	7.5	56.25	72	5	25	37.5
	$\sum x = 36$	$\sum x^2 = 202.5$		$\sum y = 36$	$\sum y^2 = 204$	$\sum xy = 177$

Since there are ties among ranks for the systolic blood pressure variable, the exact value of test statistic is calculated using the formula:

$$r_s = \frac{n\sum xy - (\sum x)(\sum y)}{\sqrt{n(\sum x^2) - (\sum x)^2}\sqrt{n(\sum y^2) - (\sum y)^2}}$$

Where r_s = rank correlation coefficient for sample paired data
n = number of pairs of sample data
x = rank of first variable
y = rank of second variable

$$r_s = \frac{8*177-(36)(36)}{\sqrt{8(202.5)-(36)^2}\sqrt{8(204)-(36)^2}} = \frac{120}{\sqrt{324}\sqrt{336}} = \frac{120}{18*18.330} = \frac{120}{329.945} = 0.364$$

Since n is less than 30, Table A-9 should be used to obtain the critical value for r_s. From Table A-9, the critical value when n=8 and $\alpha = 0.05$ is ± 0.738. Because the test statistic $r_s = 0.364$ does not exceed the critical value of 0.738, we fail to reject the null hypothesis. There does not appear to be a correlation between systolic and diastolic pressures.

15. Smoking and Nicotine

$H_0: \rho_s = 0$ There is no correlation between number of cigarettes per day and cotinine level

$H_1: \rho_s \neq 0$ There is a correlation between number of cigarettes per day and cotinine level

Cigarettes per day			Cotinine Level			
x	$x = R_x$	$x^2 = R_x^2$	y	$y = R_y$	$y^2 = R_y^2$	$xy = R_x * R_y$
60	12	144	179	7	49	84
10	6.5	42.25	283	9	81	58.5
4	2	4	75.6	5	25	10
15	9	81	174	6	36	54
10	6.5	42.25	209	8	64	52
1	1	1	9.51	2	4	2
20	10.5	110.25	350	11	121	115.5
8	4	16	1.85	1	1	4
7	3	9	43.4	4	16	12
10	6.5	42.25	25.1	3	9	19.5
10	6.5	42.25	408	12	144	78
20	10.5	110.25	344	10	100	105
	$x = 78$	$x^2 = 644.5$		$y = 78$	$y^2 = 650$	$xy = 594.5$

Since there are ties among ranks for the cigarettes per day variable, the exact value of test statistic can be calculated using the formula:

$$r_s = \frac{n \quad xy - (\quad x)(\quad y)}{\sqrt{n(\quad x^2) - (\quad x)^2}\sqrt{n(\quad y^2) - (\quad y)^2}}$$

$$= \frac{12 * 594.5 - (78)(78)}{\sqrt{12(644.5) - (78)^2}\sqrt{12(650) - (78)^2}} = \frac{1050}{\sqrt{1650}\sqrt{1716}} = \frac{1050}{40.620 * 41.425}$$

$$= \frac{1050}{1682.68} = 0.624$$

From Table A-9, the critical value is ± 0.587 (based on $\alpha = 0.05$ and $n = 12$). Because the test statistic $r_s =$ 0.624 exceeds the critical value of 0.587, we reject the null hypothesis. There is sufficient evidence to support the claim that there is a correlation between number of cigarettes smoked and the measured amounts of cotinine.

In Exercises 17 & 19, use the data from Appendix B to construct a scatterplot, find the value of the rank correlation coefficient r_s, and use a significance level of $\alpha = 0.05$ to determine whether there is a significant correlation between the variables.

17. Cholesterol and Body Mass Index

$H_0 : \rho_s = 0$ There is no correlation between cholesterol level and body mass index

$H_1 : \rho_s \neq 0$ There is a correlation between cholesterol level and body mass index

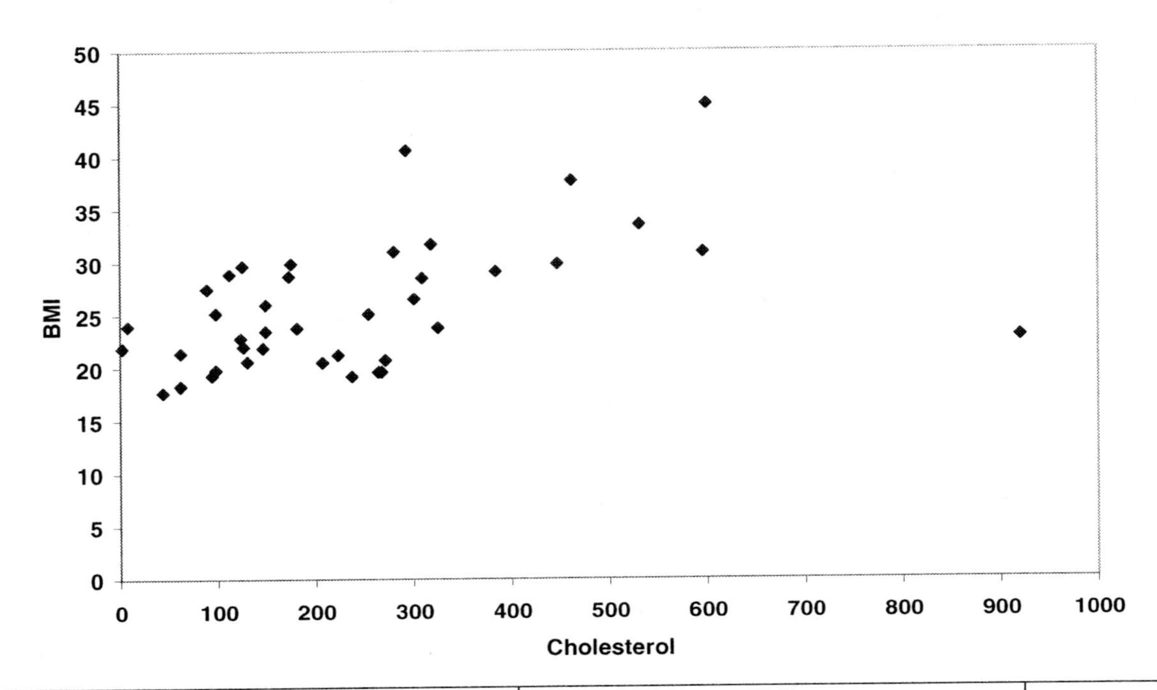

Cholesterol Level			BMI			
x	$x = R_x$	$x^2 = R_x^2$	y	$y = R_y$	$y^2 = R_y^2$	$xy = R_x * R_y$
264	25	625	19.6	5.5	30.25	137.5
181	20	400	23.8	19.5	380.25	390
267	26	676	19.6	5.5	30.25	143
384	34	1156	29.1	30	900	1020
98	8.5	72.25	25.2	23	529	195.5
62	4.5	20.25	21.4	12	144	54
126	13	169	22	15	225	195
89	6	36	27.5	26	676	156
531	37	1369	33.5	37	1369	1369
130	14	196	20.6	9	81	126
175	19	361	29.9	33	1089	627
44	3	9	17.7	1	1	3
8	2	4	24	21	441	42
112	10	100	28.9	29	841	290
462	36	1296	37.7	38	1444	1368
62	4.5	20.25	18.3	2	4	9
98	8.5	72.25	19.8	7	49	59.5
447	35	1225	29.8	32	1024	1120
125	12	144	29.7	31	961	372
318	32	1024	31.7	36	1296	1152

325	33	1089	23.8	19.5	380.25	643.5
600	39	1521	44.9	40	1600	1560
237	23	529	19.2	3	9	69
173	18	324	28.7	28	784	504
309	31	961	28.5	27	729	837
94	7	49	19.3	4	16	28
280	28	784	31	35	1225	980
254	24	576	25.1	22	484	528
123	11	121	22.8	16.5	272.25	181.5
596	38	1444	30.9	34	1156	1292
301	30	900	26.5	25	625	750
223	22	484	21.2	11	121	242
293	29	841	40.6	39	1521	1131
146	15	225	21.9	13.5	182.25	202.5
149	16.5	272.25	26	24	576	396
149	16.5	272.25	23.5	18	324	297
920	40	1600	22.8	16.5	272.25	660
271	27	729	20.7	10	100	270
207	21	441	20.5	8	64	168
2	1	1	21.9	13.5	182.25	13.5
	x=820	x^2= 22138.5		y= 820	y^2= 22138	xy= 19581.5

Since there are ties among ranks for both variables, the exact value of test statistic can be calculated using the formula:

$$r_s = \frac{n\ \ xy - (\ \ x)(\ \ y)}{\sqrt{n(\ \ x^2) - (\ \ x)^2}\sqrt{n(\ \ y^2) - (\ \ y)^2}}$$

$$= \frac{40 * 19581.5 - (820)(820)}{\sqrt{40(22138.5) - (820)^2}\sqrt{40(22138) - (820)^2}} = \frac{110860}{\sqrt{213140}\sqrt{213120}}$$

$$= \frac{110860}{461.671 * 461.649} = \frac{110860}{213129.96} = 0.520$$

Since n > 30, critical value needs to be found out from the following formula.

$$r_s = \frac{\pm z}{\sqrt{n-1}} = \frac{\pm 1.96}{\sqrt{40-1}} = \frac{\pm 1.96}{6.245} == \pm 0.314$$

Because the test statistic r_s = 0.520 exceeds the critical value of 0.314, we reject the null hypothesis. There is sufficient evidence to support the claim that there is a correlation between cholesterol levels and BMI values.

19. Chest Sizes and Weights of Bears

$H_0: \rho_s = 0$ There is no correlation between chest sizes and weights of bears

$H_1: \rho_s \neq 0$ There is a correlation between chest sizes and weights of bears

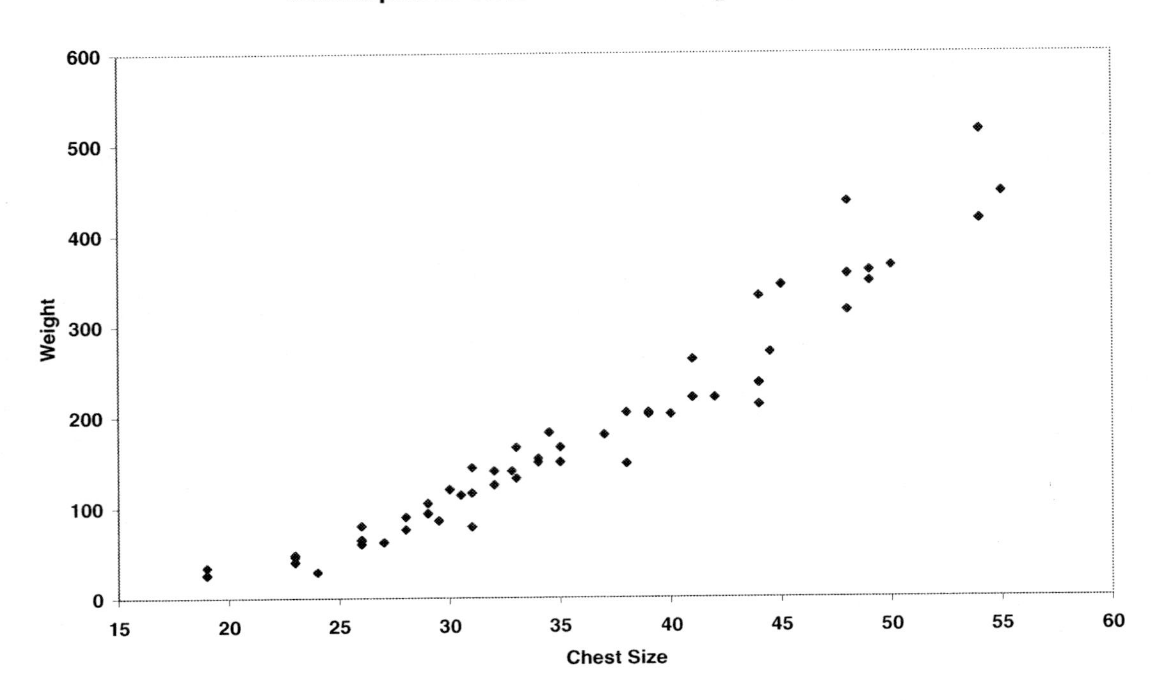

Chest Size			Weights			
x	$x= R_x$	$x^2 = R_x^2$	y	$y= R_y$	$y^2 = R_y^2$	$xy= R_x * R_y$
26.0	8.5	72.25	80	13	169	110.5
45.0	45	2025	344	46	2116	2070
54.0	52.5	2756.25	416	51	2601	2677.5
49.0	49.5	2450.25	348	47	2209	2326.5
35.0	30.5	930.25	166	30.5	930.25	930.25
41.0	38.5	1482.25	220	39.5	1560.25	1520.75
41.0	38.5	1482.25	262	42	1764	1617
49.0	49.5	2450.25	360	49	2401	2425.5
38.0	33.5	1122.25	204	36.5	1332.25	1222.75
31.0	20	400	144	25	625	500
44.0	42	1764	332	45	2025	1890
19.0	1.5	2.25	34	3	9	4.5
32.0	22.5	506.25	140	23.5	552.25	528.75
37.0	32	1024	180	32	1024	1024
29.0	14.5	210.25	105	17	289	246.5
33.0	25.5	650.25	166	30.5	930.25	777.75
39.0	35.5	1260.25	204	36.5	1332.25	1295.75
19.0	1.5	2.25	26	1	1	1.5
30.0	17	289	120	20	400	340
48.0	47	2209	436	52	2704	2444
32.0	22.5	506.25	125	21	441	472.5
33.0	25.5	650.25	132	22	484	561
28.0	12.5	156.25	90	15	225	187.5

23.0	4	16	40	4	16	16
42.0	40	1600	220	39.5	1560.25	1580
23.0	4	16	46	5	25	20
34.0	27.5	756.25	154	29	841	797.5
31.0	20	400	116	19	361	380
34.5	29	841	182	33	1089	957
34.0	27.5	756.25	150	27.5	756.25	756.25
26.0	8.5	72.25	65	10	100	85
48.0	47	2209	356	48	2304	2256
48.0	47	2209	316	44	1936	2068
29.0	14.5	210.25	94	16	256	232
29.5	16	256	86	14	196	224
35.0	30.5	930.25	150	27.5	756.25	838.75
44.5	44	1936	270	43	1849	1892
39.0	35.5	1260.25	202	34.5	1190.25	1224.75
40.0	37	1369	202	34.5	1190.25	1276.5
50.0	51	2601	365	50	2500	2550
31.0	20	400	79	12	144	240
38.0	33.5	1122.25	148	26	676	871
55.0	54	2916	446	53	2809	2862
27.0	11	121	62	8	64	88
44.0	42	1764	236	41	1681	1722
44.0	42	1764	212	38	1444	1596
26.0	8.5	72.25	60	7	49	59.5
26.0	8.5	72.25	64	9	81	76.5
30.5	18	324	114	18	324	324
28.0	12.5	156.25	76	11	121	137.5
23.0	4	16	48	6	36	24
24.0	6	36	29	2	4	12
54.0	52.5	2756.25	514	54	2916	2835
32.8	24	576	140	23.5	552.25	564
	x=1485	x^2= 53936		y= 1485	y^2=53952	xy= 53739.75

Since there are ties among ranks on both variables, the exact value of test statistic can be calculated using the formula:

$$r_s = \frac{n\ \ xy - (\ \ x)(\ \ y)}{\sqrt{n(\ \ x^2) - (\ \ x)^2}\sqrt{n(\ \ y^2) - (\ \ y)^2}}$$

$$= \frac{54 * 53739.75 - (1485)(1485)}{\sqrt{54(53936) - (1485)^2}\sqrt{54(53952) - (1485)^2}} = \frac{696721.5}{\sqrt{707319}\sqrt{708183}}$$

$$= \frac{696721.5}{841.023 * 841.536} = \frac{696721.5}{707751.13} = 0.984$$

Since n > 30, critical value needs to be found out from the following formula.

$$r_s = \frac{\pm z}{\sqrt{n-1}} = \frac{\pm 1.96}{\sqrt{54-1}} = \frac{\pm 1.96}{7.280} = \pm 0.269$$

Because the test statistic $r_s = 0.984$ exceeds the critical value of 0.269, we reject the null hypothesis. There is sufficient evidence to support the claim that there is a correlation between chest size and weight of the bears.

21. Finding Critical Values

a. $n = 8$, $\alpha = 0.05$

As per the table A-3, corresponding t value is 2.447 for $\alpha = 0.05$ and $df = n - 2 = 6$

$$r_s = \pm\sqrt{\frac{t^2}{t^2 + n - 2}} = \pm\sqrt{\frac{(2.447)^2}{(2.447)^2 + 8 - 2}} = \pm\sqrt{\frac{5.9878}{11.9878}} = \pm\sqrt{0.4995} = \pm 0.707$$

b. $n = 15$, $\alpha = 0.05$

As per the table A-3, corresponding t value is 2.160 for $\alpha = 0.05$ and $df = n - 2 = 13$

$$r_s = \pm\sqrt{\frac{t^2}{t^2 + n - 2}} = \pm\sqrt{\frac{(2.160)^2}{(2.160)^2 + 15 - 2}} = \pm\sqrt{\frac{4.6656}{17.6656}} = \pm\sqrt{0.2641} = \pm 0.514$$

c. $n = 30$, $\alpha = 0.05$

As per the table A-3, corresponding t value is 2.048 for $\alpha = 0.05$ and $df = n - 2 = 28$

$$r_s = \pm\sqrt{\frac{t^2}{t^2 + n - 2}} = \pm\sqrt{\frac{(2.048)^2}{(2.048)^2 + 30 - 2}} = \pm\sqrt{\frac{4.1943}{32.1943}} = \pm\sqrt{0.1303} = \pm 0.361$$

d. $n = 30$, $\alpha = 0.01$

As per the table A-3, corresponding t value is 2.763 for $\alpha = 0.01$ and $df = n - 2 = 28$

$$r_s = \pm\sqrt{\frac{t^2}{t^2 + n - 2}} = \pm\sqrt{\frac{(2.763)^2}{(2.763)^2 + 30 - 2}} = \pm\sqrt{\frac{7.6342}{35.6342}} = \pm\sqrt{0.2142} = \pm 0.463$$

e. $n = 8$, $\alpha = 0.01$

As per the table A-3, corresponding t value is 3.707 for $\alpha = 0.01$ and $df = n - 2 = 6$

$$r_s = \pm\sqrt{\frac{t^2}{t^2 + n - 2}} = \pm\sqrt{\frac{(3.707)^2}{(3.707)^2 + 8 - 2}} = \pm\sqrt{\frac{13.7418}{19.7418}} = \pm\sqrt{0.6961} = \pm 0.834$$

Review Exercises

In Exercises 1 – 5, use a 0.05 significance level with the indicated test. If no particular test is specified, use the appropriate nonparametric test from this chapter.

1. Temperatures and Marathons

$H_0 : \rho_s = 0$ There is no correlation between temperature and winning time

$H_1 : \rho_s \neq 0$ There is a correlation between temperature and winning time

Temperature		Winning Time			
x	R_x	Y	R_y	|Difference| (d)	d^2
55	3	145.283	2	1	1
61	5	148.717	8	3	9
49	1	148.300	7	6	36
62	6	148.100	6	0	0
70	7	147.617	5	2	4
73	8	146.400	3	5	25
51	2	144.667	1	1	1
57	4	147.533	4	0	0
					Total = 76

$r_s = 1 - \frac{6 d^2}{n(n^2 - 1)}$, where r_s = rank correlation coefficient for sample paired data

n = number of pairs of sample data

d = difference between ranks for the two values within a pair

Substituting 8 for n and 76 for d^2

$$r_s = 1 - \frac{6 d^2}{n(n^2 - 1)} = 1 - \frac{6 * 76}{8(8^2 - 1)} = 1 - \frac{456}{504} = 1 - 0.9048 = 0.0952$$

From Table A-9, the critical value is ± 0.738 (based on $\alpha = 0.05$ and $n = 8$). Because the test statistic r_s = 0.0952 does not exceed the critical value of 0.738, we fail to reject the null hypothesis. There is not sufficient evidence to support the claim that there is a correlation between temperature and winning time.

3. Do Beer Drinkers and Liquor Drinkers Have Different BAC Levels

H_0 : The populations have the same median

H_1 : The populations have different medians

$\alpha = 0.05$, two-tailed test using z statistic, CV= ±1.96

Beer Drinkers		Liquor Drinkers	
x	Rank	x	Rank
0.129	1	0.220	17
0.154	5	0.253	25
0.203	14	0.247	24
0.146	2	0.190	12.5
0.155	6	0.225	19
0.190	12.5	0.241	23
0.148	3	0.224	18
0.187	11	0.257	26
0.164	7	0.185	10
0.152	4	0.227	21
0.212	16	0.226	20
0.165	8	0.182	9
		0.205	15
		0.234	22
$n_1 = 12$ $R_1 = 89.5$		$n_2 = 14$ $R_2 = 261.5$	

$$\mu_R = \frac{n_1(n_1+n_2+1)}{2} = \frac{12(12+14+1)}{2} = \frac{324}{2} = 162$$

$$\sigma_R = \sqrt{\frac{n_1 n_2 (n_1+n_2+1)}{12}} = \sqrt{\frac{12*14(12+14+1)}{12}} = \sqrt{\frac{4536}{12}} = \sqrt{378} = 19.442$$

$$z = \frac{R-\mu_R}{\sigma_R} = \frac{89.5\text{-}162}{19.442} = \frac{-72.5}{19.442} = -3.729$$

Since it is a two tailed test and $\alpha = 0.05$ critical value of z would be ± 1.96. The test statistic of $z = -3.729$ falls within the critical region. So, we reject the claim that beer drinkers and liquor drinkers have the same median BAC levels. The liquor drinkers appear to have higher BAC levels and would be more dangerous.

5. Testing Effectiveness of Treatment

H_0 : The populations have the same before and after median weights

H_1 : The populations have different before and after median weights

Subject	A	B	C	D	E	F	G	H	I	J
Weight before treatment	700	840	830	860	840	690	830	1180	930	1070
Weight after treatment	720	840	820	900	870	700	800	1200	950	1080
Sign of difference	+	0	–	+	+	+	–	+	+	+

n= 9, 7 positive signs, 2 negative signs, 1 tie

Significance level is α= 0.05, n= 9, x (lower frequency)= 2, CV from Table A-7= 1.

Since x=2 > 1 (CV), there is not sufficient evidence to warrant rejection of the Null hypothesis, claim that there is no difference between the reported and measured heights. There is not sufficient evidence to warrant the claim that treatment was effective. The course does not appear to have an effect

Cumulative Review Exercises

For Exercises 1 – 5, refer to the accompanying table, which shows the heights of presidents matched with the candidates they beat. All heights are in inches, and only the winners and second place candidates are included. Use a 0.05 significance level for the following tests.

1. Linear correlation coefficient, Pearson r

$H_0: \rho = 0$ There is no correlation between heights of presidential winners and losers

$H_1: \rho \neq 0$ There is a correlation between heights of presidential winners and losers

Winner Height		Loser Height		
x	x^2	y	y^2	xy
76	5776	64	4096	4864
66	4356	71	5041	4686
70	4900	72	5184	5040
70	4900	72	5184	5040
74	5476	68	4624	5032
71.5	5112.25	71	5041	5076.5
73	5329	69.5	4830.25	5073.5
74	5476	74	5476	5476
x=574.5	x^2= 41325.25	y= 561.5	y^2= 39476.25	xy= 40288

$$r = \frac{n \quad xy - (\quad x)(\quad y)}{\sqrt{n(\quad x^2) - (\quad x)^2}\sqrt{n(\quad y^2) - (\quad y)^2}}$$

$$r = \frac{8 * 40288 - (574.5)(561.5)}{\sqrt{8(41325.25) - (574.5)^2}\sqrt{8(39476.25) - (561.5)^2}} = \frac{-277.75}{\sqrt{551.75}\sqrt{527.75}}$$

$$= \frac{-277.75}{23.489 * 22.973} = \frac{-277.75}{539.62} = -0.515$$

From Table A-6 the critical value is ± 0.707 (based on $\alpha = 0.05$ and $n = 8$). Because the absolute value of test statistic $r = -0.515$ is not less than the critical value of -0.707, we conclude that there is no significant correlation between the height of winners and those that were their matched losers.

3. Sign test comparing heights

H_0 : There is no difference between median heights of presidential winners and losers

H_1 : There is a difference between median heights of presidential winners and losers

Winner	76	66	70	70	74	71.5	73	74
Runner-up	64	71	72	72	68	71	69.5	74
Sign of difference	+	–	–	–	+	+	+	0

n= 7, 4 positive signs, 3 negative sign, 1 tie

Significance level is α= 0.05 with one-tailed test, n= 7, x (lower frequency)= 3, CV from Table A-7, T= 0. Since x=3 > 0 (CV), there is not sufficient evidence to warrant rejection of the claim that there is no difference between the presidential winner and loser's heights. The results support the claim that median height of presidential winners and losers or runner-ups is not different.

5. Parametric t Test

$H_0: \mu_d = 0$ There is no difference between mean heights of presidential winners and losers

$H_1: \mu_d \neq 0$ There is a difference between mean heights of presidential winners and losers

t test statistic for matched pairs

$$t = \frac{\bar{d} - \mu_d}{\frac{s_d}{\sqrt{n}}}$$

where d = individual difference between the two values in a single matched pair

μ_d = mean value of the differences d for the population of all matched pairs

$\bar{d}$ = mean value of the differences d for the paired sample data

s_d = standard deviation of the differences d for the paired sample data

n = number of pairs of data

Winner	Runner-up	Difference (d)	$\bar{d}$	$(d - \bar{d})$	$(d - \bar{d})^2$
76	64	12	1.625	10.375	107.641
66	71	-5	1.625	-6.625	43.891
70	72	-2	1.625	-3.625	13.141
70	72	-2	1.625	-3.625	13.141
74	68	6	1.625	4.375	19.141
71.5	71	0.5	1.625	-1.125	1.266
73	69.5	3.5	1.625	1.875	3.516
74	74	0	1.625	-1.625	2.641
					$(d - \bar{d})^2$ =204.378

$$\bar{d} = \frac{d}{n} = \frac{13}{8} = 1.625$$

$$s_d = \sqrt{\frac{(d-\bar{d})^2}{n-1}} = \sqrt{\frac{204.378}{7}} = \sqrt{29.197} = 5.403$$

$$t = \frac{\bar{d} - \mu_d}{\frac{s_d}{\sqrt{n}}} = \frac{1.625 - 0}{\frac{5.403}{\sqrt{8}}} = \frac{1.625}{1.910} = 0.851$$

From Table A-3, the critical value of *t*, using the column for 0.05 and two-tailed, with *df*= n – 1= 7 is ± 2.365. Since the test statistic is smaller than critical value, there is not sufficient evidence to support the claim that there is a difference between the heights of the winning and losing candidates.

Chapter 13. Life Tables

13-2 Elements of a Life Table

1. **Types of Life Tables**
A cohort life table is a record of the actual mortality experience of a particular group, such as all people born in 1970. This would include data of the mortality of this group at each year. A period life table, on the other hand, the mortality conditions and results for different ages are based on data from one particular year.

3. **Population Size**
The probability of dying during the first interval would remain the same, as would the expected remaining lifetime value. However, the number surviving to the beginning of the interval, number of deaths during the interval, person-years lived, and total person years lived would all be half of the stated values.

In Exercises 5 & 7, refer to the accompanying life table for white females in the United States for the year 2000.

5. **Probability of Dying**
There were 99,487 white females alive on their first birthday, and there were 41 deaths during the interval. The probability of dying in that interval was 41/99,487 = 0.000412.

7. **Finding Survival Rate**
There were 100,000 white females at birth, and there were 99,446 that survived until their second birthday. The probability of surviving until the second birthday is 99,446/100,000 = 0.99446

9. **Probability of Surviving**
The probability of dying between the 20th birthday and the 21st birthday is 0.000888, so the probability of surviving that same period is 1 – 0.000888 = 0.999112. If 250 people live to their 20th birthday, we expect that 250 × 0.999112 = 249 to survive to their 21st birthday.

11. **Probability of Surviving**
There were 98,943 people alive on their 16th birthday, and 80,812 people alive on their 66th birthday. The probability of surviving is 80,812/98,943 = 0.81675308.

13. **Abridged Table**
a. The probability of dying in the age interval 0-5 is found by first finding the probability of surviving. There were 100,000 people to begin with. There were 99,177 alive at their 5th birthday. The probability of surviving is 99,177/100,000 = 0.99177, and so the probability of dying is 1 – 0.99177 = 0.00823.
b The probability of dying in the age interval 5-10 is found by first finding the probability of surviving. There were 99,177 people to begin with. There were 99,095 alive at their 10th birthday. The probability of surviving is 99,095/99,177 = 0.999173195, and so the probability of dying is 1 – 0.999173195 = 0.000826805.
The results for the two intervals is so different due to infant mortality. Comparably, the probability of dying before the first birthday is very high, and so the probability of dying in the age interval 0-5 is correspondingly high also.

15. **Abridged Table**
The values would be:

0-5	0.00823	100000	822	496,308	7996958	76.9

13-3 Applications of Life Tables

In Exercises 1 & 3, assume there are 4,453,000 births in the United States this year (based on data from the U.S. National Center for Health Statistics). Use Table 13-1 to find the given value.

1. Future Drivers

From Table 13-1, we see that among 100,000 births, 98,943 are expected to make it to their 16th birthday, so the probability of surviving until the 16th birthday is 0.98943. If each of the 4,453,000 people have a 0.98943 probability of surviving to their 16th birthday, the expected number of such survivors is 4,453,000×0.98943 = 4,405,931. These results are important as a practical matter to help plan for driving patterns and number of cars on the road in the future.

3. Future Health Care

From Table 13-1, we see that among 100,000 births, 87,498 are expected to make it to their 60th birthday, so the probability of surviving until the 60th birthday is 0.87498. If each of the 4,453,000 people have a 0.87498 probability of surviving to their 60th birthday, the expected number of such survivors is 4,453,000×0.87498 = 3,896,285.

5. Hypothesis Test

From Table 13-1, we see that the probability of dying during the age interval 20-21 is 0.000888. The proportion of actual deaths was $\hat{p} = x/n = 15/12500 = 0.0012$. The significance level is not mentioned, so we use the common value of $\alpha = 0.05$. Using the methods of Section 7-3,

H_0: $p = 0.000888$

H_1: $p > 0.000888$

The test statistic is $z = \dfrac{\hat{p}-p}{\sqrt{\dfrac{pq}{n}}} = \dfrac{0.0012-0.000888}{\sqrt{\dfrac{0.000888\times 0.999112}{12500}}} = 1.171$

In a right-tailed test at the 0.05 significance level, the critical value is $z_{\alpha/2} = z_{.05} = 1.645$.

Since this is a right-tailed test, the P-value is the area to the right of the test statistic. Using Table A-2, we find that the $P-value = (1-0.8790) = 0.1210$.

We fail to reject the null hypothesis.

There is not sufficient evidence to support the claim that this was an unusually high number of deaths.

7. Hypothesis Test

From Table 13-1, we see that there were 98,943 people alive on their 16th birthday, and 97,696 people alive on their 30th birthday, so the probability of dying during that interval is 1 – 97,696/98,943 = 0.01260322. The proportion of actual deaths was $\hat{p} = x/n = 147/8774 = 0.01675405$. The significance level is not mentioned, so we use the common value of $\alpha = 0.05$. Using the methods of Section 7-3,

H_0: $p = 0.01260322$

H_1: $p > 0.01260322$

The test statistic is $z = \dfrac{\hat{p}-p}{\sqrt{\dfrac{pq}{n}}} = \dfrac{0.01675405-0.01260322}{\sqrt{\dfrac{0.01260322\times 0.98739678}{8774}}} = 3.485$

In a right-tailed test at the 0.05 significance level, the critical values are $z_{\alpha/2} = z_{.05} = 1.645$.

Since this is a right-tailed test, the P-value is the area to the right of the test statistic. Using Table A-2, we find that the $P-value = (1-0.9998) = 0.0002$.

We reject the null hypothesis.

There is sufficient evidence to support the claim that this was an unusually high number of deaths.

9. Life Insurance Rates

The break-even cost can be found using the following simple formula:

$$\text{Break - even cost of policy} = (\text{Amount of policy}) \times \frac{P(\text{dying})}{1-P(\text{dying})}.$$

Using Table 13-1, we see that the probability of dying in the age category 49-50 is 0.004138. So the break even cost for a policy of $100,000 is

$$(\text{Amount of policy})\times\frac{P(\text{dying})}{1-P(\text{dying})}=(100{,}000)\times\frac{0.004138}{0.995832}=415.50.$$

(Similar to the text, we round at the fourth significant digit, as the probability of dying has four significant digits)

11. Life Insurance Rates

The break-even cost can be found using the following simple formula:

$$\text{Break - even cost of policy} = (\text{Amount of policy})\times\frac{P(\text{dying})}{1-P(\text{dying})}.$$

Using Table 13-1, we see that there are 87,498 people alive at their 60th birthday, and 74,561 alive on their 70th birthday. The probability of dying during that time is 1 – 74,561/87,498 = 0.14785481. So the break even cost for a policy of $50,000 is

$$(\text{Amount of policy})\times\frac{P(\text{dying})}{1-P(\text{dying})}=(50{,}000)\times\frac{0.14785481}{0.85214519}=8{,}675.45.$$

13. Life Insurance Rates

The break-even cost can be found using the following simple formula:

$$\text{Break - even cost of policy} = (\text{Amount of policy})\times\frac{P(\text{dying})}{1-P(\text{dying})}.$$

The probability of a black female dying in the age category 49-50 is 0.005829. So the break even cost for a policy of $100,000 is

$$(\text{Amount of policy})\times\frac{P(\text{dying})}{1-P(\text{dying})}=(100{,}000)\times\frac{0.005829}{0.994171}=586.30.$$

(Similar to the text, we round at the fourth significant digit, as the probability of dying has four significant digits)

The result for a person, absent gender and race specification, is $415.50. This suggests that the mortality rate for black females in the age category 49-50 is higher than that of people in general.

Review Exercises

In Exercises 1-5, refer to the accompanying life table for black females in the United States for the year 2000.

1. Probability of Dying

There were 98,733 black females alive on their first birthday, and there were 78 deaths during the interval. The probability of dying in that interval was 78/98,733 = 0.00079001.

3. Finding Survival Rate

There were 100,000 black females at birth, and there were 98,655 that survived until their second birthday. The probability of surviving until the second birthday is 98,655/100,000 = 0.98655.

5. Expected Remaining Lifetime

The expected remaining lifetime for a black female who was just born is 74.9 years. It is the same for a black female who has reached her first birthday. It would appear that once a black female survives to her first birthday, she is likely to live an extra year. Mathematically, this is due to the very high number of deaths before the first birthday is reached.

Cumulative Review Exercises

In Exercises 1 & 3, use the same life table used for Review Exercises 1-5.

1. Hypothesis Test of Experimental Results

From the life table in the Review Exercises, we see that the probability of a black female dying before her first birthday is 0.012672. The proportion of actual deaths in the community following the implementation of the program was $\hat{p} = x/n = 4/786 = 0.00508906$. The significance level is not mentioned, so we use the common value of $\alpha = 0.05$. Using the methods of Section 7-3,

H_0: $p = 0.012672$

H_1: $p < 0.012672$

The test statistic is $z = \dfrac{\hat{p}-p}{\sqrt{\dfrac{pq}{n}}} = \dfrac{0.00508906-0.012672}{\sqrt{\dfrac{0.012672 \times 0.987328}{768}}} = -1.901$

In a left-tailed test at the 0.05 significance level, the critical value is $-z_{\alpha} = -z_{.05} = -1.645$.

Since this is a left-tailed test, the P-value is the area to the left of the test statistic. Using Table A-2, we find that the $P-value = 0.0287$.

We reject the null hypothesis.

There is sufficient evidence to support the claim that this program was effective in reducing the mortality rate of black females before their first birthday.

3. Probability of Surviving

The probability of a black female surviving to her first birthday is 0.987328. Assuming that the survival for the three black females is independent, the probability that all three survive to their first birthday is $(0.987328)^3 = 0.96246$.